Metal Plasticity and Fatigue at High Temperature

Metal Plasticity and Fatigue at High Temperature

Special Issue Editors

Denis Benasciutti
Luciano Moro
Jelena Srnec Novak

MDPI • Basel • Beijing • Wuhan • Barcelona • Belgrade

Special Issue Editors

Denis Benasciutti
University of Ferrara
Italy

Luciano Moro
University of Udine
Italy

Jelena Srnec Novak
University of Udine
Italy

Editorial Office
MDPI
St. Alban-Anlage 66
4052 Basel, Switzerland

This is a reprint of articles from the Special Issue published online in the open access journal *Metals* (ISSN 2075-4701) from 2018 to 2020 (available at: https://www.mdpi.com/journal/metals/special_issues/metal_plasticity_Fatigue).

For citation purposes, cite each article independently as indicated on the article page online and as indicated below:

LastName, A.A.; LastName, B.B.; LastName, C.C. Article Title. *Journal Name* **Year**, *Article Number, Page Range.*

ISBN 978-3-03928-770-3 (Pbk)
ISBN 978-3-03928-771-0 (PDF)

Cover image courtesy of John Wiley & Sons, Inc.

Contents

About the Special Issue Editors

Denis Benasciutti (Prof.) is Associate Professor of Machine Design at the University of Ferrara, Italy. In 2001, he graduated cum laude in Materials Engineering at the University of Ferrara where, in 2005, he also received his Ph.D. In 2006, he earned a Master in Welding Engineering and diploma in European/International Welding Engineering (EWE/IWE). From 2006 to 2015 he was Assistant Professor at the University of Udine, Italy, where he lectured on Machine Design. At the University of Ferrara, he now teaches Mechanical Drawing and Computer Assisted Design (Finite Element Method). His main research interests involve the structural durability of mechanical components subjected to stochastic uniaxial and multiaxial loadings (with special focus on frequency-domain methods) and cyclic plasticity behavior of metals. He has co-authored over 100 scientific papers in the topics of metal fatigue and structural durability, finite element method, energy harvesting, and cyclic plasticity of metals.

Luciano Moro (Prof.) is Assistant Professor at the Polytechnic Department of Engineering and Architecture (DPIA) at the University of Udine, Italy. He graduated cum laude at the University of Udine where, in 2013, he received his Ph.D. degree in Industrial and Information Engineering. From 2013 to 2016, he was a Postdoctoral Researcher at the University of Udine, Italy. His research mainly focused on multiphysics analysis, non-linear finite element simulations, and failure analysis. He co-authored over 30 scientific papers, in the fields of energy harvesting, finite element methods, cyclic plasticity, and thermomechanical problems.

Jelena Srnec Novak (Dr.) is a Postdoctoral Researcher at the Polytechnic Department of Engineering and Architecture (DPIA) at the University of Udine, Italy. In 2012, she graduated in Mechanical Engineering at the University of Rijeka, Croatia. In 2016, she received her Ph.D. degree in Industrial and Information Engineering at the University of Udine, Italy. Her main research interests include the mechanical behavior of materials, cyclic plasticity, material characterization, low-cycle fatigue prediction, and non-linear finite element simulations. She is co-author of over 20 scientific papers in the topics of cyclic plasticity of metals, material calibration, damage analysis, and finite element method.

Editorial

Metal Plasticity and Fatigue at High Temperature

Denis Benasciutti [1],*, Luciano Moro [2] and Jelena Srnec Novak [2],*

[1] Department of Engineering, University of Ferrara, via Saragat 1, 44122 Ferrara, Italy
[2] Politechnic Department of Engineering and Architecture (DPIA), University of Udine, via delle Scienze 208, 33100 Udine, Italy; luciano.moro@uniud.it
* Correspondence: denis.benasciutti@unife.it (D.B.); jelena.srnec@uniud.it (J.S.N.); Tel.: +39-(0)532-974-976 (D.B.); +39-0432-558-297 (J.S.N.)

Received: 10 February 2020; Accepted: 26 February 2020; Published: 1 March 2020

1. Introduction and Scope

The situation in which a component or structure is maintained at high temperature under the action of cyclic thermal and/or mechanical loadings represents, perhaps, one of the most demanding engineering applications—if not, in fact, the most demanding one. Examples can be found in many industrial fields, such as automotive (cylinder head, engine, disk brakes), steel-making (hot rolling), machining (milling, turning), aerospace (turbine blades), and fire protection systems (fire doors).

The presence of high temperatures usually induces some amount of material plasticity or creep deformation in the most stressed regions of the structure. Plasticity, if combined with the action of cyclic loading variation, may lead to low-cycle fatigue (LCF) failure.

In order to estimate the component fatigue life in such demanding operative condition, it is often necessary to characterize the high-temperature material behavior under cyclic loading, in terms, for example, of cyclic stress–strain response, strain hardening or softening, creep behavior, experimental fatigue strength under isothermal and/or non-isothermal conditions. Moreover, it is also necessary to develop a reliable structural durability approach that is able to include experimental results in numerical and/or predictive models (e.g., plasticity models, fatigue strength curves).

The choice of the most appropriate material model to be used in simulations, or even calibrating the model to experimental data, often represents the most critical step in the whole design approach. Experimental techniques and modeling have to be properly managed to guarantee the reliability of the estimated fatigue life.

2. Contributions

As a part of this Special Issue, researchers were invited to submit their innovative research papers aimed at providing a state-of-the-art knowledge on the topic of metal plasticity, creep deformation and fatigue strength of metals operating at high temperatures, with emphasis on both experimental characterization and numerical modeling of material behavior. A total of eleven research papers were published [1–11].

Problems correlated with the creep phenomenon have been investigated in [1,2]. The paper by Liu et al. [1] presents an explicit fatigue-creep model which develops one formulation that covers the full range of conditions: from pure fatigue, to creep-fatigue up to pure creep. Kloc et al. [2] study a complex phenomenological creep model with particular attention focused on transient effects in the creep behavior of the Sanicro 25 steel.

In the paper by Aigner et al. [3] a model based on the $\sqrt{\text{area}}$ concept proposed by Murakami is extended to elevated temperature by introducing an additional exponent.

Poulain et al. [4] examine the dependence and interaction between the effect of pressurized water reactors environment, strain rate and shape of loading waveform on the LCF resistance of a 304L steel in terms of stress-strain response, crack growth, fatigue life and fracture surfaces morphology.

Thermo-mechanical fatigue (TMF) is studied in [5–7]. The paper written by Szmytka et al. [5] provides an overview of a TMF design protocol through the analysis of the specific case of a cylinder head, which summarizes the studies led by the authors over the last five years. Four typical issues in high-temperature design—loading identification, aging and constitutive models, TMF criteria, and validation tests—are addressed and critically analyzed, while some improvements are proposed. In [6], Wagner et al. work on the identification of a criterion for crack initiation to provide the basis for a nondestructive quality control of TMF-loaded porous components with improved statistical safety. The paper by Ghodrat et al. [7] presents an adapted version of Paris' fatigue crack-growth law where cyclic plastic strains at the crack tip are considered as the parameter that controls crack growth. Finally, it is found that the modified Paris' law is able to assess TMF lifetimes of spheroidal graphite iron (SGI) very well for all constraint levels with a single set of parameters. Crack initiation life models are studied in [8], where three approaches (one based on Weibull distributed crack initiation life, the other two based on probabilistic Schmid factor) are presented.

The effect of ageing temperature, initial stress levels and pre-strains on the stress-relaxation ageing behavior of aluminium alloy AA7150-T7751 are investigated through a series of experiments and presented in the work of Cai et al. [9]. The authors discuss a stress relaxation constitutive model with the ability to reproduce stress relaxation curves under different process conditions. The paper by Testa et al. [10] describes a model to estimate the yield stress at different strain rates and temperatures for metals with body-centered-cubic (bcc) structure. Srnec Novak et al. [11] develop a new isotropic model to describe the cyclic hardening/softening plasticity behavior of metals. The proposed model is described with three parameters which were calibrated based on LCF experimental data of CuAg alloy. An improvement is observed with respect to another nonlinear isotropic model generally used in the literature.

Papers in this Special Issue addressed different types of material, from stainless steel studied by [1,2,4,10], to aluminum alloys investigated by [3,6,9], Ni-based supper alloy in [8], spheroidal graphite iron by [7], and copper alloy discussed in [11].

The developed constitutive methods have, in most cases, been compared to experimental data obtained with tensile tests [3], fatigue tests [3,4,8], TMF tests [6,7], creep tests [2], and stress relaxation, aging tests presented in [9].

3. Conclusions

This Special Issue presents a collection of research articles covering the relevant topics in the field of metals plasticity, creep and fatigue at high temperature. As Guest Editors, we hope that these articles may be useful to scientists working in this field to deepen or widen their research.

Acknowledgments: As Guest Editors, we thank all authors who contributed to this Special Issue with high-quality manuscripts that shared results of their research activities. Thanks to Assistant Editor Kinsee Guo for his valuable and inexhaustible support during the preparation of this issue, and thanks also to all staff members of Metals Editorial Office for their management and practical support in the publication process of this issue.

Conflicts of Interest: The authors declare no conflict of interest.

Dedication: This Special Issue is dedicated to the memory of Luciano Moro. He was a promising researcher, a colleague, a friend.

References

1. Liu, D.; Pons, D.J. An explicit creep-fatigue model for engineering design purposes. *Metals* **2018**, *8*, 853. [CrossRef]
2. Kloc, L.; Sklenicka, V.; Dymacek, P. Transient effect in creep of Sanicro 25 austenitic steel and their modelling. *Metals* **2019**, *9*, 245. [CrossRef]
3. Aigner, R.; Garb, C.; Leitner, M.; Stoschka, M.; Grun, F. Application of a $\sqrt{area}$-approach for fatigue assessment of cast aluminum alloys at elevated temperature. *Metals* **2018**, *8*, 1033. [CrossRef]

4. Poulain, T.; de Baglion, L.; Mendez, J.; Henaff, G. Influence of strain rate and waveshape on environmentally-assisted cracking during low-cycle fatigue of a 304L austenitic stainless steel in a PWR water environment. *Metals* **2019**, *9*, 197. [CrossRef]

5. Szmytka, F.; Osmond, P.; Remy, L.; Masson, P.D.; Forre, A.; Hocke, F.X. Some recent advances on thermal-mechanical fatigue design and upcoming challenges for the automotive industry. *Metals* **2019**, *9*, 794. [CrossRef]

6. Wagner, M.; Mosenbacher, A.; Eiber, M.; Hoyer, M.; Riva, M.; Christ, H.J. Thermomechanical fatigue of lost foam cast Al-Si cylinder heads-assessment of crack origin based on the evaluation of pore distribution. *Metals* **2019**, *9*, 821. [CrossRef]

7. Ghodrat, S.; Kalra, A.; Kestens, L.A.I.; Riemslag, T.A.C. Thermo-mechanical fatigue lifetime assessment of spheroidal cast iron at different thermal constraint levels. *Metals* **2019**, *9*, 1068. [CrossRef]

8. Engel, B.; Made, L.; Lion, P.; Moch, N.; Gottschalk, H.; Beck, T. Probabilistic modeling of slip system-based shear stresses and fatigue behavior of coarse-grained Ni-based superalloy considering local grain anisotropy and grain orientation. *Metals* **2019**, *9*, 813. [CrossRef]

9. Cai, Y.; Zhan, L.; Xu, Y.; Liu, C.; Wang, J.; Zhao, X.; Xu, L.; Tong, C.; Jin, G.; Wang, Q.; et al. Stress relaxation aging behavior and constitutive modelling of AA7150-T7751 under different temperatures, initial stress levels and pre-strains. *Metals* **2019**, *9*, 1215. [CrossRef]

10. Testa, G.; Bonora, N.; Ruggiero, A.; Iannitti, G. Flow stress of bcc metals over a wide range of temperature and strain rates. *Metals* **2020**, *10*, 120. [CrossRef]

11. Srnec Novak, J.; De Bona, F.; Benasciutti, D. An isotropic model for cyclic plasticity calibration on the whole shape of hardening/softening evolution curve. *Metals* **2019**, *9*, 950. [CrossRef]

Article

Some Recent Advances on Thermal–mechanical Fatigue Design and Upcoming Challenges for the Automotive Industry

Fabien Szmytka [1],*, Pierre Osmond [2], Luc Rémy [3], Pierre-Damien Masson [2], Agathe Forré [4] and François-Xavier Hoche [2]

[1] Institute for Mechanical Sciences and Industrial Applications (IMSIA-UME), ENSTA Paris, Institut Polytechnique de Paris, 91120 Palaiseau, France
[2] Chassis and Engine Department, Groupe PSA, 78955 Poissy, France
[3] Centre des Matériaux, MINES Paristech, PSL University UMR CNRS 7633, 91003 Evry CEDEX, France
[4] Science and Future Technologies Department, Groupe PSA, 78140 Vélizy-Villacoublay, France
* Correspondence: fabien.szmytka@ensta-paristech.fr; Tel.: +331-6931-9840

Received: 12 June 2019; Accepted: 12 July 2019; Published: 17 July 2019

Abstract: Automotive industry faces numerous evolutions regarding environment regulations and parts reliability. Through the specific case of a cylinder head, actual and forthcoming challenges for low-cycle and thermal–mechanical fatigue design in an industrial context are presented. With a description of current design approaches and highlighting limitations, this work focuses on variable loadings, constitutive models and their interaction with the environment, fatigue criteria, and structure validations, the four major steps to meet a reliable design. The need to carry out extended experimental databases for different complexity levels is emphasized to provide a better understanding of loadings and their impact on the strength of materials and structures, as well as the production of more physically-based models that are easier to identify and lead to higher levels of reliability in the thermal–mechanical design process.

Keywords: thermal–mechanical fatigue; probabilistic design; constitutive models; fatigue criterion; experimental set-ups

---

1. Introduction

In the automotive industry, fatigue-critical parts most of the time exhibit complex geometries and severe mechanical environments. They are subject to loads from various sources that systematically show high levels of randomness. These structures are usually designed for enduring a vehicle usage described as "standard" (thus designating the mean conditions of use plus some standard deviations) as well as for a prescribed number of exceptional or accidental events. The largest car manufacturers produce several million vehicles per year and it is therefore necessary to guarantee a high level of reliability for each of the components (engine, suspensions, etc.) to avoid any risk of mechanical failures. The automotive market is also highly competitive and manufacturers are compelled to optimize the produced parts by constantly seeking the best balance between mass, cost, mechanical resistance, and production times. These constraints then reduce drastically the number of design options. For instance, Engine component development requires multiple optimizations for combustion strategies and emission control systems. At the same time, the applied thermal loads become increasingly severe in terms of maximum temperatures and spatial thermal gradients. These loading conditions lead to the use of materials often at the edge of their capacity and make these parts subject to significant risks of damage by thermal–mechanical fatigue (TMF) [1,2].

For about 20 years, the automotive industry has thus contributed to the development of numerous calculation methods and test protocols to determine the service life of automotive structures subject to coupled thermal–mechanical loadings. They commonly focused on the relevant analysis of boundary conditions and mainly on the continuous improvement of non-linear constitutive models and fatigue criteria. At the heart of the developed methods, there are several fundamental hypotheses that have been put forward for the automotive industry by Constantinescu and co-workers [3] and have since been taken up by several authors [4–8]. A total decoupling between behavior and damage evolution is thus first of all assumed. Viscoplastic shakedown at each point of the structure is then guessed after only a few loading cycles the stabilized hysteresis loop is usually taken as a reference to characterize the damage per cycle. A fatigue criterion based on the density of energy dissipated per stabilized cycle is then commonly used [9]. Figure 1 summarizes common protocols for cylinder heads and emphasizes (in red) on assumptions that tend to simplify the mechanical problem and to jeopardize the reliability of the fatigue lifetime estimation. These drawbacks, that will be detailed, become a real concern as operating temperatures increase and design constraints become more severe. Moreover, recent evolutions in engines due to new usages or regulations induce the employment of different materials and processes for which common methods are therefore inefficient or lead to very conservative designs.

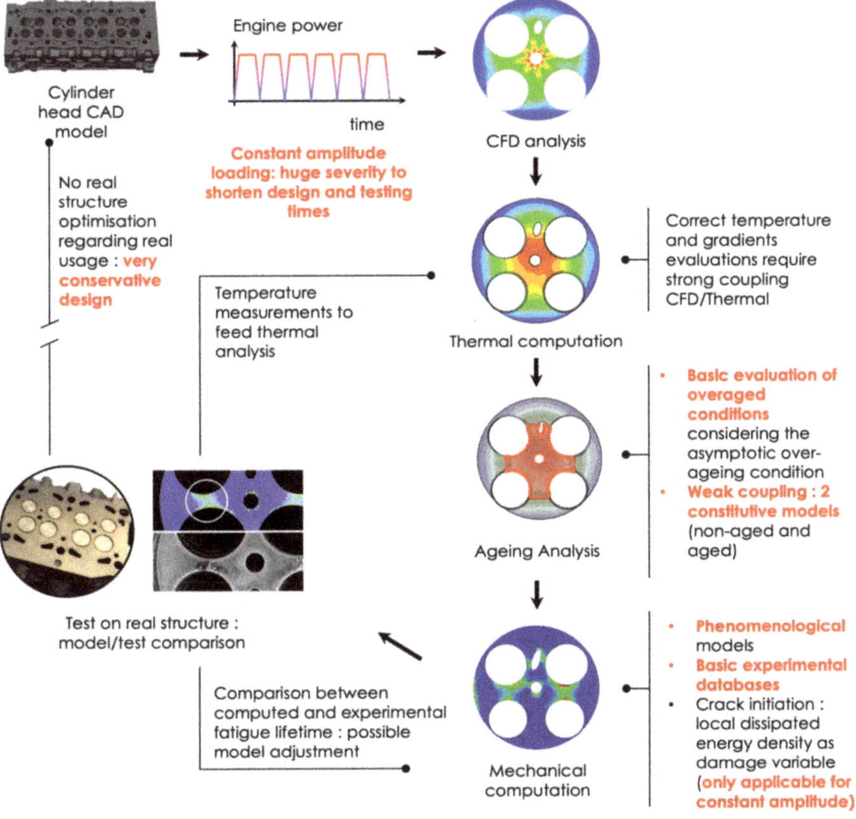

Figure 1. Global strategy for thermal–mechanical fatigue (TMF) design protocols and main weakness (in red) with focus on the fire deck as the main critical zone for TMF (CAD is for Computer Aided Design and CFD is for Computed Fluid Dynamics).

In order to deal with these limitations and improve design processes, car manufacturers, and more particularly Groupe PSA, have regularly investigated model and methods evolutions since

Constantinescu and co-workers developed the first protocol against TMF for automotive parts, beneficiating a better understanding of the structure behavior, enhanced computational capacities, and novel experimental investigation techniques. This article aims therefore at providing a complete upgrade of the TMF design protocol based on studies led by the authors during the last 5 years and of the upcoming challenges mainly through the example of the cylinder head. As depicted in Figure 1, four steps of design—loading identification, aging and constitutive models, TMF criteria, and validation tests—will be addressed, criticized, and improvements will be proposed.

Research on TMF resistance has most often focused on deterministic methods to ensure the strength of structures subjected to thermal and mechanical coupled loads. However, fatigue is a probabilistic phenomenon by nature: manufacturing processes, geometric tolerances, or operating conditions are some examples of variability sources while estimating the lifetime expectancy of an industrial structure. Car usage, even if it is limited by its pure performance, is unique because of the types of used roads, the weather conditions or even its driver's behavior. As a result, thermal–mechanical loads applied to a cylinder head, an exhaust manifold or a piston become probabilistic while the intrinsic fatigue strength of the structure itself is variable (machining, process, distribution of foundry defects). A complete protocol for analyzing the risk of failure of a structure subject to TMF has been developed by some of the authors based on the Strength-Stress Interference Method [10]. In this method, a detailed analysis of vehicle use, through measurements and customer surveys, is carried out to measure the thermal–mechanical loads variability while numerical methods link a succession of macroscopic loads (engine speed and torque) to a local damage in TMF. We will here quickly come back to this method and stress the importance of considering variable amplitude loadings, particularly for aluminum alloys used for cylinder heads, which are sensitive to thermal aging. We will therefore focus on the impact of this type of method while considering the influence of aging as part of a design protocol.

In a second time, constitutive models using inspired-by-physic equations relying on dislocations densities are proposed as a substitute to usual phenomenological model. Main objectives are to cope with a wide variety of loading conditions (temperature, strain rates, and amplitudes) and to facilitate model parameters identification. Micro-crack growth laws and defects population are also introduced in fatigue criterion in place of simple crack initiation lifetime estimation, once again to deal with variable amplitude loading. The impact of such evolutions on the fatigue design protocol is then evaluated and discussed. The reliability of these two latter elements is indeed presented as crucial for proposing non-linear damage accumulation strategies, more appropriate for probabilistic design protocols, particularly when dealing with material exhibiting thermal aging. Simplified methods for numerical integration of such models leading to a more flexible use of complex equations are jointly presented.

Finally, a specific attention is dedicated to an experimental set-up based on a simplified but representative structure (single cylinder head), half the way from specimens to real industrial structures. They reproduce the complexity of real loadings on structures for which the strain and stress analyses are far easier and could quickly lead both to geometries and models improvements. Example of comparison between models and experiment is detailed to underline the previously discussed evolutions.

2. Materials

Cylinder heads are nowadays mainly produced by die casting from aluminum alloys containing silicon, magnesium, and often copper due to their high strength to weight ratio, high thermal conductivity, good casting properties as well as machinability. These alloys are widely used and strengthened by precipitation (T7 heat treatment). Copper can be often added in different amounts in order to enhance alloys properties, especially thermal stabilities. Two primary cast aluminum alloys are mentioned in this article: an AlSi7Cu0.5Mg0.3 and an AlSi7Cu3.5Mg0.1, close to A356 and A319 alloys respectively.

The design protocol used today has been established for AlSi7Cu0.5Mg0.3 T7 aluminum alloy that is widely used for Groupe PSA cylinder heads [11]. Such components can experience operating temperatures that can exceed 250 °C in some critical areas and thus the precipitation treatment temperature. implying that over-aging occurs during service. As seen on Figure 2, these high-temperature incursions cause changes in nature and morphology of hardening precipitation population and consequently the alloys mechanical behavior, resulting in a loss of strength and of fatigue resistance. Vickers microhardness tests are commonly used to represent the effect of thermal aging. Here, samples of material are exposed to different temperature levels for different periods of time. The hardness is then measured inside dendrites (about 20 measurement points per test condition) and an average value is obtained. Figure 3a,d represent these values as a function of temperature and exposure time for the two considered alloys, while typical morphologies of precipitates in T7 and overaged conditions are shown in Figure 3b,c and Figure 3e,f for alloys AlSi7Cu0.5Mg0.3 and AlSi7Cu3.5Mg0.1 respectively. A complete description of microstructure is given in [12].

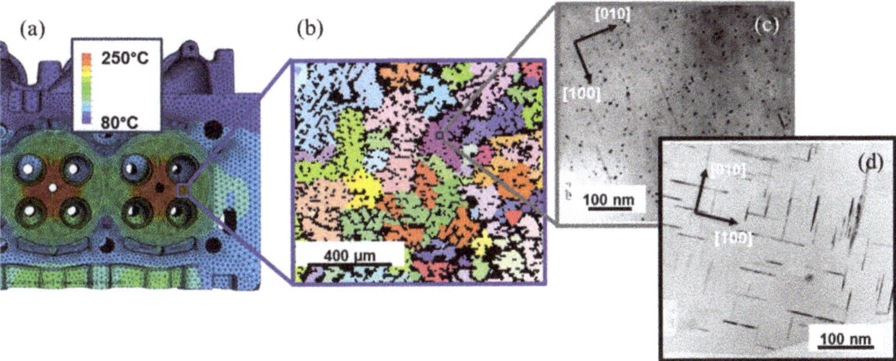

Figure 2. (**a**) Temperature map on the cylinder head fire deck. (**b**) Electron Backscatter Diffraction (EBSD) Inverse pole figure showing the grain microstructure in the inter-valve bridge. (**c**) Alloy AlSiCu0.5Mg0.3 T7, initial hardening precipitates microstructure, (**d**) AlSi7Cu3.5Mg0.1 T7, initial hardening precipitates microstructure.

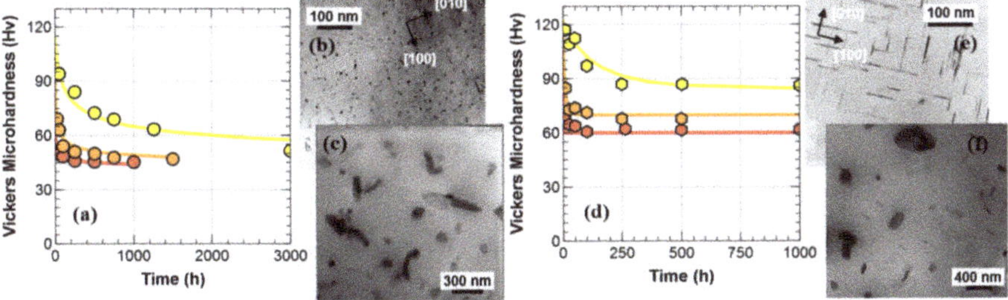

Figure 3. Alloy AlSi7Cu0.5Mg0.3-T7: (**a**) Microhardness evolution with heat treatment time and temperature; (**b**) TEM dark field image of semi-coherent precipitates in initial condition, negative contrast (both image plane and thin foil plane normal directions close to [001] (Al) zone axis); (**c**) TEM bright field image of incoherent precipitates in over-aged condition. Alloy AlSi7Cu3.5Mg0.1 T7: (**d**) Microhardness evolution with heat treatment time and temperature; (**e**) TEM dark field image of semi-coherent precipitates in initial condition, negative contrast (both image plane and thin foil plane normal directions close to [001] (Al) zone axis); (**f**) TEM bright field image of stable precipitates in over-aged condition treated at 300 °C.

Hardening precipitate populations in T7 condition are composed of thin, semi coherent precipitates with morphologies depending on the Cu amount in the alloy. For AlSi7Cu0.5Mg0.3, lath and rod-shaped precipitates, from β and Q precipitates families are observed and shown in Figure 3b in edge view. With addition of Cu, hardening precipitates are plate shaped θ family with semi coherent $\theta\prime$ precipitate in T7 condition, as shown in Figure 3e in side view. With aging, these precipitates evolve to their relative coarse and stable forms, as presented in Figure 3c,f. The effect of Cu composition on the nature and morphologies of hardening precipitates induce significant evolutions of the aging process, as shown in Figure 3a,d. While the stabilized overaged state is almost insensitive to temperature in the alloy AlSi7Cu0.5Mg0.3, this state strongly depends on aging temperature for the AlSi7Cu3.5Mg0.1 alloy.

Its mechanical behavior and the evolution of its mechanical characteristics then explicitly depends on the thermal history, which has a significant influence on the cylinder head design. Finally, the impact on the lifetime analysis is considerable and must be considered in the design protocol. In the light of these elements, it seems obvious that the precise analysis of the loads on the structure to be designed becomes a central point of TMF life analysis methods.

Meanwhile, the lost foam casting process was recently introduced [13] to reduce the processing cost while opening new opportunities for geometry optimization. However, the slower cooling rates during solidification process combined with partial contaminations of liquid aluminum by foam sublimation-induced gas result in coarser microstructures and larger porosity as presented for an AlSi7Cu0.5Mg0.3 alloy on Figure 4. This Figure compares the microstructures obtained by die and lost foam casting. Without necessarily quantitatively analyzing the microstructures—perfectly representative of the processes—the size of the secondary inter-dendritic spaces appears considerably larger for lost foam casting material. This coarser microstructure is accompanied by the presence of large shrinkage pores. It then seems obvious that such microstructures will have a deleterious impact on the lifetime of the structures, even though the chemical composition of the alloys is the same. The methods commonly used to describe fatigue strength can only be challenged, confronted with these new materials to be improved.

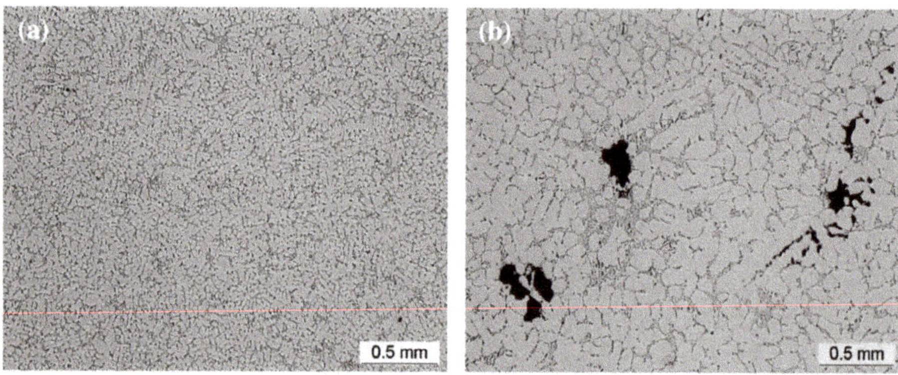

Figure 4. A356-T7 alloy microstructure for die casting (**a**) and lost foam casting (**b**).

3. Loadings Characterization and Its Impact on the Design Process

Developments of TMF design protocol have mainly been achieved within a deterministic framework where loading is well-known and controlled. To put it differently, design loading is chosen to be severe to cover both the majority of loading cases and to reduce the time required for validation tests. The immediate consequence of this choice is the very conservative design of cylinder heads that are too fatigue-resistant for almost all cases of use, expensive and not optimized as underlined on Figure 1. Car usage induces however a very wide variety of loading conditions, implying several degrees of damage and should thus be treated within a necessarily probabilistic framework. In order to reach an optimal design with regard to fatigue lifetime for an automotive component while guaranteeing a high and quantified level of reliability, it is therefore generally necessary to develop numerical and experimental tools to measure the variability of the severity of use. Combined with a precise knowledge of the variability of the mechanical strength of all the parts produced, it then calculates a failure risk that leads to a probabilistic design of the structures while optimizing both cost and mass. This kind of protocol was introduced in 2013 for thermal–mechanical cylinder head fatigue [10], then extended to exhaust manifolds and finally to brake discs [14].

Precise knowledge of loading and its statistical variability must then become the core of any design protocol. One of the central points, in addition to having the most accurate behavior and fatigue models, is thus the analysis and understanding of the random loads to which the structure to be designed is subjected. Estimating the variability of use of an automotive part used in uncontrolled situations by a very diverse population of customers is however a complex task. It is then essential to carry out as many measurements as possible in order to estimate the different moments of the observed probability densities assigned to the different loading conditions. A detailed description of the methods used to build this experimental database is given in [10]. The first step consists in obtaining reliable information on drivers' driving habits by using statistical studies-surveys-on a fairly large panel of representative drivers of the overall customer population. The use of the vehicle can thus be easily categorized by the type of used road (as "highway", "road", "city", and "mountain" [15]). These first categories can then be modulated by the vehicle load or any other parameter likely to modify the fatigue behavior of the component under study. The result is a finite number of elementary life situations. Surveys must then be regularly repeated to estimate, for each of the users, the percentage of time or the number of kilometers that correspond to each of these life situations and thus obtain a statistical measure of the variability in the use of the vehicle. This step makes it possible to build representative usage sequences. It is then a question of quantifying the loading amplitude for each usage and then the damage they cause to the structure according to the severity of the customer's use.

In order to answer to this problem, vehicles are equipped with sensors to be subsequently used either by professional drivers on calibrated paths representative of the identified life situations. A database containing the temporal monitoring of the vehicle's use is then collected. TMF damage to the cylinder head results from the temporal variation in its temperature field and the thermal gradient created during this variation both on the surface and in the core of the material. Depending on the severity of the thermal–mechanical loading, the local behavior in the structure exhibits non-linearities related to viscoplasticity, which makes it impossible to quickly convert the imposed global loading into a local loading for the most TMF critical area. The second step in the analysis of load sequences variability then relies on signal processing for a variable wisely chosen to characterize the severity of loading at the local level. The thermal–mechanical stresses for the engine are controlled in the first order by the thermal exchanges with the combustion gases. The surface heat flux on the fire deck is therefore naturally chosen as a measure of the severity for the cylinder head.

Figure 5 shows an example of the time evolution of the heat flow for the cylinder head. These raw signals cannot be used directly to characterize the overall severity of a time recording. Indeed, they are too noisy to be introduced as boundary conditions for a finite element calculation to compute the local damage. A simplifying treatment must be carried out to consider them as a succession of elementary loading cycles. The elementary load is a simplified signal (often symmetrical, linear per part, and parameterized) calibrated for finite element computation. Signal processing therefore focuses on displaying sequences corresponding to heating, high temperature dwell times, cooling and then cold temperature dwell times, taking care to measure as accurately as possible the transition rates and duration of holding times, which are decisive parameters for describing the viscous behavior for the material. Figure 5 also shows the result of such signal processing, which must integrate the inertial effects of the component, the thermal properties of the materials and the sensitivities of the mechanical response to such loads. We then obtain here loading cycle sequences representative of the use of the cylinder head. This raises the question of how to use this statistical data and how to integrate it in a way that is relevant to the design protocol

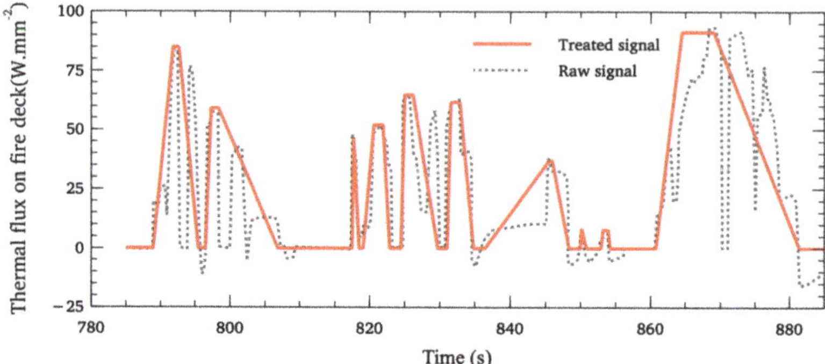

Figure 5. Evolution of thermal flux on cylinder head fire deck with time and signal processing for TMF design analysis.

A strong hypothesis makes each cycle as independent and thus eliminates the influence of the loading order. Each cycle is treated as stabilized and can then be directly related to a damage quantum. At first glance, this hypothesis seems conservative because it cancels out the cumulative plasticity effect which, on a volume element, tends to make it less severe in terms of damage to cycles of low amplitudes following very severe cycles of amplitudes. This also has the advantage of easily generating loading sequences to be used during design without having to worry about the history. If it is validated on some cylinder heads where the variability of the load remains moderate and the thermal aging evolution quite simple [10], it can be questioned in the case of significant local gradients or very high variabilities of the load. This protocol also raises questions when it comes to designing structures with constituent material subjected to thermal aging and when the latter presents distinct overaged states as a function of the history of thermal loading.

In this case, coupling strategies between aging and mechanical behavior will have to be developed. Two simple examples are then proposed in an illustrative form in Figure 6. This step requires in any case a relevant consideration of aging within the mechanical behavior modelling itself, for which some possibilities will be given in the following section. The relevant definition of the loading sequences to be introduced in such numerical methods also remains a largely open question: if effective signal processing is available, it is necessary to consider the statistical representativeness of the sequences and their link with the calculation of local damage. This step will be a key point for improvement in methods in the coming years.

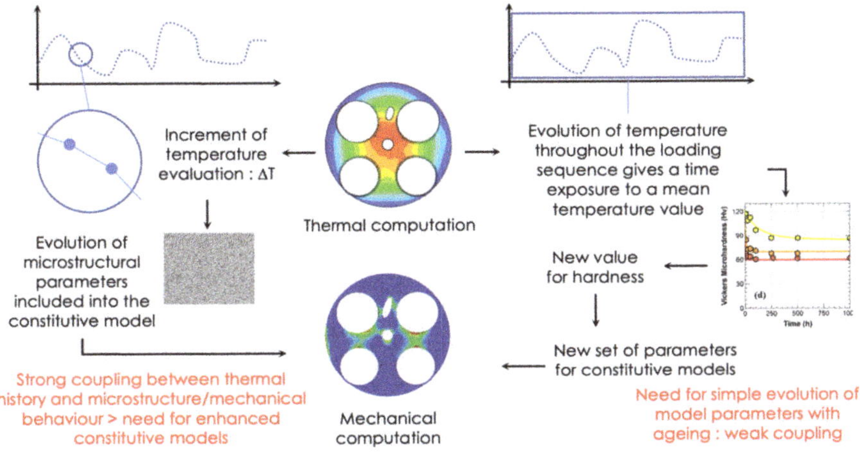

Figure 6. Coupling strategies: some possibilities.

4. Constitutive Models and Aging Conditions

The commonly used design protocol and particularly the way over-aging is considered in the constitutive model, was defined for AlSi7Cu0.5Mg0.3 T7 aluminum alloy that was widely used for PSA cylinder heads [11]. To evaluate the relevance of a constitutive model and establish its choice, it is mandatory here to re-examine thermal aging. Operating temperatures for cylinder heads fire decks can exceed 250 °C in critical areas and for specific loading conditions, a thermal state that can thus exceed the precipitation heat treatment temperature and induce over-aging of precipitates during service, a problem that has received attention in high temperature components [16]. This gives rise to a loss in strength and thus fatigue resistance that is often simply hinted by a drop in local microhardness as easily observed on Figure 3 [17]. For AlSi7Mg0.3 T7, exposure to high temperature during a sufficiently long time results in a micro-hardness that is almost insensitive to maximum temperature. The design protocol therefore usually considers two different constitutive behavior models—more precisely two parameter sets for the same model—for low and high temperature areas in components as exposed in Figure 1. A viscoplastic Chaboche-type model is most of the time chose to represent the aluminum alloy mechanical evolution [18]. Parameters are identified for T7 condition from cyclic hardening tests [19] and the corresponding model is used for simulate low temperature areas. Another set of parameters is then identified for an asymptotic over-aging condition and is used for high temperature areas. Figure 7b highlights the zone corresponding to each identified behavior for a specific cylinder head experiencing severe loading conditions. This procedure may yield conservative predictions of fatigue life depending upon alloy composition or thermo-mechanical fatigue loading conditions. However, it seems limited in the context of random loading sequences and particularly when the stabilized aging state is intrinsically dependent on the temperature history.

A detailed investigation of over-aging conditions and variation in composition has been done during the last ten years [12,17,20]. It combines micro-hardness measurements that are rather sensitive to precipitation conditions and transmission electron microscopy (TEM) to measure precipitate distribution parameters at micro-scale: nature of precipitates, size and volume fraction. Figure 3 illustrates the variation of micro-hardness with temperature and duration of exposure for alloy AlSi7Cu0.5Mg0.3 T7 and AlSi7Cu3.5Mg0.1 T7 (Figure 3a,d). The initial microstructure and after long time over-aging at high temperatures are shown in Figure 3b,c for alloy AlSi7Cu0.5Mg0.3 T7 and Figure 3e,f for alloy AlSi7Cu3.5Mg0.1 T7. A complete description of microstructure evolution is given in [12].

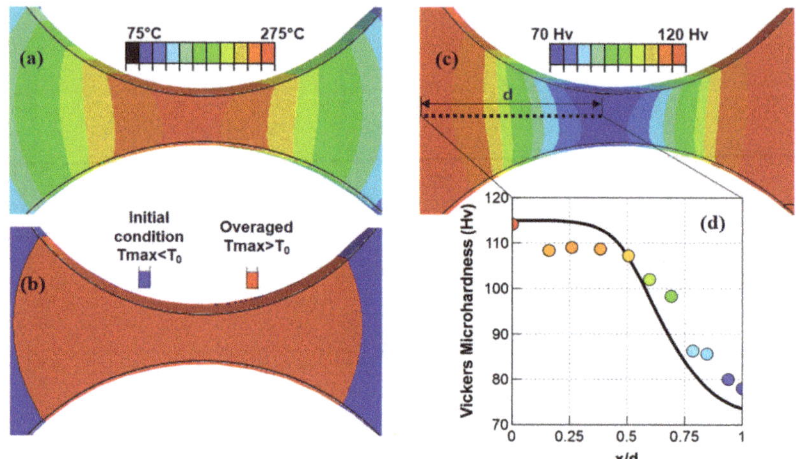

Figure 7. Alloy AlSi7Cu0.5Mg0.3-T7. Detail analysis of aging of a simplified structure of cylinder head submitted to a full thermal load (**a**) thermal map in the inter-valve bridge. (**b**) Temperature areas and aging conditions considered in historical design protocol. The blue area corresponds to low temperature area where initial aging conditions is used, the red one to high temperature area where an asymptotic over-aging conditions is used. (**c**) Microhardness gradient predicted in inter-valve bridge after 10,000 thermal cycles. (**d**) Comparison between experiment (symbols) and simulated (solid curve) microhardness gradient along a half of the inter-valve bridge. The location of microhardness analysis is presented in (**c**).

From this microstructure study, specific over-aging conditions are chosen (for different times and temperatures) and for each condition, the tensile and cyclic stress–strain behavior is characterized at different temperatures. This investigation highlights that micro-hardness can be used as a generalized variable to characterize the thermal aging on both tensile and cyclic yield strength. A microstructural model has thus been developed to predict micro-hardness evolution as a function of exposure time and temperature. The modeled curves are compared with isothermal data in Figure 3a,d, underlying a good correlation.

Once this has been established and the microstructural model validated, it becomes clear that conventional constitutive models cannot accurately reproduce such behavior without clear improvement. An alternative is to move towards microstructure-sensitive models, able to cope with complex thermal aging evolutions. and to establish the link between microstructure variations quantified from TEM observations. The global scheme of this modeling is detailed in Figure 8 and distinguishes between isotropic and kinematic hardening as Chaboche-type model at macroscopic level. As a first contribution to isotropic hardening, this mean-field model uses dislocation density as an internal variable, and it considers dislocation multiplication and dynamic recovery as proposed by Kocks and Estrin [21–23]. The basic Equation is:

$$d\rho = d\gamma_p \frac{1}{b}\left[\frac{1}{L} - y\rho\right] \tag{1}$$

where ρ is stored dislocation density, b Burgers vector, $d\gamma_p$ shear strain increment, L slip distance and y annihilation distance. Static recovery is easily introduced and this type of model formalism with adapted equations was previously shown to be very effective for stainless steels used in exhaust manifolds [24]. In over-aged aluminum, alloys precipitates are non-shearable: dislocations by-pass precipitates by Orowan mechanism and leave loops around them. Equations similar to Equation (1) could be used for the variation of loops density at precipitates and special care is taken of cyclic slip reversibility due to the presence of the loops [25,26]. This mechanism causes additional contributions to isotropic and kinematic hardening, σ_{ppt} and σ_{kin} respectively. The first one varies as the inverse of the distance between precipitates λ_{ppt}, the second one increases as the precipitate volume fraction f_{ppt}, the height of precipitate h_{ppt}, and the number of Orowan's loops n [27,28].

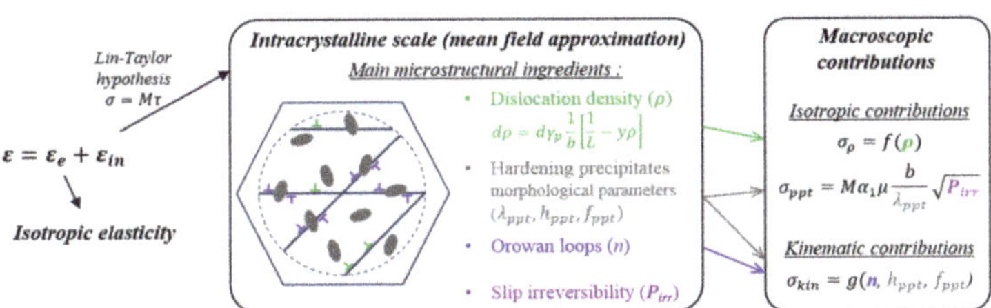

Figure 8. Global scheme for the proposed constitutive model.

A reduced and optimized database combining tensile and cyclic hardening tests [19] with specific observations is proposed. The model parameters are identified from initial and extreme overaged conditions by combining physical constants from literature, average microstructure parameters from TEM observations, and optimized values from experimental database. The model can then be used on a range of alloy compositions and over-aging conditions while keeping almost the entire parameter set initially identified [26]. A few examples, revealing excellent matches between simulation and experiments, are given in Figure 9. The crucial point is to evaluate the real impact of long-term operation on a cylinder head and the differences between the established design procedure and the proposition relying on the real aging conditions. Instead of using a complete engine, a simplified structure described as single-cylinder specimen and detailed in Section 6, is tested under a cyclic thermal load with constant amplitude. Figure 7a illustrates the thermal map of the inter-valve bridge where TMF cracks can occur. We recall that Figure 7b shows the two areas defined in the usual design procedure, corresponding to the application of initial T7 constitutive model and long-term overaged condition.

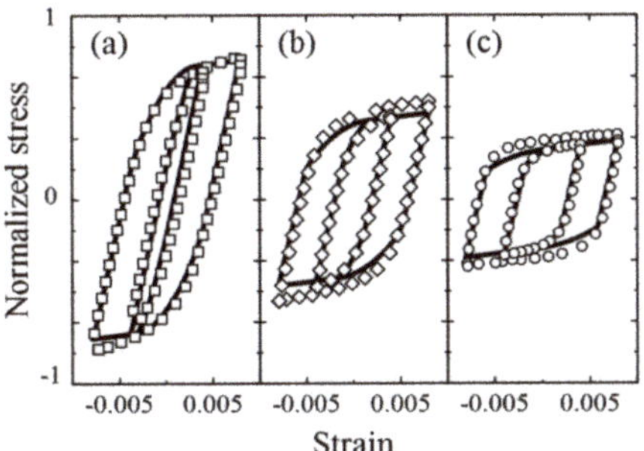

Figure 9. Comparison between experiment (empty symbols) and simulated (solid curve) cyclic stress–strain response obtained for three aging conditions of AlSi7Cu0.5Mg0.3 alloy with a unique identification for our model. Stresses are normalized for confidentiality reasons.

Our microstructural model is first used to predict the micro-hardness for the chosen component and Figure 7c shows the gradient obtained by post-processing of the thermal map. Figure 7d finally shows the comparison between the micro-hardness model predictions and experimental measurements and highlight that model predictions are quite accurate. A major result obtained here is that the hardness in the TMF critical area is proven to be significantly higher than assumed in the current design procedure, with an immediate over-aged condition. Let us now focus more specifically on the structure local behavior and its influence on the lifetime estimation for alloy AlSi7Cu0.5Mg0.3

Figure 10 shows the hysteresis strain-stress loops assumed in the design procedure and estimated combining hardness model and cyclic stress strain data for established and proposed models. The current design process yields a dissipated plastic energy, here qualitatively represented as the area of the hysteresis loop and used as cyclic damage indicator as proposed by Constantinescu et al. [3], about twice higher which results in a lifetime smaller within a factor of 5. This clearly shows the gain in lifetime prediction that can be achieved using a more realistic estimate of over-aging under service conditions as shown in Table 1.

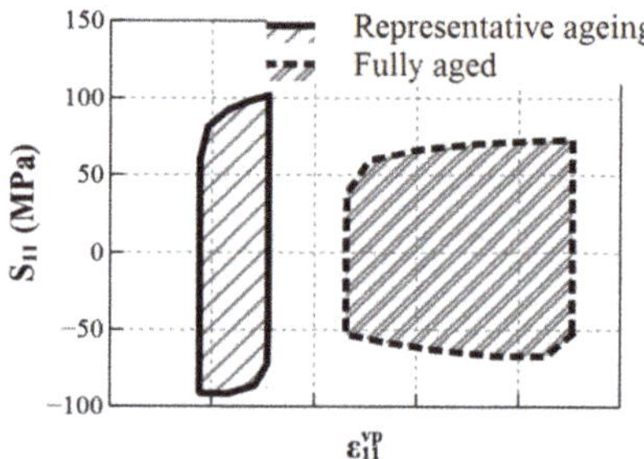

Figure 10. Stress–plastic strain hysteresis loops in critical area for representative aging and asymptotic over-aging conditions, highlighting significative differences between established and proposed model.

Table 1. Effect of aging conditions used for simulating high-temperature areas on dissipated plastic energy and lifetime.

Condition	Inelastic Dissipated Energy Densities per Stabilized Cycle (mJ/mm^3)	Lifetime in Cycles
Representative aging	0.20	27,250
Fully aged	0.38	4750

Once this observation has been made, the bases exist for applying this metallurgical model to loads of variable amplitudes that are much more realistic in terms of actual use. By defining the most statistically representative loading sequences and estimating the associated hardness changes, the estimated lifetime gains could be substantial and allow a more efficient optimization of the cylinder heads.

5. Fatigue Criteria Evolutions

Fatigue criteria are then the last item in design protocols that can be adapted and improved by considering more realistic loading sequences and variable amplitudes. Fatigue data are most of the time obtained on over-aged materials to ensure stable mechanical behavior. However, progressive aging coupled with variable amplitude loading may lead, in the future, to an evolution of fatigue tests protocols and analyses to propose proper fatigue damage estimation throughout robust fatigue criterion. Therefore, in the following we will focus on a proposition on the formalism of models without specifically addressing the theme of thermal aging.

Until recently, and as previously mentioned, one of the main hypotheses commonly used in the automotive industry in TMF design was the decoupling between behavior and damage. Viscoplastic shakedown at each point of the structure is thus assumed after a few loading cycles and it is the "stabilized" cycle that then serves as a reference to characterize the cyclic damage. Constantinescu and co-workers [3] made the choice, based on previous work [9,29], to propose a fatigue criterion relying on the energy density dissipated per stabilized cycle. This choice often leads to quite correct estimation of fatigue lifetime if damage is governed by micro-cracks behavior at the microstructure scale [30]. Despite the recognized ability of this type of criterion to provide a good estimate of fatigue life under a given cyclic load at constant amplitude, Figure 11 reveals for the AlSi7Cu3.5Mg0.1 aluminum alloy a significant dispersion on the correlation between the experimental life and that estimated by the criterion with an average deviation of 41%.

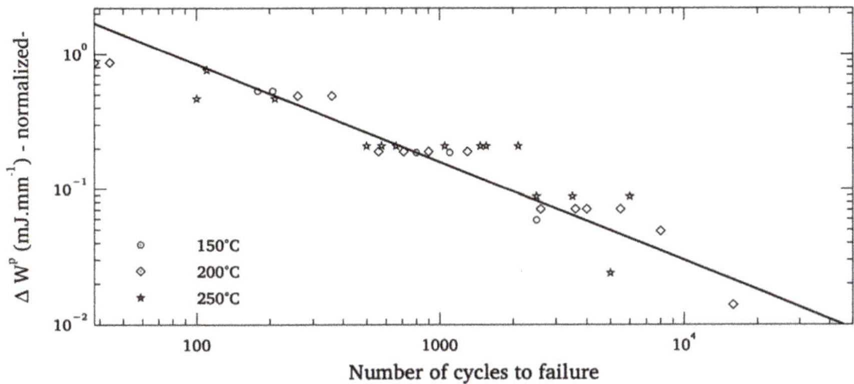

Figure 11. Dispersion observed on aluminum alloys Low-Cycle Fatigue criterion.

Fatigue resistance is a naturally dispersed property for metallic materials and the analysis of the previous Figure 11 obviously does not make it possible to decide on the origin of the observed dispersion (inadequate choice of the criterion mathematical form or damage variable, influence of the material heterogeneities, etc.) and it is maybe simply too ambitious to want to combine under the same formalism and with only two adjustment parameters tests carried out over wide ranges of temperatures and strain amplitude. Indeed, it seems very difficult to characterize the robustness of such criteria without a more detailed analysis of damage mechanisms at the microstructure scale and by mixing experimental and numerical approaches to better understand the physical origin of dispersion. It is also crucial to note that the formalism commonly used and recalled above is not only unable to describe dispersion -and therefore unsuitable for probabilistic design approaches- but also unable to efficiently consider variable amplitude loading. It then appears necessary to build a criterion formalism explicitly incorporating microstructural defect statistics suspected of being responsible for the initiation and micro-propagation mechanisms for foundry metal fatigue: such formalism can easily be integrated into a probabilistic framework but can above all allow a progressive evolution of the damage linked to loads of variable amplitude. Foundry alloys indeed systematically display microstructural heterogeneities which are the consequence of interactions between alloy elements during casting and cooling phases. Aluminum alloys alloyed with silicon, magnesium and copper thus contain many intermetallic compounds [31,32] with size and shape that vary according to the cooling rate of the alloy [33,34].

The complexity of the castings as well as the parameters of the processes controlling the cooling rates can also generate shrinkage or pores [35]. These different elements are randomly distributed in the material with particular statistical distributions of size and shape. The study of their role in fatigue damage has often been studied in the high cycle fatigue regime [36–41] and it seems relevant to evaluate their role in the case of TMF which is generally synonymous with generalized plasticity. Shadan Tabibian and co-worker's work on aluminum alloys obtained by lost foam casting has recently opened up reflection on the role of defects in TMF damage mechanisms and introduced a formalism to take into account of the defect population statistics in the basic evaluation of the fatigue lifetime variability [42]. The main idea developed here is therefore to consider some populations in the material as initial defects, seen as notches from which fatigue damage would develop on a microstructural scale before causing ruin on an engineer's scale as proposed by Charkaluk. A full 3D characterization of the microstructure through X-ray tomography [43] can then be performed in order to measure the size and shape distribution of pores in a Representative Elementary Volume as seen on Figure 12.

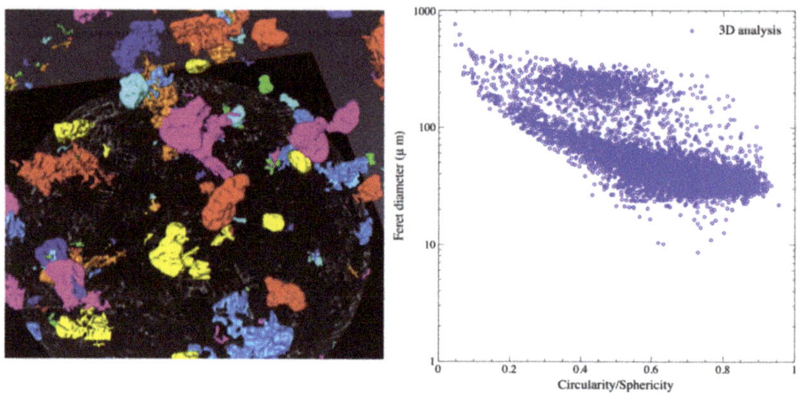

Figure 12. Example of pore detection by X-ray tomography for the studied AlSi7Cu0.5Mg0.3 lost foam cast alloy and pore population ranked by size and sphericity.

A quantitative description of the pore population can be obtained through the measurement of their Feret diameter and sphericity, which could reveal the presence of gassing porosities for lost foam casting processes which are nearly not observed in standard die casting processes. The largest and sharpest pores have to be accurately observed as they play an important part in the fatigue process by acting as crack initiation zone. The pores size distributions could then be introduced in the computational approach used to design engine parts against TMF. Under a basic assumption of growth equation of pores from an initial size a_0 to a final size a_f at N_f characterizing a macroscopic crack initiation (a 2 mm-long crack is usually considered here), a fatigue criterion can be developed under a probabilistic form as proposed by [42] but with an improved formalism regarding the micro-propagation model. Different variable can here be taken as the driving forces for micro-crack-propagation, as proposed by [30,44–47]. The proposition can be summed up in the following improved form of the fatigue criterion:

$$N_f = \int_{a_o}^{a_f} \frac{1}{\left[\left[\frac{w^d}{k_1}\right]^{n_1} a^{n_1} + \left[\frac{w^e}{k_2}\right]^{n_2} a^{n_2}\right]} da \tag{2}$$

$$N_f = \int_{a_o}^{a_f} \frac{1}{\left[\left[\frac{w^d}{k_1}\right]^{n_1} a^{n_1} + \left[\frac{w^e}{k_2}\right]^{n_2} a^{n_2}\right]} da \tag{3}$$

$$w^e = \int_{cycle} sup\left(tr\underline{\underline{\sigma}}(x), 0\right) : \underline{\underline{\varepsilon}}^e(x) dt \tag{4}$$

where w^d and w^e denotes respectively inelastic and elastic dissipated energy densities per stabilized cycle, and $k_1, n_1, k_2,$ and n_2 are material constants. a is a random variable that consists in the observed statistical distribution of pore size in the tested volume and finally to the risk of finding a pore of a given size. The observed then analyzed volumes are however often of different sizes from those of fatigue specimens. Statistical treatments then correct the observed distribution in order to and to become independent of the size of the observed areas [48]. Corrected pore size distribution can be directly integrated in the semi-probabilistic fatigue criterion as probability of occurrence while material parameters are identified from the standard macroscopic low-cycle fatigue and TMF experiments.

This criterion better reflects the effects of mean stresses, which are very important for uniaxial fatigue tests as well as for crack propagation tests, especially under conditions where plasticity is widespread. The particular mathematical form of Equation (2) forbids a conventional lifetime analysis. The number of cycles to failure is therefore linked to an energy density couple and a defect-size distribution. When crack propagation is investigated on laboratory specimens, a direct access to the parameters is possible. In other cases, the simple numerical identification of the parameters of such a model is quite complex because no analytical description is possible. While experimental data are missing, treatment is then carried out for classes of pore size. For a given pore size and for a given set of parameters $(k_1; n_1; k_2; n_2)$, the estimated lifetime is easily calculated. A histogram and a complete distribution of the number of cycles to failure are then obtained. For a given experimental couple $\left(w^d; w^e\right)$, the number of cycles to failure with the highest probability is then finally compared to the experimental lifetime. A least-squares optimization is then performed to obtain the most relevant parameter set.

Lifetime prediction results can then be compared with laboratory specimens. Following this simple principle, good matching is obtained between estimated and experimental lifetimes as seen on Figure 13. A micro-propagation crack growth law, closer to physics, therefore makes it possible to obtain a more realistic estimate of the fatigue lifetime scattering and certainly more in line with the experiment [42]. A formalism that introduces both micro-propagation cracks growth law with driving forces composed from macroscopic energies combined with statistics of microstructural features taken as initial defects yields promising results. The introduction of the role of microstructure in damage evolution seems unavoidable for foundry materials, but the proposed protocol is most of time faced with the difficulty of obtaining reliable experimental results. It therefore seems mandatory, in view of these initial results, to add crack propagation tests on SEN-type specimens [30,47–50] to have access to reliable laws and to carry out test campaigns systematically proposing repetitions of one or more tests to obtain first estimates, even partial, of the effective fatigue lifetime scattering of the couple material/process to be evaluated. In the case of materials sensitive to thermal aging, the choice of relevant tests to be carried out is all the more crucial, as is the need to have a large database. The development of strategies to maximize the information obtained on fatigue resistance by minimizing the number of tests to be performed remains an important challenge for the coming years.

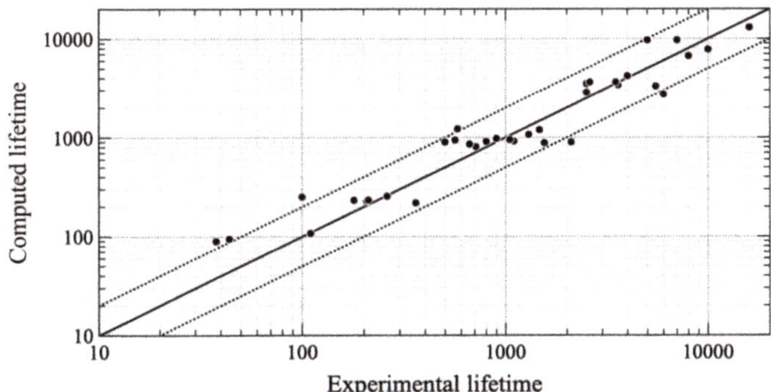

Figure 13. Computed lifetimes vs experimental lifetimes in number of cycles to failure for an AlSi7Cu3.5Mg0.1 alloy.

The use of micro-propagation laws also opens the way to a new formalism for criteria and introduces a notion that until now has often been set aside from design protocols for TMF: non-linear accumulation of damage. For the sake of simplicity, the accumulation of damage is generally described by Miner's rule [51] without considering the role of the loading history. This hypothesis, which is often conservative, is also perfectly adapted to the standard formalism of the criteria. A propagation law, because it incorporates an initial crack size and varies the propagation rate according to the crack size, is more suitable for estimating the historical effects of loading and also variations related to thermal aging. However, it has the major drawback, for a design process that has to be the fastest as possible, of having to compute more loading cycles. The analyses of a realistic sequence of loading cycles as seen on Figure 5 however undoubtedly underlines the real need to change the paradigm from a hypothesis of almost systematic structural shakedown under a single type of loading cycle, assumed representative and involving a linear accumulation of damage, to longer evolutions based on statistically representative sequences and non-linear cumulative damage. It will then be a question of estimating as well as possible the sequences of cycles to be analyzed in a compromise to be found between the computation times and the confidence given to the estimation of the mechanical resistance

6. From Experimental Specimens to Real Structures: A Validation Strategy

The validation strategy of an industrial structure submitted to thermal–mechanical loads is usually basic. Constitutive models and fatigue criterion are validated on laboratory specimen with tests close to the ones that are used to identify their parameters [2,48]. These tests, due to their ease of implementation, deeply simplify real structural loading conditions. Thermal gradients, which are most often responsible for TMF damage, are suppressed in order to provide a stress/strain direct analysis. When these gradients become very severe and when variable amplitude loadings existence is well-established, it is mandatory to include them in an objective comparison of material/process solutions. Final tests on real structure are also performed but due, to their difficult analyses, are unsuitable to really optimize models.

In these cases, it becomes necessary to build specimens closer to the structure to be designed and therefore often more complex in terms of geometry and loading conditions. In the automotive world, a first solution has been proposed by Constantinescu and his co-authors [3]: a clamped heated specimen that reproduces loading conditions observed on a cylinder head but with gradients often too severe. The last part of the validation strategy against TMF for cylinder heads uses an experimental analysis of a "single-cylinder" specimen as seen on Figure 14 which provides loading conditions as close as possible to the structure ones while respecting both the microstructure and the surface roughness in the critical zone. A blowtorch is used to heat the fire deck and is first designed to reproduce the thermal loading generated by the engine start and stop. The single-cylinder stands in front of the blowtorch during the heating phase then draw aside from the blowtorch to reproduce a natural cooling (like a real cylinder head, the specimen is cooled by a dedicated water circuit maintained at 20 °C). During cooling, Damage evolution on the surface is monitored thank to a camera that detects cracks macroscopic initiation. To correctly reproduce real cylinder heads rigidity, the single-cylinder specimen includes seats and valves guides both shrink-fitted on the structure. A support is added to represent the real interface with cylinder block. In order to evaluate the representativeness of the thermal loadings, welded thermocouples carry out a real-time thermal loading control.

Figure 14. Single-cylinder specimens and experimental set-up with blowtorch heating.

By controlling the blowtorch power, it could be possible to modulate the loading amplitude but so far unfortunately impossible to apply complex sequences of variable loadings. However, the set-up evaluates the evolution of the macroscopic damage on the structure, and then, thanks to a comprehensive finite element model, to validate the design strategy, including constitutive models and damage approaches. The numerical part of the validation strategy is based on the sequence described in Figure 1. All the improvements that have been previously proposed can be easily integrated into this protocol.

The first step of the calculation is a transient thermal analysis performed by Abaqus as seen on Figure 15, corresponding to the heating and cooling phases of the test on the single-cylinder. The calculation is adapted to fits with the experimental temperature measurements. As previously stressed, an accurate temperature field during a cycle is mandatory to analyze thermal aging and then to obtain the correct mechanicals stresses. As presented in Section 4, the maximum temperature field during heating determines the aging condition of the structure. Then, material properties are updated for each integration point. The mechanical calculation considers the evolution of the temperature field during the loading cycle as boundary conditions (weak coupling) to obtain the structure local mechanical evolution: a complete shakedown is usually observed after 3 to 10 loading cycles at constant temperature amplitude.

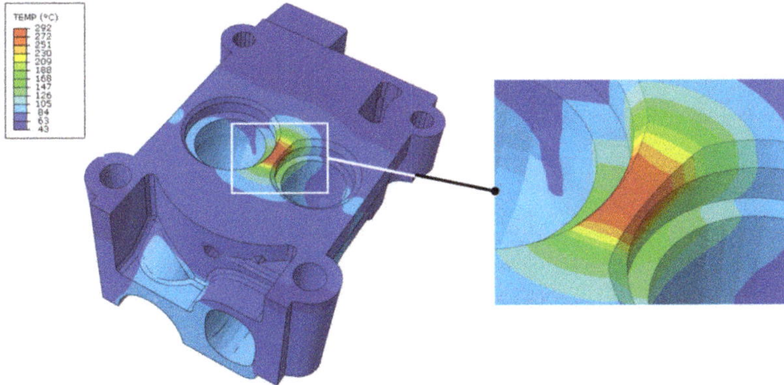

Figure 15. Single-cylinder specimen thermal computation and thermal gradient localization.

Finally, a TMF criterion is applied to estimate the lifetime of the component and detect the critical zones as seen on Figure 16. These results highlight that the value of the dissipated energy density is of course dependent on the mesh size. It is therefore mandatory to adjust the probability of finding pores taken as initiation defects in an element volume V_c to obtain reliable results when using our proposed formalism for fatigue criteria. More precisely, one must compute the probability of the presence of at least one pore of a given size in the critical volume V_c (as explained in [49]) and estimate the associated lifetime using Equation (2). This method, applied on AlSi7Cu3.5Mg0.1 single cylinder specimens, provides encouraging results as highlighted on Figure 17. A good estimation of the lifetime mean value is obtained and a first approximation of the standard deviation could also be underlined.

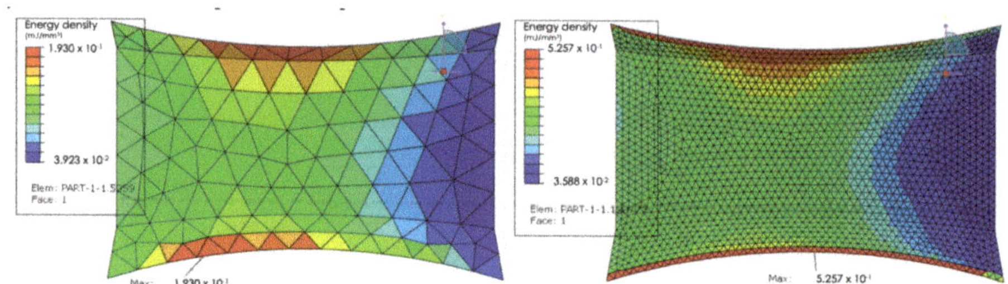

Figure 16. Damage localization for different kind of meshes.

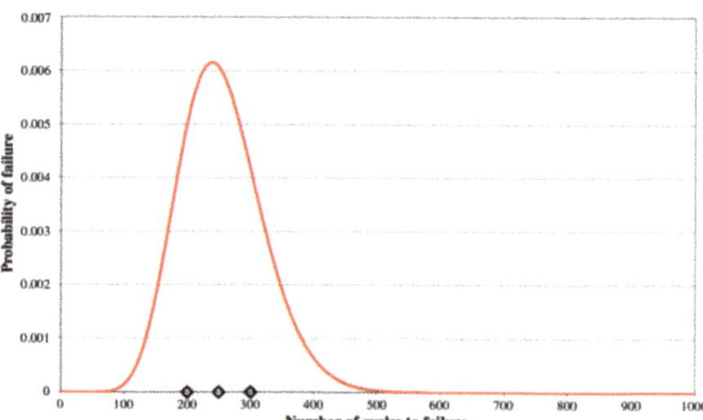

Figure 17. Comparison between experimental results (fatigue crack initiation on single-cylinder specimens represented by diamonds) and the foreseen scattering for fatigue lifetime.

Finally, it is crucial to underline that the use of elasto-viscoplastic constitutive models combining specific equations such as those presented above, induces a major difficulty for finite element analyses. Such complex models are usually not integrated in industrial solvers as Abaqus which however offer to code a user model. This step requires the necessary complex writing of a numerical integration scheme, widely documented in the literature [52] and already applied for physically-based models in the automotive industry [19,53]. The integration procedures are based on a fully implicit integration scheme associated with a radial return, which has been proved unconditionally stable and faster for temperature dependent yield limit and viscous flow. Different schemes could then be used with strong differences in complexity, time computation and precision [54]. If in the past, it was mandatory to write and then code the entire integration scheme in this case, generally including the calculation and then the inversion of a Jacobian matrix, tools are now available to facilitate this step and thus facilitate access to the use of more advanced constitutive models such as MFRONT [55] or Zset [56]. This simplification of numerical integration is a major step in facilitating the use of more reliable models that are essential for representing complex loads with variable amplitudes.

7. Conclusions

Since the late 1990s, the automotive industry has developed design protocols against thermomechanical fatigue for engine parts. These protocols were based on design and validation loads, generally identical, and deliberately chosen as severe to ensure the necessary safety margins and avoid any failure in use. These protocols make it difficult to optimize the designed structures and tend to provide a high level of conservatism.

The improvement of knowledge on foundry materials, coupled with the possibility of making measurements more and more easily and the increase in the capacity of computing means, opens the way to probabilistic design protocols. The latter make it possible to evaluate a risk of failure but above all to introduce, in the protocols, the necessarily random nature of car loads. This point is thus the key to new design protocols and above all the gateway to many model improvements, which will allow a real optimization of the designed structures and therefore a net decrease in conservatism. Figure 18 summarize the proposed improvements made to well-established TMF design protocol.

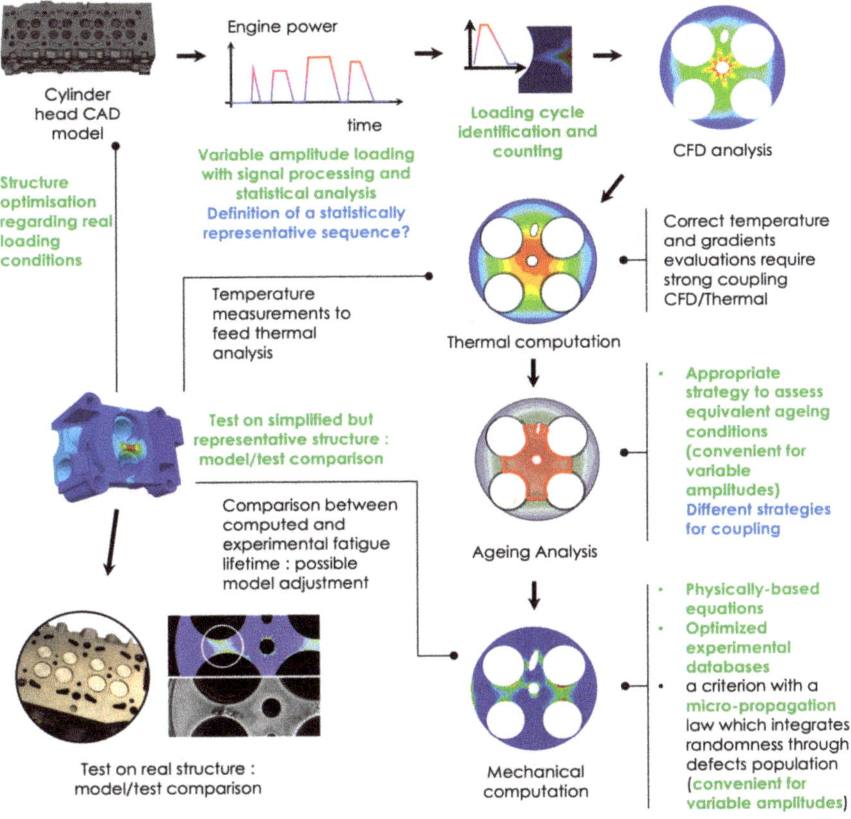

Figure 18. Summary of design protocol improvements (in green) and questions still to be answered (in blue). (CAD is for Computer Aided Design and CFD is for Computed Fluid Dynamics).

At the origin of the reconsideration of the usual protocols, we underlined the introduction of aluminum alloys richer in copper and therefore better adapted to the increasing temperatures of use. These materials have a different sensitivity to thermal aging than those observed on previous alloys with a dependence of the over-aged state on the aging temperature. This observation directly challenges the computation strategies based on the definition of two different stable metallurgical zones from the beginning of the calculation. In addition, it also fails to use a very severe design cycle that will then induce a high and unacceptable level of conservatism to achieve true optimization for the cylinder head. The precise analysis of automotive loads also highlights their high variability with thermal and mechanical amplitudes that are often lower than those used in conventional design protocols. We therefore proposed to establish loading sequences statistically representative of actual use based on complete measurements of the various usage situations and signal processing appropriate for thermomechanical fatigue.

Considering the reality of the thermal and mechanical loadings and their variability necessary modifies the way of designing structures against thermomechanical fatigue. The use of simplifying assumptions to consider the behavior of aluminum alloys and their over-aging is now incompatible with the need to produce reliable and robust design protocols. In fact, the in-depth analysis of loading conditions and the development of models that are as close to physical reality as possible for both behavior and damage are nowadays indispensable ways of ensuring reliable fatigue design.

We therefore proposed an original solution to model the behavior of aluminum alloys using a metallurgical model linking microhardness to the thermal aging condition. By introducing a dislocation density as an internal variable in the constitutive model and using an adapted experimental database, we propose a reliable constitutive model that successfully reproduces the alloy behavior for many aging conditions without increasing the number of parameters to be identified and by introducing physical constants from the literature, the evolution of which with temperature is well known. This model is perfectly adapted to loads of variable amplitudes while the associated numerical strategy still needs to be refined and in particular the strength of the coupling to be introduced to obtain the best compromise between the accuracy of the results and the duration of the calculation. In a complementary way and once again to follow behaviors that evolve over different amplitude cycles, we propose to introduce fatigue criteria once again related to the physics of damage. Based on the probability of finding pores of different sizes in a given volume, we introduce a probabilistic criterion based on a micro-propagation model. This proposal is perfectly in line with the variable nature of the load and allows the thermomechanical problem to be dealt with by means of an accumulation of non-linear damage.

Examining these results with the search of an always better knowledge of thermal loads and the possibility of implementing more complex tests also pave the way for "structure" type tests, which are proving to be essential in future design strategies. At the heart of the digital strategy, these will first of all validate and fault models and then necessarily make them evolve. These models will then make it possible, in a virtuous loop, to improve experimental protocols in order to produce more reliable and instructive tests. By concentrating all the richness of the real loads in a single test, the way also opens up to the systematic study of experiential lifetime scattering, essential for probabilistic design. These tests can also lead to systematic optimization of cylinder heads, in line with the real usage of these structures.

The search for intermediate specimens between the industrial structure subjected to often complex use constraints and the simple volume element of the fatigue specimen will therefore undoubtedly constitute a field of research in the coming years, accelerated by the emergence of experimental techniques for fatigue such as image correlation or infrared thermography. Today, it is a necessary step to improve constitutive models and fatigue criteria while accessing to a fully integrated consideration of the role of microstructure in thermo-mechanical fatigue damage mechanisms.

Finally, it is necessary to stress that all these developments will have to be accompanied by appropriate numerical strategies. If computing powers tend to increase regularly, it seems illusory to think that numerical models will see their degree of freedom increase forever. Digital development will have to be more agile and adapted to design protocols evolutions: considering more variability sources and introducing more physics in models must be accompanied with reasoned calculations with limited size and that are really useful to design processes.

Author Contributions: Investigation, F.S., P.O., L.R., P.-D.M., A.F. and F.-X.H.; Methodology, F.S., P.O., L.R., P.-D.M., A.F. and F.-X.H.; Writing–original draft, F.S., P.O., L.R. and A.F.; Writing–review & editing, F.S., P.O., L.R. and A.F.

Funding: This research received no external funding excepted from the part dealing with thermal ageing that was partially funding by a CIFRE grant from French National Association of Research and Technology.

Conflicts of Interest: The authors declare no conflict of interest.

References

1. Charkaluk, E.; Rémy, L. Thermal Fatigue. In *Fatigue of Materials and Structures*; John Wiley & Sons, Ltd.: Hoboken, NJ, USA, 2013; pp. 271–338.
2. Rémy, L. Thermal-mechanical fatigue (including thermal shock). In *Comprehensive Structural Integrity*; Elsevier: Amsterdam, The Netherlands, 2003; Volume 5, pp. 113–199.
3. Constantinescu, A.; Charkaluk, E.; Lederer, G.; Verger, L. A computational approach to thermomechanical fatigue. *Int. J. Fatigue* **2004**, *26*, 805–818. [CrossRef]
4. Su, X.; Zubeck, M.; Lasecki, J.; Engler-Pinto, C.; Tang, C.; Sehitoglu, H.; Allison, J. *Thermal Fatigue Analysis of Cast Aluminum Cylinder Heads*; SAE Technical Paper 2002-01–0657; SAE International: Warrendale, PA, USA, 2002.
5. Zhao, W.M.; Zhang, W.Z. Study on Thermal Fatigue Life Prediction of Cylinder Head. *Adv. Mater. Res.* **2012**, *433–440*, 3–8. [CrossRef]
6. Thalmair, S.; Thiele, J.; Fischersworring-Bunk, A.; Ehart, R.; Guillou, M. *Cylinder Heads for High Power Gasoline Engines—Thermomechanical Fatigue Life Prediction*; SAE Technical Paper 2006-01–0541; SAE International: Warrendale, PA, USA, 2006.
7. Trampert, S.; Gocmez, T.; Pischinger, S. Thermomechanical Fatigue Life Prediction of Cylinder Heads in Combustion Engines. *J. Eng. Gas Turbines Power* **2008**, *130*, 012806. [CrossRef]
8. Nagode, M.; Längler, F.; Hack, M. A time-dependent damage operator approach to thermo-mechanical fatigue of Ni-resist D-5S. *Int. J. Fatigue* **2011**, *33*, 692–699. [CrossRef]
9. Skelton, R.P. Energy criterion for high temperature low cycle fatigue failure. *Mater. Sci. Technol.* **1991**, *7*, 427–440. [CrossRef]
10. Szmytka, F.; Oudin, A. A reliability analysis method in thermomechanical fatigue design. *Int. J. Fatigue* **2013**, *53*, 82–91. [CrossRef]
11. Verger, L.; Constantinescu, A.; Charkaluk, E. Thermomechanical fatigue design of aluminium components. In *European Structural Integrity Society*; Rémy, L., Petit, J., Eds.; Elsevier: Amsterdam, The Netherlands, 2002; Volume 29, pp. 309–318.
12. Osmond, P.; Rémy, L.; Nazé, L.; Köster, A.; Hoche, F.-X. Effect of thermal ageing of precipitation and mechanical behaviour of two industrial particle hardened cast aluminium alloys. *Mater. Sci. Eng. A* **2019**. to be submitted.
13. Shivkumar, S.; Wang, L.; Apelian, D. The lost-foam casting of aluminum alloy components. *JOM* **1990**, *42*, 38–44. [CrossRef]
14. Augustins, L.; Billardon, R.; Hild, F. Constitutive model for flake graphite cast iron automotive brake discs: From macroscopic multiscale models to a 1D rheological description. *Contin. Mech. Thermodyn.* **2016**, *28*, 1009–1025. [CrossRef]
15. Thomas, J.J.; Perroud, G.; Bignonnet, A.; Monnet, D. Fatigue design and reliability in the automotive industry. In *European Structural Integrity Society*; Marquis, G., Solin, J., Eds.; Elsevier: Amsterdam, The Netherlands, 1999; pp. 1–11.
16. Chateau, E.; Rémy, L. Oxidation-assisted creep damage in a wrought nickel-based superalloy: Experiments and modelling. *Mater. Sci. Eng. A* **2010**, *527*, 1655–1664. [CrossRef]
17. Tabibian, S.; Charkaluk, E.; Constantinescu, A.; Guillemot, G.; Szmytka, F. Influence of process-induced microstructure on hardness of two Al–Si alloys. *Mater. Sci. Eng. A* **2015**, *646*, 190–200. [CrossRef]
18. Chaboche, J.L. A review of some plasticity and viscoplasticity constitutive theories. *Int. J. Plast.* **2008**, *24*, 1642–1693. [CrossRef]
19. Szmytka, F.; Rémy, L.; Maitournam, H.; Köster, A.; Bourgeois, M. New flow rules in elasto-viscoplastic constitutive models for spheroidal graphite cast-iron. *Int. J. Plast.* **2010**, *26*, 905–924. [CrossRef]
20. Hoche, F.-X.; Nazé, L.; Rémy, L.; Köster, A.; Osmond, P. Effect of thermo-mechanical fatigue on precipitation microstructure in two particle hardened cast aluminum alloys. *Metall. Mater. Trans. A* **2019**. to be submitted.
21. Kocks, U.F. The relation between polycrystal deformation and single-crystal deformation. *Metall. Mater. Trans. B* **1970**, *1*, 1121–1143. [CrossRef]
22. Kocks, U.F. Laws for Work-Hardening and Low-Temperature Creep. *J. Eng. Mater. Technol.* **1976**, *98*, 76–85. [CrossRef]

23. Estrin, Y.; Braasch, H.; Brechet, Y. A Dislocation Density Based Constitutive Model for Cyclic Deformation. *J. Eng. Mater. Technol.* **1996**, *118*, 441–447. [CrossRef]

24. Rémy, L.; Szmytka, F.; Bucher, L. Constitutive models for bcc engineering iron alloys exposed to thermal–mechanical fatigue. *Int. J. Fatigue* **2013**, *53*, 2–14. [CrossRef]

25. Han, W.Z.; Vinogradov, A.; Hutchinson, C.R. On the reversibility of dislocation slip during cyclic deformation of Al alloys containing shear-resistant particles. *Acta Mater.* **2011**, *59*, 3720–3736. [CrossRef]

26. Osmond, P.; Rémy, L.; Nazé, L. Microstructure sensitive model of cyclic stress-strain behavior of particle-hardened cast aluminium alloys: Effect of non-shearable precipitate microstructure at ambient temperature. *Mater. Sci. Eng. A* **2019**. to be submitted.

27. Orowan, E. Discussion on internal stresses. In Proceedings of the Symposium on Internal Stresses in Metals and Alloys, London, UK, 15–16 October 1948.

28. Brown, L.M.; Stobbs, W.M. The work-hardening of copper-silica. *Philos. Mag. J. Theor. Exp. Appl. Phys.* **1971**, *23*, 1185–1199.

29. Garud, Y.S. A New Approach to the Evaluation of Fatigue Under Multiaxial Loadings. *J. Eng. Mater. Technol.* **1981**, *103*, 118–125. [CrossRef]

30. Rémy, L.; Alam, A.; Haddar, N.; Köster, A.; Marchal, N. Growth of small cracks and prediction of lifetime in high-temperature alloys. *Mater. Sci. Eng. A* **2007**, *468–470*, 40–50. [CrossRef]

31. Closset, B.; Gruzleski, J.E. Structure and properties of hypoeutectic Al-Si-Mg alloys modified with pure strontium. *Metall. Trans. A* **1982**, *13*, 945–951. [CrossRef]

32. Gustafsson, G.; Thorvaldsson, T.; Dunlop, G.L. The influence of Fe and Cr on the microstructure of cast Al-Si-Mg alloys. *Metall. Trans. A* **1986**, *17*, 45–52. [CrossRef]

33. Murali, S.; Raman, K.S.; Murthy, K.S.S. Effect of magnesium, iron (impurity) and solidification rates on the fracture toughness of Al7Si0.3Mg casting alloy. *Mater. Sci. Eng. A* **1992**, *151*, 1–10. [CrossRef]

34. Tabibian, S.; Charkaluk, E.; Constantinescu, A.; Oudin, A.; Szmytka, F. Behavior, damage and fatigue life assessment of lost foam casting aluminum alloys under thermo-mechanical fatigue conditions. *Procedia Eng.* **2010**, *2*, 1145–1154. [CrossRef]

35. Ammar, H.R.; Samuel, A.M.; Samuel, F.H. Porosity and the fatigue behavior of hypoeutectic and hypereutectic aluminum–silicon casting alloys. *Int. J. Fatigue* **2008**, *30*, 1024–1035. [CrossRef]

36. Murakami, Y.; Endo, M. Effects of defects, inclusions and inhomogeneities on fatigue strength. *Int. J. Fatigue* **1994**, *16*, 163–182. [CrossRef]

37. Le, V.-D.; Morel, F.; Bellett, D.; Saintier, N.; Osmond, P. Multiaxial high cycle fatigue damage mechanisms associated with the different microstructural heterogeneities of cast aluminium alloys. *Mater. Sci. Eng. A* **2016**, *649*, 426–440. [CrossRef]

38. Le, V.-D.; Morel, F.; Bellett, D.; Saintier, N.; Osmond, P. Simulation of the Kitagawa-Takahashi diagram using a probabilistic approach for cast Al-Si alloys under different multiaxial loads. *Int. J. Fatigue* **2016**, *93*, 109–121. [CrossRef]

39. Wang, Q.G.; Apelian, D.; Lados, D.A. Fatigue behavior of A356-T6 aluminum cast alloys. Part I. Effect of casting defects. *J. Light Met.* **2001**, *1*, 73–84. [CrossRef]

40. Mayer, H.; Papakyriacou, M.; Zettl, B.; Stanzl-Tschegg, S.E. Influence of porosity on the fatigue limit of die cast magnesium and aluminium alloys. *Int. J. Fatigue* **2003**, *25*, 245–256. [CrossRef]

41. Dezecot, S.; Brochu, M. Microstructural characterization and high cycle fatigue behavior of investment cast A357 aluminum alloy. *Int. J. Fatigue* **2015**, *77*, 154–159. [CrossRef]

42. Charkaluk, E.; Constantinescu, A.; Szmytka, F.; Tabibian, S. Probability density functions: From porosities to fatigue lifetime. *Int. J. Fatigue* **2014**, *63*, 127–136. [CrossRef]

43. Dezecot, S.; Buffiere, J.; Koster, A.; Maurel, V.; Szmytka, F.; Charkaluk, E.; Dahdah, N.; el Bartali, A.; Limodin, N.; Witz, J. In situ 3D characterization of high temperature fatigue damage mechanisms in a cast aluminum alloy using synchrotron X-ray tomography. *Scr. Mater.* **2016**, *113*, 254–258. [CrossRef]

44. Tabibian, S.; Charkaluk, E.; Constantinescu, A.; Szmytka, F.; Oudin, A. TMF criteria for Lost Foam Casting aluminum alloys. *Fatigue Fract. Eng. Mater. Struct.* **2013**, *36*, 349–360. [CrossRef]

45. Maurel, V.; Rémy, L.; Dahmen, F.; Haddar, N. An engineering model for low cycle fatigue life based on a partition of energy and micro-crack growth. *Int. J. Fatigue* **2009**, *31*, 952–961. [CrossRef]

46. Bourbita, F.; Rémy, L. A combined critical distance and energy density model to predict high temperature fatigue life in notched single crystal superalloy members. *Int. J. Fatigue* **2016**, *84*, 17–27. [CrossRef]

47. Rémy, L. Low cycle fatigue of alloys in hot section components: Progress in life assessment. *Procedia Struct. Integr.* **2019**, *14*, 3–10. [CrossRef]
48. Grison, J.; Remy, L. Fatigue failure probability in a powder metallurgy Ni-base superalloy. *Eng. Fract. Mech.* **1997**, *57*, 41–55. [CrossRef]
49. Dezecot, S.; Rambaudon, M.; Koster, A.; Szmytka, F.; Maurel, V.; Buffiere, J.-Y. Fatigue crack growth under large scale yielding condition in a cast automotive aluminum alloy. *Mater. Sci. Eng. A* **2019**, *743*, 87–97. [CrossRef]
50. Merhy, E.; Rémy, L.; Maitournam, H.; Augustins, L. Crack growth characterisation of A356-T7 aluminum alloy under thermo-mechanical fatigue loading. *Eng. Fract. Mech.* **2013**, *110*, 99–112. [CrossRef]
51. Miner, M.A. Cumulative damage in fatigue. *J. Appl. Mech.* **1945**, *12*, 159–164.
52. Simo, J.C.; Hughes, T.J.R. *Computational Inelasticity*; Springer: New York, NY, USA, 1998.
53. Szmytka, F.; Forré, A.; Augustins, L. A time increment control for return mapping algorithm applied to cyclic viscoplastic constitutive models. *Finite Elem. Anal. Des.* **2015**, *102–103*, 19–28. [CrossRef]
54. Forré, A.; Helfer, T.; Derouiche, K. Comparison of different implicit integration procedures for an elasto-viscoplastic model. In Proceedings of the 6th International Conference on Material Modelling, Lund, Sweden, 26–28 June 2019.
55. Helfer, T.; Michel, B.; Proix, J.-M.; Salvo, M.; Sercombe, J.; Casella, M. Introducing the open-source mfront code generator: Application to mechanical behaviours and material knowledge management within the PLEIADES fuel element modelling platform. *Comput. Math. Appl.* **2015**, *70*, 994–1023. [CrossRef]
56. Chaboche, J.L.; Cailletaud, G. Integration methods for complex plastic constitutive equations. *Comput. Methods Appl. Mech. Eng.* **1996**, *133*, 125–155. [CrossRef]

Article

Thermomechanical Fatigue of Lost Foam Cast Al–Si Cylinder Heads—Assessment of Crack Origin Based on the Evaluation of Pore Distribution

Martin Wagner [1],*, Andreas Mösenbacher [1], Marion Eiber [1], Martin Hoyer [2], Marco Riva [2] and Hans-Jürgen Christ [3]

[1] Method Development & Fatigue Strength, IABG mbH, 85521 Ottobrunn, Germany
[2] Simulation Structural Mechanics Powertrain, BMW AG, 80788 München, Germany
[3] Institut für Werkstofftechnik, Universität Siegen, 57076 Siegen, Germany
*   Correspondence: Wagnerm@iabg.de; Tel.: +49-89-6088-2847

Received: 30 June 2019; Accepted: 23 July 2019; Published: 26 July 2019

Abstract: In automotive cylinder heads, thermomechanical fatigue (TMF) leads to crack initiation within the critical loaded sections. This effect becomes even more relevant in lost foam cast cylinder heads since its system-dependent porosity shows a significant influence on the lifetime under TMF loading. This work covers the identification of a criterion for crack initiation in order to provide the basis for an effective quality control with improved statistical safety by nondestructive testing. Specimens extracted from lost foam cylinder heads were investigated by uniaxial TMF tests, X-ray micro computer tomography (µCT), and scanning electron microscopy (SEM). Due to pore analyses on a global and local scale, it is concluded that pore networks are crucial for crack initiation. Thus, a tool for computation of pore accumulations from µCT data containing interaction criteria by Murakami was developed in order to assess the crack origin. The consideration of pore accumulations significantly improves the predictive accuracy compared to the consideration of single pores.

Keywords: aluminum-silicon cylinder head; lost foam; pore accumulation; pore distribution; thermomechanical fatigue; X-ray micro computer tomography

---

1. Introduction

In automotive engine components such as cylinder heads, aluminum alloys are widely spread due to their good mechanical and physical properties, along with suitable manufacturing characteristics and the excellent suitability for lightweight design [1]. In cylinder heads, temperature cycles (engine start-stop, full load, partial load) in combination with high temperature gradients lead to thermomechanical fatigue (TMF). The result is crack initiation within the critical section on the combustion chamber side [2] and integrated exhaust manifolds. In the past, the thermomechanical fatigue behavior of cylinder heads was investigated in detail [2,3].

Cylinder heads are increasingly manufactured by lost foam rather than gravity die casting due to process cost reduction, geometry optimization, and consumption control [4,5]. However, the result of systemic lower cooling rates is a coarser microstructure as well as an increase in number, size, and volume fraction of pores [6,7]. Besides rather spherical gas pores, the solidification of the melt causes extensive shrinkage pores with small sphericities [7] associated with high stress intensities. A few studies [7,8] deal with the 3D in situ analysis of the crack initiation and growth caused by shrinkage pores in lost foam cast components. The TMF life of lost foam cast cylinder heads can be assessed by using energy based damage models with additional consideration of the local damage [5]. There is a significant variation of the lifetime of identically loaded sections on the combustion chamber side of cylinder heads. It is well studied that the interaction of defects plays an important role with regard to crack initiation [9–11]. Due to the presence of large pore numbers and extensive shrinkage pores, the interaction might be the main reason for lifetime variation.

It is common to use statistical methods to evaluate pore distributions because of their suitability for volume extrapolation [12,13]. Romano et al. [14] worked on the statistical evaluation of single defects within additive manufactured parts using exponential and extreme value distributions. Furthermore, extreme value models for estimating defect size in clean steels were developed for multiple types of large inclusions [15]. In the past, many authors studied the influence of material defects on the LCF lifetime and the fatigue limit in the HCF regime. Charkaluk et al. [16] determined a lifetime probability density function from a probability density function of defects using a growth law. Rödling et al. [17] investigated the influence of critical non-metallic inclusions in high-strength steels on HCF design properties by means of statistics of extremes. Regarding an aluminum alloy, doubling the initial defect diameter causes an approximate 30% decrease in run-out stress [18]. These studies [17,18] underlie the recognition that, when defects become the fracture origin, crack initiation occurs at the largest defect or the critical combination of defects within the volume [9].

However, there exists a strong need to systematically and effectively qualify large sample sizes of TMF-loaded lost foam cast components for industrial applications, by means of nondestructive testing instead of time-consuming and cost-intensive destructive testing. The application of X-ray micro computer tomography (μCT) enables the nondestructive detection of defects within the volume of components, even inside inaccessible sections. The detection of defects by μCT requires a significant difference in density to the surrounding material and is limited by the resolution of the μCT system. In this study, the application of μCT is well suited as the porosities in lost foam cast components have many times larger expansions than the resolution of the used μCT system (10 μm).

The aim of this work is the identification of a criterion for crack initiation based on μCT defect data. Due to the presence of large pore networks caused by the lost foam casting process, defect interaction criteria by Murakami [9] are used. Subsequently, a correlation between the criterion for crack initiation and the lifetime has to be derived in order to assess the lifetime variation of cylinder heads as a function of the variation of the criterion in specimen geometry.

An extensive experimental program comprising TMF testing, scanning electron microscopy (SEM) of the fracture surface, and μCT scans before and after TMF testing is conducted. On a global scale, the pore volume fraction and number of pores are analyzed. On a local scale, the characteristic parameters of single pores are statistically evaluated by means of probability distributions. A tool for the computation of pore accumulations is developed in order to assess the crack origin.

2. Materials and Methods

2.1. Material

Lost foam cast cylinder heads were taken from the series process to consider the actual present casting conditions and the resulting quality of the material. The cylinder heads are manufactured from the precipitation-hardening aluminum-silicon alloy AlSi7MgCu0.5 with heat treatment T5. The specimens were extracted from the region between the intake and exhaust port of the cylinder heads (Figure 1) to consider the influence of the lost foam casting process on the lifetime under TMF loading. Thus, the specimens have small dimensions. The total specimen length is 80 mm, while the mount and test diameter read 10 mm and 7 mm, respectively.

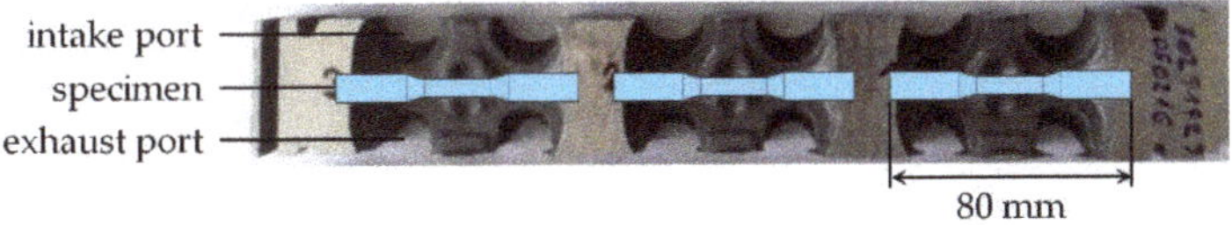

Figure 1. Specimen extraction.

2.2. TMF Tests

Strain controlled uniaxial TMF tests were performed with a servo-hydraulic test rig (MTS Systems GmbH, Berlin, Germany). For strain measurement a high-temperature extensometer with a gauge length of 12 mm was used. The strain control was realized by the software MTS 793 (software version 5.9, MTS Systems GmbH, Berlin, Germany). The heating of the specimen was conducted by a 10 kW high-frequency generator and an induction coil. In order to be able to perform TMF tests with small specimens (Section 2.1), the induction coil consists of two pipes: an outer water-cooled pipe and an inner compressed air pipe with nozzles for air transportation to the specimen. A flattened thermocouple (type K) with 0.1 mm thickness was used for temperature measurement (Figure 2). The test procedure and temperature control were realized by the software LabVIEW (software version 2014, National Instruments Germany GmbH, Munich, Germany).

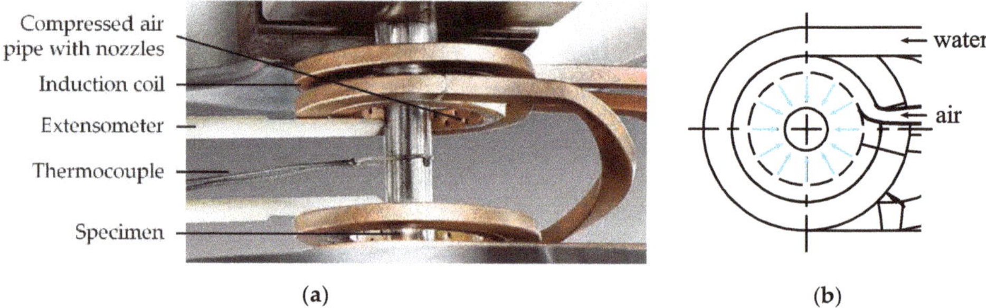

Figure 2. (**a**) Thermomechanical fatigue (TMF) test setup. (**b**) Induction coil with outer water-cooled pipe and inner compressed air pipe with nozzles for air transportation to the specimen.

Twenty-one TMF tests of specimens extracted from two batches were performed with a minimum temperature of 50 °C and two different maximum temperatures (225 °C and 250 °C) with heating and cooling rates of 5 °C/s using a triangle shaped signal. Three mechanical strain range values $\Delta\varepsilon_{mech,1}$, $\Delta\varepsilon_{mech,2}$, and $\Delta\varepsilon_{mech,3}$ were applied based on the engine service conditions. The strains are scaled with respect to $\Delta\varepsilon_{mech,2}$ according to $\Delta\varepsilon_{mech,1} = 1.2\,\Delta\varepsilon_{mech,2}$ and $\Delta\varepsilon_{mech,3} = 0.8\,\Delta\varepsilon_{mech,2}$. The maximum mechanical strain of each cycle was defined as $\varepsilon_{mech,max} = 0$. A hold time of 60 s was applied at the maximum temperature in order to cover a worst case automotive usage profile with regard to creep damage, i.e., during the TMF tests the specimen is exposed to the same amount of creep damage as the cylinder head under worst case service conditions. The TMF tests were carried out in out-of-phase (OP) mode, i.e., with a phase shift of 180° between the thermal and mechanical strain (Figure 3).

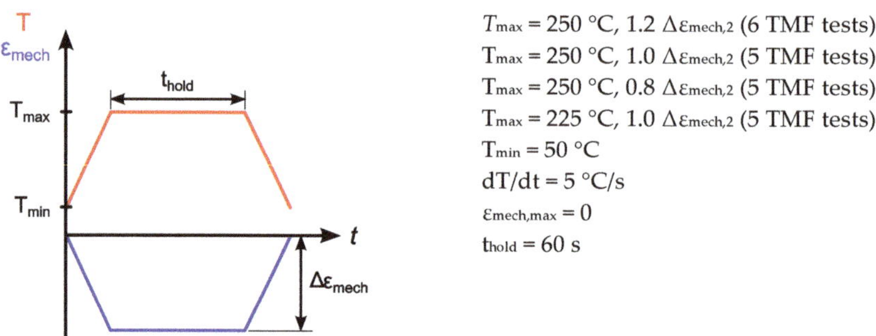

Figure 3. Loading conditions of performed TMF tests.

2.3. Microstructure Analyses

After TMF testing, the fracture surfaces were investigated by SEM (Carl Zeiss AG, Oberkochen, Germany) in order to identify the crack origin of each specimen. All fracture surfaces show striations (Figure 4a), and fracture propagation lines are detected near material defects. Figure 4a shows freely solidified surfaces which occur due to shrinkage of the material during the casting process and result in shrinkage pores. The system-dependent, slow cooling rate of the lost foam casting process increases the trend to the evolution of freely solidified surfaces and thus shrinkage pores.

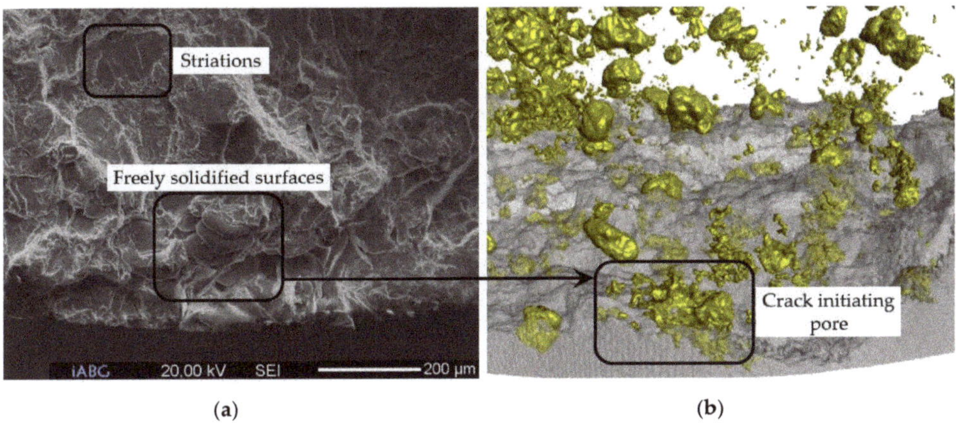

Figure 4. (a) Crack origin detected by scanning electron microscopy (SEM); (b) Crack-initiating pore detected by X-ray micro computer tomography (µCT).

µCT scans with a resolution of 10 µm were performed by means of a X-ray inspection system (GE Sensing & Inspection Technologies GmbH, Wunstorf, Germany) before and after the TMF tests. The µCT scan before TMF testing was performed across the entire gauge length of the specimen to consider all material defects, including pores open to the specimen surface. Moreover, the pore volume fraction and the number of pores within the volume were determined. The scan after TMF testing focused on the fracture area. For µCT data analysis the software VGSTUDIO MAX (software version 2.2, Volume Graphics GmbH, Heidelberg, Germany) was used. The volumes before and after TMF testing were virtually overlapped. In order to achieve this, both volumes were manually positioned to each other by aligning the surfaces of conspicuous pores, followed by an automatic alignment of both volumes by the evaluation software. Subsequently, the fracture surfaces scanned by SEM and µCT were visually compared. The crack-initiating pores were detected in the µCT scan looking on the location of the actual crack origin, i.e., freely solidified surfaces, detected by SEM (Figure 4b) [19].

Additionally, all pores located on the fracture surface were detected by the following method: after the described virtual overlap of the volumes before and after TMF testing, some pores detected in the µCT scan before TMF testing did not touch the fracture surface, even though SEM investigations obviously showed that they are located on the fracture surface. It is assumed that this small shift between the pores and the fracture surface is caused by the plastic deformation of the material close to the fracture surface during the TMF test. In order to consider these pores the smallest distance of each pore to the fracture surface was computed. All pores closer than 50 µm to the fracture surface were defined as pores on the fracture surface. The value of 50 µm (five times the scan resolution) was visually adjusted. By means of this procedure, the influence of inner pores on the crack initiation behavior can be evaluated.

3. Experimental Results

3.1. TMF Tests

In Figure 5, the relation between the mechanical strain range (scaled) and lifetime (number of cycles to a stress drop of 20%) is shown for all TMF tests performed. There is no influence of the batch on the lifetime. However, the lifetime increases with increasing maximum temperature. There are lower maximum stresses, but a larger plastic strain range with increasing temperature. Thus, it is assumed that the maximum stresses have a greater influence on the lifetime than the plastic strain range.

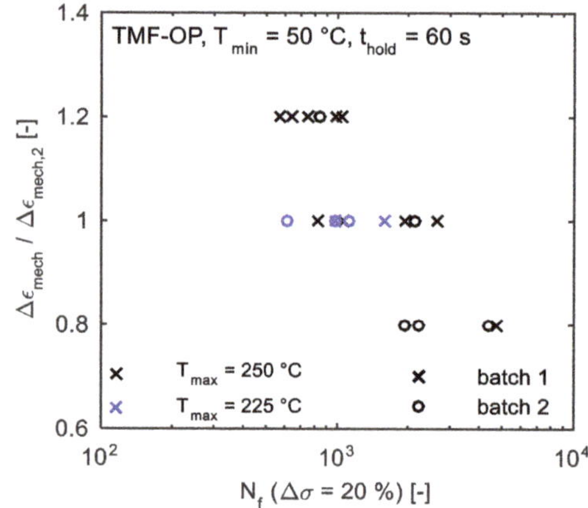

Figure 5. Relation between mechanical strain range (scaled) and lifetime for all TMF tests performed.

Figure 6 provides the cyclic hardening–softening curves and the stress–strain hystereses of cycle N = 500 for the performed TMF tests with a mechanical strain range of $\Delta\varepsilon_{mech}$ = 1.2 $\Delta\varepsilon_{mech,2}$ and a maximum temperature of T_{max} = 250 °C. All stresses are scaled to the mean stress range $\Delta\sigma_2$ at half number of cycles to failure for the TMF tests with a mechanical strain range of $\Delta\varepsilon_{mech,2}$. Each TMF test shows a positive mean stress due to out-of-phase loading. The nominal maximum and minimum stresses of each cycle decrease with increasing cycle number due to ageing effects. The nominal maximum and minimum stresses lie within a scatter band of 10% and the lifetime varies from about 500 to 1000 cycles (Figure 6a). Additionally, the shape and area of the stress–strain hystereses as well as the stress relaxation during the hold time are almost identical (Figure 6b). Thus, there is no appreciable influence of the nominal maximum and minimum stresses as well as the characteristics of the stress–strain hystereses on the lifetime.

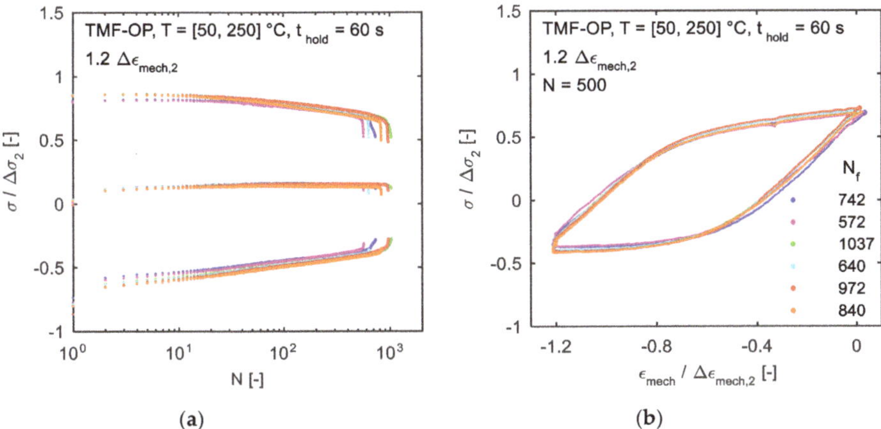

(a)   **(b)**

Figure 6. Performed TMF tests with $\Delta\varepsilon_{mech} = 1.2\ \Delta\varepsilon_{mech,2}$ and $T_{max} = 250\ °C$: (**a**) Cyclic hardening-softening curves; (**b**) Stress–strain hystereses of cycle N = 500.

3.2. Microstructure Results

SEM investigations of the fracture surface showed that, in all specimens, cracks initiate from macro-scale porosities. Thus, there is no influence of the applied mechanical strain range ($\Delta\varepsilon_{mech,1}$, $\Delta\varepsilon_{mech,2}$ and $\Delta\varepsilon_{mech,3}$) and the maximum temperature (225 °C and 250 °C) on the type of crack initiation. Except for a few specimens where cracks initiate from an extensive single pore (Figure 4), mostly a pore network consisting of several pores was identified at the crack origin. It is assumed that adjacent pores coalesce during a few cycles and therefore act as crack initiator. In most specimens, pore networks consist of shrinkage pores, but also a combination of shrinkage and gas pores was detected at the crack origin (Figure 7a). The crack-initiating pores and pore networks have a large projected area perpendicular to loading direction and a small distance to the specimen surface. The latter observation may be caused by higher stress intensities of pores near to the specimen surface compared to pores within the material [9].

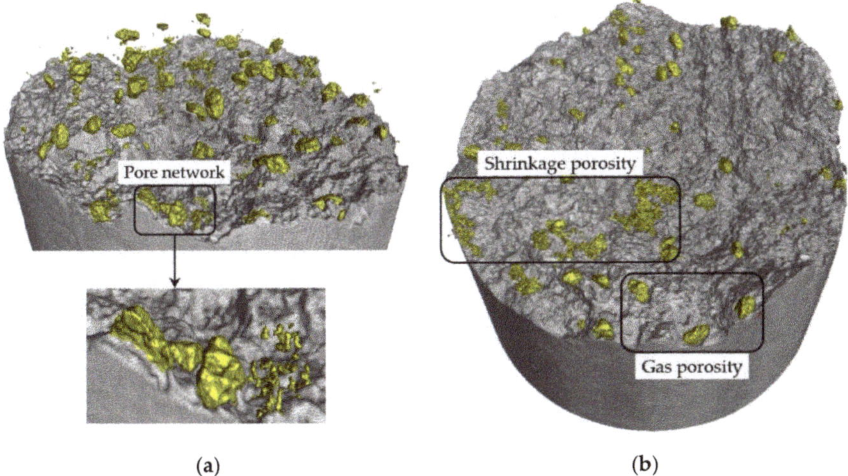

(a)   **(b)**

Figure 7. Fracture surface scanned by μCT: (**a**) Discontiguous crack-initiating pores (crack origin detected by SEM, specimen No. 09); (**b**) Exclusively pores on the fracture surface. (specimen No. 01).

Additionally, the influence of inner pores on the crack initiation behavior was evaluated by analyzing the pores on the fracture surface (Section 2.3). A combination of shrinkage and gas pores is present on the fracture surface (Figure 7b). Shrinkage pores are assumed to be more crucial for crack initiation due to their lower sphericity and larger projected area perpendicular to the loading direction.

The above observations indicate that extensive single pores and pore networks caused by the low cooling rate of the lost foam casting process are crucial for crack initiation. It is concluded that fatigue damage is the predominant failure mode compared to creep damage as large pores cause high stress intensities under cyclic loading, intensified by their interaction.

3.3. Pore Analysis on a Global Scale

Figure 8 shows the pore volume fraction ϕ_{Vol} and the number of pores n_{pores} with a maximum Feret diameter $d_{Feret,max} > 100$ µm over the lifetime (number of cycles to a stress drop of 20%). The maximum Feret diameter $d_{Feret,max}$ is equal to the diameter of the outer sphere of the pore. There is a trend towards lower lifetimes for higher pore volume fractions, as consistently observed under each test condition. The second batch shows a higher average pore volume fraction (1.17%) than the first batch (0.99%). The number of pores within the scanned volume does not correlate with the lifetime. The second batch shows a lower average number of pores (1489) than the first batch (2150). Thus, the second batch contains fewer but larger defects caused by the casting process of the cylinder heads.

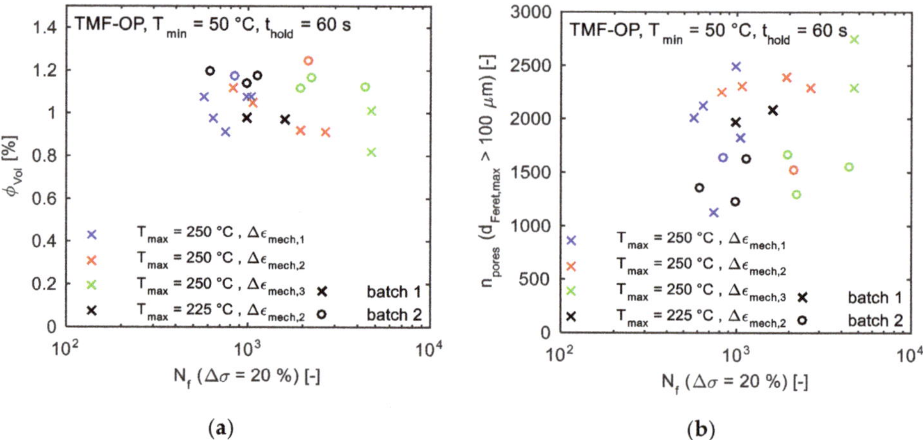

(a) (b)

Figure 8. Global pore parameters vs. lifetime: (**a**) Pore volume fraction; (**b**) Number of pores with a maximum Feret diameter larger than 100 µm.

3.4. Pore Analysis on a Local Scale

Two characteristic parameters of single pores are considered: the maximum Feret diameter of the pore $d_{Feret,max}$ (Section 3.3) and the projected area of the pore to the plane perpendicular to loading direction A_{proj} (Figure 9). Both parameters were derived from the µCT scans. It should be noted that by using $d_{Feret,max}$ and A_{proj}, the shape and orientation of the pore are not taken into account. Though the evaluation software provides the sphericity of the pore, there is no available information about the orientation of the pore.

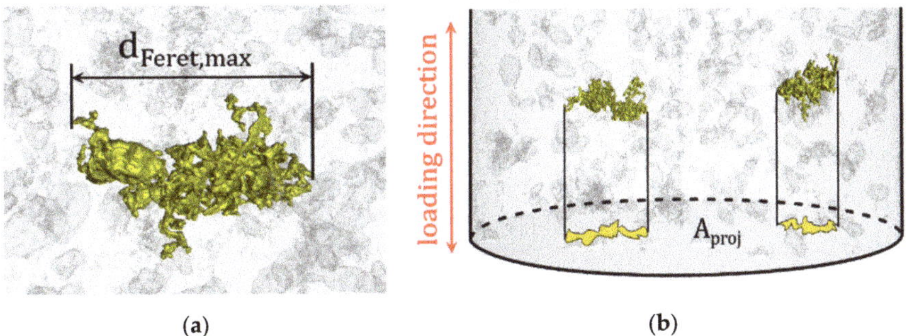

(a) (b)

Figure 9. Characteristic parameters of single pores: (**a**) $d_{Feret,max}$; (**b**) A_{proj}.

For the statistical evaluation of the characteristic parameters, the exponential distribution is used [14] because of the presence of many small pores and less and less larger pores regarding the size of $d_{Feret,max}$ and A_{proj}. The exponential distribution is well suited to rank the actual crack-initiating pores detected by SEM as well as the pores on the fracture surface in the totality of pores within the volume. The equation for the exponential distribution is as follows:

$$d_{Feret,max,i} = -\ln(1 - P_i) \quad \text{and} \quad A_{proj,i} = -\ln(1 - P_i) \tag{1}$$

P_i is the probability that the characteristic parameters are lower than the value of the corresponding $d_{Feret,max,i}$ and $A_{proj,i}$.

In Figure 10, $d_{Feret,max}$ and A_{proj} are shown in the probability plot of the exponential distribution for specimens which were TMF tested at 1.2 $\Delta\varepsilon_{mech,2}$ and $T_{max} = 250$ °C. For both $d_{Feret,max}$ and A_{proj}, the data lie approximately on a straight line, thus demonstrating the suitability of the exponential distribution. The data of the four specimens in Figure 10 are hardly discernable. It is concluded that there is no correlation between the course of the probability distribution of $d_{Feret,max}$ and A_{proj} with the lifetime. Furthermore, the crack-initiating pores detected by SEM are marked with crosses. Notably, the actual crack-initiating pores of each specimen are not the pores with the largest $d_{Feret,max}$ and A_{proj}.

Additionally, Figure 11 depicts the characteristic parameter A_{proj} of pores located on the fracture surface (Section 2.3) in the probability plot of the exponential distribution. For several specimens the pores with the largest A_{proj} within the entire volume lie on the fracture surface (e.g., specimen No. 01, Figure 11a). On the other hand, there are some specimens where pores with the largest A_{proj} within the entire volume do not lie on the fracture surface (e.g., specimen No. 07, Figure 11b). The same observations are made for the characteristic parameter $d_{Feret,max}$.

These investigations show that a proper criterion for crack initiation cannot be identified by exclusively studying single pores. There are two explanations for this observation: firstly, pores are present as pore networks, coalesce during a few cycles, and therefore act as crack initiator (see Section 3.2); secondly, shrinkage pores have discontiguous structures and therefore cannot be detected as one single pore by μCT. For these reasons, the sum of the projected areas A_{proj} of discontiguous single pores at the crack origins detected by SEM and μCT (Figure 7a) is used as the projected area $A_{proj,sum}$ of the crack origin. All available projected areas $A_{proj,sum}$ are plotted in the probability plot of the logit distribution (Figure 12) [17]. The equation for the logit distribution is:

$$A_{proj,sum,i} = \frac{\sqrt{3}}{\pi} \cdot \ln\left(\frac{P_i}{1 - P_i}\right) \tag{2}$$

Since almost all data points lie within the 90% confidence region, the application of pore accumulations is considered appropriate.

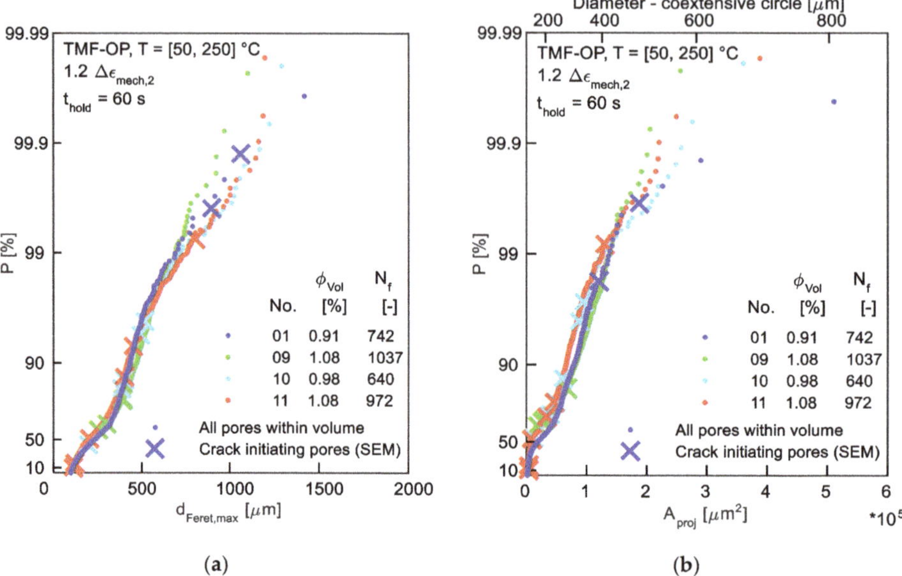

Figure 10. Characteristic parameters in probability plot of exponential distribution for specimens which were TMF tested at 1.2 $\Delta\varepsilon_{\text{mech},2}$ and $T_{\text{max}} = 250\ ^\circ$C—crack-initiating pores detected by SEM are shown by the crosses. (**a**) $d_{\text{Feret,max}}$; (**b**) A_{proj}.

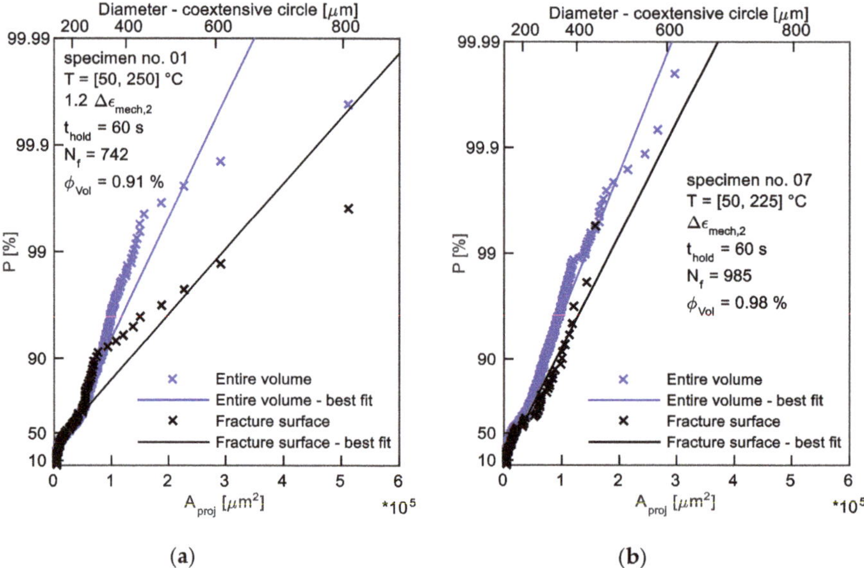

Figure 11. A_{proj} in probability plot of exponential distribution—pores within the entire volume (blue) and exclusively on the fracture surface (black). (**a**) Specimen No. 01; (**b**) Specimen No. 07.

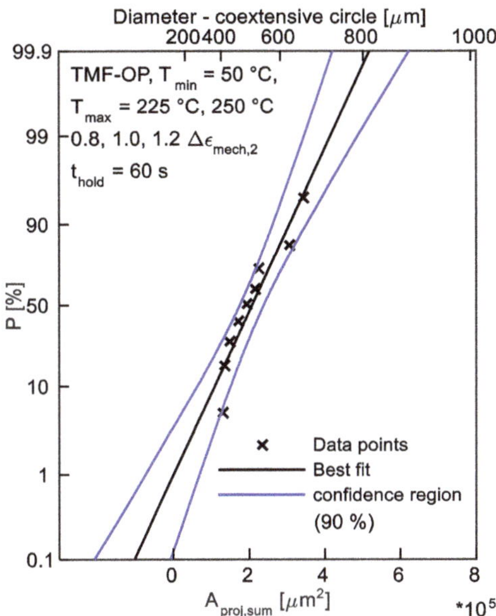

Figure 12. Sum of A_{proj} of discontiguous single pores at crack origins detected by SEM and µCT in probability plot of logit distribution for all available specimens.

4. Computation of Pore Accumulations

4.1. Method of Computation

In order to assess the actual crack origin, a tool for the computation of pore accumulations was developed with the software MATLAB (software version R2015a, The MathWorks GmbH, Ismaning, Germany). For each pore, all surrounding pores within a defined inspection volume were detected. The inspection volume was defined in cylinder coordinates by r_{iv} in radial direction, φ_{iv} in angular direction, and z_{iv} in axial direction (Figure 13b). In order to reach the best crack origin assessment by the computation of pore accumulations, different sizes of the inspection volume were manually tested (0.25 mm to 1 mm in radial direction, 10° to 30° in angular direction and 0.1 mm to 1 mm in axial direction). The tested sizes of the inspection volume were derived from the size of the crack origins detected by SEM. Additionally, the SEM investigations of the fracture surfaces showed that in all cases, crack initiation occurred near the specimen surface (see Section 3.2). Therefore, only inspection volumes touching the specimen surface were considered.

The projected areas A_{proj} of the collected pores were summed up to $A_{proj,acc}$ for each inspection volume. Subsequently, the pore accumulations were sorted in descending order. The pore accumulation with the largest computed $A_{proj,acc}$ is marked green. The crack-initiating pore accumulation is marked red. All crack-initiating pores detected by SEM lie within the inspection volume of the corresponding pore accumulation. The projected areas are presented as area equivalent circles (Figure 13).

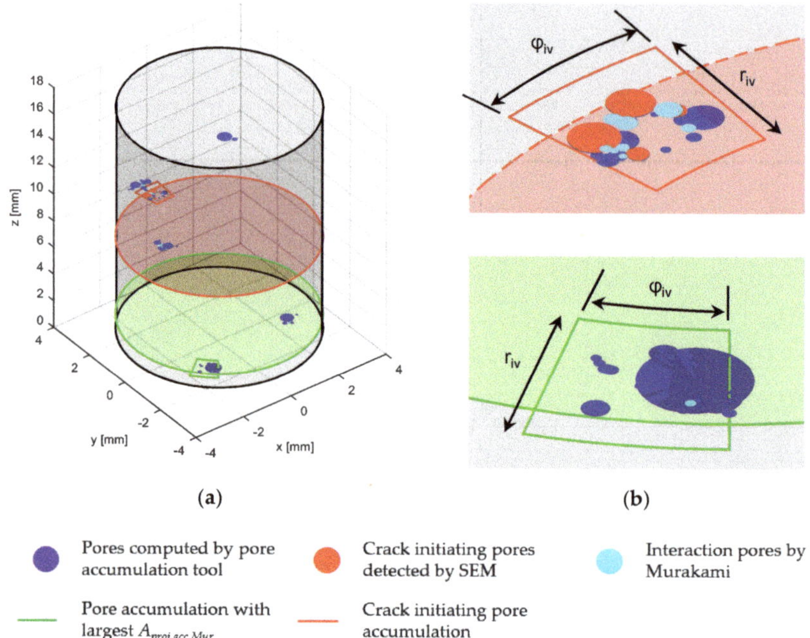

Figure 13. (**a**) Computed pore accumulations in terms of A_{proj} for specimen No. 04; (**b**) Detail view of the crack-initiating pore accumulation including crack-initiating pores detected by SEM (top) and the pore accumulation with the largest computed $A_{proj,acc,Mur}$ (bottom).

Additionally, the described procedure is expanded using interaction criteria by Murakami [9]: for two pores having a smaller distance between each other than the diameter of the smaller one an additional pore is generated between the two pores (Figure 13). This method is based on the theory that the section between the pores can be interpreted as initial crack because two adjacent pores coalesce during a few cycles.

It should be noted that the provided method is based on the projected areas of the pores. Therefore, different pore shapes and orientations with their resulting higher stress intensities are not considered. Using the developed tool, spherical pores are evaluated as more and flat pores less dangerous than they are in reality.

4.2. Results of Computation of Pore Accumulations

In order to provide a measure for the suitability of computing pore accumulations with respect to assessing the actual crack origin, the computed pore accumulations are sorted in descending order according to the size of their accumulated projected areas $A_{proj,acc}$. From this ranking, the placement of the actual crack-initiating pore accumulation detected by SEM is determined (Table 1, column "$A_{proj,acc}$"). On the other hand, the single pores are sorted in descending order according to the size of their projected areas A_{proj}. From this ranking, the placement of the largest actual crack-initiating pore detected by SEM is determined (Table 1, right column). Comparing the computation of pore accumulations with the exclusive consideration of single pores, a smaller number indicates a better placement and thus a better predictive accuracy of crack origins. The approach with the best placement is marked grey for each specimen. Number '1' means that the actual crack origin is exactly matched. (Table 1).

For most specimens, the computation of pore accumulations is suitable for crack origin assessment because it improves the predictive accuracy compared to the exclusive consideration of single pores.

However, for two specimens (No. 07 and 08) the tool failed. There is no significant improvement of the results by the consideration of interaction pores by Murakami (Table 1, column "$A_{proj,acc,Mur}$"). A possible reason for that is the much larger projected area of the accumulated pores compared to the area between the pores. The inspection volume providing the optimal results has the dimensions $r_{iv} = 1$ mm, $\varphi_{iv} = 15°$, $z_{iv} = 0.5$ mm.

Table 1. Evaluation of the suitability of computing pore accumulations for crack origin assessment (a smaller number indicates a better placement and thus a better predictive accuracy of crack origins; number '1' means that the actual crack origin is exactly matched).

Specimen No.	Pore Accumulations			Single Pores
	Placement of Crack-Initiating Pore Accumulation in the Descending Order of the Computed Pore Accumulations			Placement of Largest Crack-Initiating Pore Detected by SEM in the Descending Order of Single Pores
	$A_{proj,acc}$	$A_{proj,acc,Mur}$		
01	1	1	<	2
02	6	5	<	6
04	3	3	<	79
06	7	9	<	37
07	32	35	>	8
08	2	4	>	1
09	26	12	<	149
10	4	4	<	25
11	4	6	<	6

5. Conclusions

The aim of this work was the identification of a criterion for crack initiation in order to provide the basis for a nondestructive quality control of TMF loaded porous components with improved statistical safety. TMF tests under realistic service conditions were performed on specimens extracted from automotive cylinder heads manufactured by lost foam casting. By performing µCT scans before and after the TMF tests, as well as comparing the fracture surfaces from SEM and µCT scans, the characteristics of crack-initiating pores were detected. Several characteristic pore parameters on the global and local scale were statistically examined in order to identify their influence on lifetime and crack initiation. Furthermore, a tool for computation of pore accumulations was developed. The following main conclusions can be drawn:

1. The positive mean stress in the TMF tests is caused by the out-of-phase loading condition and cyclic softening by ageing of the material.
2. Large shrinkage and gas pores, caused by the lost foam production process, are crucial for crack initiation.
3. Crack-initiating pores and pore networks have a small distance to the specimen surface and a large projected area to the plane perpendicular to loading direction.
4. On a global scale, the pore volume fraction and number of pores do not correlate with the lifetime.
5. On a local scale, statistical evaluation of single pores implied that pore accumulations are crucial for crack initiation.
6. For most specimens, the computation of pore accumulations is suitable for crack origin assessment because it improves the predictive accuracy compared to the exclusive consideration of single pores.
7. There is no further improvement of the predictive accuracy by consideration of interaction of pores by Murakami [9] because the projected area of the accumulated pores is much larger than the area between the pores.

A criterion for crack initiation in lost foam cast cylinder heads consisting of the application of pore accumulations under consideration of pore interaction was found. Applying the developed tool for the computation of pore accumulations to µCT scans of lost foam cast components enables the early-stage assessment of critical sections within the material. In industrial applications, the statistical power can be improved because of time-saving and cost-effective nondestructive testing (application of µCT) instead of time-consuming and cost-intensive destructive testing (application of TMF tests).

Even though, for most specimens, the computation of pore accumulations provides a better predictive accuracy of crack origins compared to the consideration of single pores, it can still be improved. A possible way is expanding the pore accumulation tool with the consideration of the edge distance of pores [10] since pores near to the specimen surface have higher stress intensities than pores within the material [9]. Furthermore, even though SEM investigations showed that crack initiation occurred near the specimen surface, it is planned to extend the applied inspection volume to the entire cross-section of the specimen to consider the possible influence of inner pores on crack initiation. Subsequently, a correlation between the final criterion and the lifetime can be derived in order to assess the lifetime variation of cylinder heads as a function of the variation of the criterion.

Author Contributions: Conceptualization, M.W., M.H., M.R. and H.-J.C.; formal analysis, M.W.; investigation, M.W.; methodology, M.W. and M.E.; project administration, M.W.; resources, A.M., M.E., M.H. and M.R.; software, M.W.; supervision, A.M., M.H., M.R. and H.-J.C.; validation, M.W., M.H., M.R. and H.-J.C.; visualization, M.W.; writing—original draft, M.W.; writing—review & editing, A.M., M.E., M.H., M.R. and H.-J.C.

Acknowledgments: The authors thank BMW AG for providing series cylinder heads as well as Manfred Hück, Stefan Slaby and Fabian Sartorius for their general support and discussions.

Conflicts of Interest: The authors declare no conflict of interest.

References

1. Hirsch, J. Aluminium in Innovative Light-Weight Car Design. *Mater. Trans.* **2011**, *52*, 818–824. [CrossRef]
2. Thalmair, S.; Thiele, J.; Fischersworring-Bunk, A.; Ehart, R.; Guillou, M. Cylinder Heads for High Power Gasoline Engines—Thermomechanical Fatigue Life Prediction. *SAE Tech. Pap.* **2006**, 548–555. [CrossRef]
3. Grieb, M. *Untersuchung von Aluminiumgusslegierungen auf Thermomechanische und HCF Ermüdung und Verbesserung der Lebensdauervorhersage*; VDI-Verlag: Düsseldorf, Germany, 2010; ISBN 978-3-18-374205-9.
4. Geffroy, P.-M.; Lakehal, M.; Goñi, J.; Beaugnon, E.; Heintz, J.-M.; Silvain, J.-F. Thermal and mechanical behavior of Al-Si alloy cast using magnetic molding and lost foam processes. *Metall. Mater. Trans. A* **2006**, *37*, 441–447. [CrossRef]
5. Tabibian, S.; Charkaluk, E.; Constantinescu, A.; Oudin, A.; Szmytka, F. Behavior, damage and fatigue life assessment of lost foam casting aluminum alloys under thermo-mechanical fatigue conditions. *Procedia Eng.* **2010**, *2*, 1145–1154. [CrossRef]
6. Boileau, J.M.; Allison, J.E. The effect of solidification time and heat treatment on the fatigue properties of a cast 319 aluminum alloy. *Metall. Mater. Trans. A* **2003**, *34*, 1807–1820. [CrossRef]
7. Wang, L.; Limodin, N.; El Bartali, A.; Witz, J.-F.; Seghir, R.; Buffiere, J.-Y.; Charkaluk, E. Influence of pores on crack initiation in monotonic tensile and cyclic loadings in lost foam casting A319 alloy by using 3D in-situ analysis. *Mater. Sci. Eng. A* **2016**, *673*, 362–372. [CrossRef]
8. Dezecot, S.; Maurel, V.; Buffiere, J.-Y.; Szmytka, F.; Koster, A. 3D characterization and modeling of low cycle fatigue damage mechanisms at high temperature in a cast aluminum alloy. *Acta Mater.* **2017**, *123*, 24–34. [CrossRef]
9. Murakami, Y. *Metal Fatigue: Effects of Small Defects and Nonmetallic Inclusions*; Elsevier: Amsterdam, The Netherlands, 2002; ISBN 978-0-08-049656-6.
10. Masuo, H.; Tanaka, Y.; Morokoshi, S.; Yagura, H.; Uchida, T.; Yamamoto, Y.; Murakami, Y. Effects of Defects, Surface Roughness and HIP on Fatigue Strength of Ti-6Al-4V Manufactured by Additive Manufacturing. *Procedia Struct. Integr.* **2017**, *7*, 19–26. [CrossRef]
11. Åman, M.; Okazaki, S.; Matsunaga, H.; Marquis, G.B.; Remes, H. The effect of interacting small defects on the fatigue limit of a medium carbon steel. *Procedia Struct. Integr.* **2016**, *2*, 3322–3329. [CrossRef]

12. Beretta, S.; Murakami, Y. Statistical Analysis of Defects for Fatigue Strength Prediction and Quality Control of Materials. *Fatigue Fract. Eng. Mater. Struct.* **1998**, *21*, 1049–1065. [CrossRef]
13. Wilson, P.; Saintier, N.; Palin-Luc, T.; Bergamo, S. Isothermal fatigue damage mechanisms at ambient and elevated temperature of a cast Al-Si-Cu aluminium alloy. *Int. J. Fatigue* **2019**, *121*, 112–123. [CrossRef]
14. Romano, S.; Brandão, A.; Gumpinger, J.; Gschweitl, M.; Beretta, S. Qualification of AM parts: Extreme value statistics applied to tomographic measurements. *Mater. Des.* **2017**, *131*, 32–48. [CrossRef]
15. Beretta, S.; Anderson, C.; Murakami, Y. Extreme value models for the assessment of steels containing multiple types of inclusion. *Acta Mater.* **2006**, *54*, 2277–2289. [CrossRef]
16. Charkaluk, E.; Constantinescu, A.; Szmytka, F.; Tabibian, S. Probability density functions: From porosities to fatigue lifetime. *Int. J. Fatigue* **2014**, *63*, 127–136. [CrossRef]
17. Rödling, S.; Fröschl, J.; Hück, M.; Decker, M. Influence of non-metallic inclusions on acceptable HCF design properties. *Mater. Test.* **2011**, *53*, 455–462. [CrossRef]
18. Boileau, J.M.; Allison, J.E. *The Effect of Porosity Size on the Fatigue Properties in a Cast 319 Aluminum Alloy*; SAE International: Warrendale, PA, USA, 2001.
19. Wagner, M.; Decker, M.; Hoyer, M.; Riva, M.; Christ, H.-J. *Experimental and Computational Investigations of a Lost Foam Cast Aluminium-Silicon Alloy under Thermomechanical Loading*; German Association for Materials Research and Testing E.V.: Dresden, Germany, 2017; pp. 405–410.

Article

Thermo-Mechanical Fatigue Lifetime Assessment of Spheroidal Cast Iron at Different Thermal Constraint Levels

Sepideh Ghodrat [1,*], Aakarshit Kalra [2,3], Leo A.I. Kestens [2,4] and Ton (A.C.) Riemslag [2]

[1] Department of Design Engineering, Faculty of Industrial Design Engineering, Delft University of Technology, Landbergstraat 15, 2628 CE Delft, The Netherlands
[2] Department of Materials Science and Engineering, Delft University of Technology, Mekelweg 2, 2628 CD Delft, The Netherlands; A-Aakarshit.Kalra@DAFTRUCKS.com (A.K.); leo.kestens@ugent.be (L.A.I.K.); A.C.Riemslag@tudelft.nl (T.A.C.R.)
[3] DAF Trucks N.V., Hugo van der Goeslaan 1, 5643 TW Eindhoven, The Netherlands
[4] Metal Science and Technology Group, EEMMeCS Department, Ghent University, Technologiepark 46, B9052 Ghent, Belgium
*   Correspondence: s.ghodrat@tudelft.nl; Tel.: +31-1527-81655

Received: 27 July 2019; Accepted: 26 September 2019; Published: 1 October 2019

Abstract: In previous work on the thermo-mechanical fatigue (TMF) of compacted graphite iron (CGI), lifetimes measured under total constraint were confirmed analytically by numerical integration of Paris' crack-growth law. In current work, the results for CGI are further validated for spheroidal cast iron (SGI), while TMF tests at different constraint levels were additionally performed. The Paris crack-growth law is found to require a different C_{Paris} parameter value per distinct constraint level, indicating that Paris' law does not capture all physical backgrounds of TMF crack growth, such as the effect of constraint level. An adapted version of Paris' law is developed, designated as the local strain model. The new model considers cyclic plastic strains at the crack tip to control crack growth and is found to predict TMF lifetimes of SGI very well for all constraint levels with a single set of parameters. This includes not only full constraint but also over and partial constraint conditions, as encountered in diesel engine service conditions. The local strain model considers the crack tip to experience a distinct sharpening and blunting stage during each TMF cycle, with separate contributions to crack-tip plasticity, originating from cyclic bulk stresses in the sharpening stage and cyclic plastic bulk strains in the blunting stage.

Keywords: thermo-mechanical fatigue; spheroidal cast iron; partial constraint; crack growth models; crack-tip cyclic plasticity; crack-tip blunting and sharpening

---

1. Introduction

Cast iron finds widespread application in the automotive industry. Spheroidal (or nodular) cast iron is a grade of cast iron frequently used in engine components. It is often preferred over flake and compacted cast irons for load-bearing applications. Higher strength of spheroidal cast iron stems from the spheroidal shape of graphite particles. However, the spheroidal shape of the graphite particles also leads to lower thermal conductivity. In SiMo spheroidal cast iron, silicon and molybdenum are added to the material to compensate for the lower thermal conductivity by providing strength to the material at high temperatures.

As a result of sequential start-up and shutdown, these engine components are subjected to repeated thermal cycling, resulting in a phenomenon known as thermo-mechanical fatigue (TMF). The extent of resulting TMF damage depends on the amount of constraint during the thermal expansion.

This constraint emanates from the spatial temperature gradients that develop in the material during start-up and shutdown.

For the manufacturers of engine components, it is imperative to predict the thermo-mechanical fatigue lifetime of these components under the constraint conditions during service. Various approaches are available for TMF lifetime prediction as summarized by Gocmez et al. [1]. According to this reference, three main types of models can be distinguished: (i) phenomenological models, (ii) cumulative damage models, and (iii) crack-growth models.

Rémy et al. [2] successfully calculated lifetimes under high cycle fatigue for a powder metallurgy (PM) material containing defects, using the hypothesis that a defect can be considered a crack. To calculate the high cycle fatigue (HCF) lifetime, they employed Paris' fatigue crack-growth equation, which describes the crack-growth rate (da/dN) as a function of the stress intensity range (ΔK), taking a representative dimension of particles as the initial crack size.

Fatigue-crack initiation and growth in graphitic cast irons is largely affected by the presence of graphite particles. Because of the likely fast initiation of TMF cracks in cast irons as a result of delamination at the graphite–metal interface, Ghodrat et al. [3–5] evaluated TMF lifetime using a crack-growth model, and they proved that, in tension, the graphite particles can be considered as internal notches or defects, from which TMF cracks start to grow during the very first TMF cycles. In the presence of an external notch, the notch depth can be considered as the initial crack length. The mechanical graphite/matrix interaction of CGI is demonstrated by Ghodrat and Kestens [6], showing a weak mechanical bonding. This was confirmed by a recent work [7], studying the mechanical behavior of the graphite/matrix interface for cycling load conditions at room temperature, for three types of cast iron, including spheroidal cast iron. The measured macroscopic cyclic stress–strain behavior was validated using both micromechanical calculations with the finite element method (FEM) and microstructural strain measurements by digital image correlation (DIC). De-bonding of the graphite/matrix interface was found to develop during the initial load cycles, resulting in an interface free of bonding forces. Consistently, both the FEM calculations and the DIC observations showed a pronounced increase in strain levels at the graphite particle boundaries. Therefore, the results of Reference [7] confirmed that graphite particles can be considered as internal notches, as also argued in the current work.

Despite the fact that TMF loading is often a case of low cycle fatigue involving bulk cyclic plasticity, Ghodrat et al. [3] proved the applicability of the Paris' crack-growth law in successfully predicting the TMF lifetime of compacted graphite iron (CGI) under total constraint conditions, although the Paris law is based on a linear elastic fracture mechanics (LEFM) approach, which ignores low cycle fatigue conditions. However, total constraint conditions are rarely encountered in service. A fracture-mechanical approach, capable of determining TMF lifetime under any possible constraint condition, is lacking. Considering the diesel engine background of this research, this work focuses on out-of-phase TMF loading, signifying that there is a 180° phase difference between temperature and mechanical cycling.

This work aims to understand and model crack growth in spheroidal cast iron for any TMF constraint condition. To this purpose, TMF tests were performed at different constraint levels, and results were analyzed using two fatigue crack-growth models. Both models were developed using the same set of experimental results, with their backgrounds covered in Section 3.

It is acknowledged that, apart from TMF constraint levels, other factors also influence TMF lifetime, most notably high temperature effects such as creep and oxidation. However, the additional variation of test conditions identifying high temperature effects, such as using prolonged holding times [5], was considered to be beyond the scope of this research.

2. Experimental Set-Up and Methods

This section covers details about the materials used, as well as the experimental set-up, and specifies the definition of test conditions.

2.1. Material and Specimens

The material under investigation for this study was a type of SiMo spheroidal graphite cast iron, also known as ductile cast iron, with a ferritic matrix, and Si and Mo as the major alloying elements. The microstructure of the material under investigation is represented in Figure 1, showing graphite nodules with an average size of 30 µm ± 8 µm. The chemical composition is listed in Table 1. The coefficient α of linear thermal expansion of the material was obtained by measurements of the strain during heating from 50 to 550 °C in free expansion. By plotting the axial strain measured as a function of temperature, linear expansion was observed, of which the slope, α, was found to be 13.6×10^{-6} °C^{-1}.

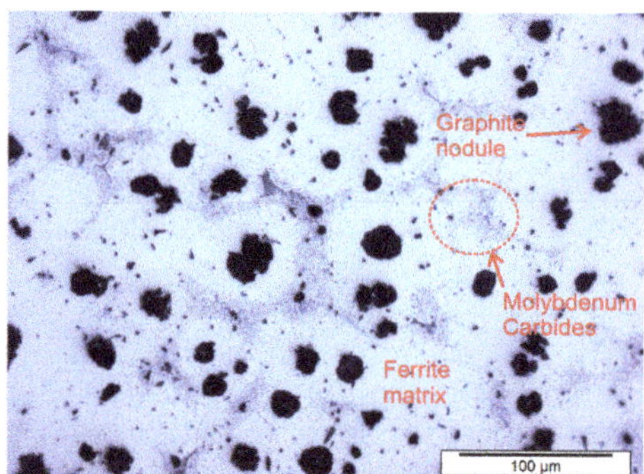

Figure 1. Typical microstructure of the material under consideration.

Table 1. Chemical composition (in wt%) of the spheroidal cast iron under consideration.

C	Si	Mo	Mn	S	P	Fe
3.40	4.20	0.80	0.50	0.05	0.02	Balance

Cylindrical dog bone-shaped specimens were used for TMF testing. A sharp circumferential notch was machined in the center of the gauge section of some of the samples, using a lath and a sharp chisel. The machined notch depth used for each TMF test is reported in Table 2. The geometrical specifications of the sample and notch are given in Figure 2. A dedicated extensometer with a gauge length of 12 mm was used to measure the total strain (extensometer model 632.53 F14, MTS systems, Eden Prairie, MN, USA).

Table 2. Typical values for thermo-mechanical (TMF) crack-growth parameters for all TMF tests performed, with calculated lifetimes according to the local stress model (Paris), with the C_{Paris} coefficient based on units for $\frac{da}{dn}$ and ΔK of m(cycle)$^{-1}$ and MPa m$^{0.5}$, respectively.

γ (%)	a_o (mm)	Z [1]	Experimental Values			Local Stress Model Calculations (Paris' Law)		
			S [2] (MPa)	N_{10} [2] (–)	Δe_{mech} (%)	$N_{\Delta K}$ [2] (–)	C_{paris}	m (–)
125	0.15	3	836 ± 3%	29 ± 41%	0.84	24 ± 12%	9.0×10^{-11}	3.58
	0.40	2	808 ± 3%	8 ± 0%		8 ± 12%		
100	0.03	6	718 ± 1%	157 ± 55%	0.67	287 ± 4%	8.5×10^{-11}	3.58
	0.15	3	772 ± 3%	48 ± 23%		41 ± 10%		
	0.40	2	710 ± 2%	13 ± 8%		14 ± 7%		
75	0.15	3	654 ± 4%	168 ± 19%	0.51	193 ± 16%	3.3×10^{-11}	3.58
	0.40	2	631 ± 0%	56 ± 7%		55 ± 0%		
50	0.15	3	480 ± 2%	803 ± 28%	0.34	990 ± 8%	1.8×10^{-11}	3.58
	0.40	4	480 ± 7%	284 ± 20%		276 ± 21%		
	0.60	3	453 ± 2%	244 ± 14%		144 ± 7%		

[1] Number of replicate tests; [2] standard deviation (SD), given as a percentage of the average value.

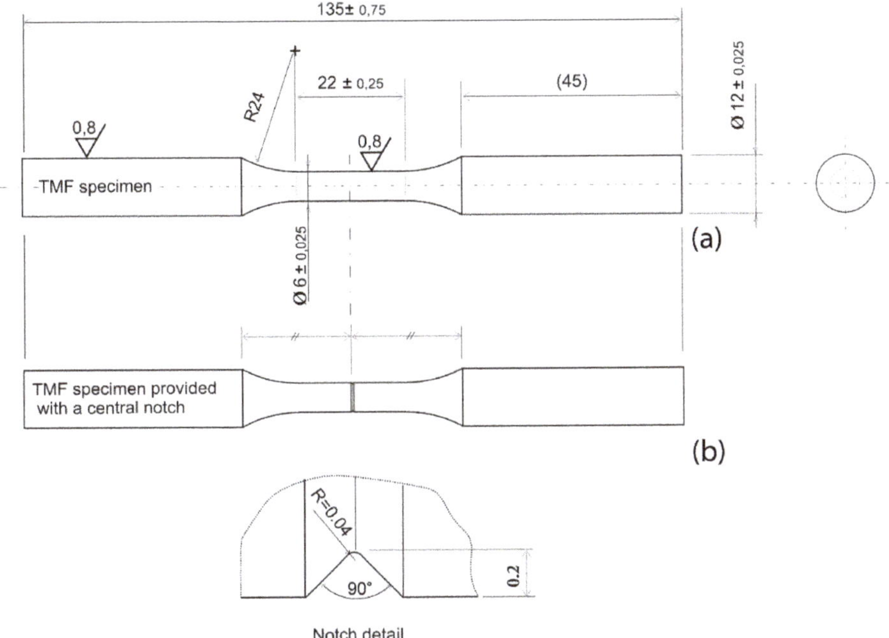

Figure 2. Geometry of thermo-mechanical fatigue (TMF) dog-bone specimen for (**a**) the case of unnotched (smooth) specimens, and (**b**) for circumferentially notched specimens with notch depth of 0.2 mm (as example) and a 0.04-mm notch-tip radius (dimensions are given in mm).

2.2. Experimental Set-up

A TMF test set-up, capable of independently imposing temperature and strain profiles on the specimen, was employed for TMF testing identical to the set-up that was as used by Ghodrat et al. For a detailed description of the TMF test set-up, the reader is referred to this previous work [3,4].

The material was subjected to temperature cycling between minimum and maximum values of 50 °C and 550 °C, respectively. Holding times of 30 and 140 s were applied at maximum and minimum temperatures, respectively, whereas the heating and cooling rates were 7 and 6 °C·s^{-1}, respectively. Out-of-phase TMF tests were performed at the following constraint levels: 125% (over constraint), 100% (total constraint), 75% (partial constraint), and 50% (partial constraint). The definition of the constraint levels is given in Section 2.4. It is acknowledged that 100% is the maximum theoretical constraint level possible, as a result of a thermal mismatch, in product service situations. However, the enforced 125% constraint level, enabled by using a TMF test machine, was chosen to broaden the range of TMF conditions, so as to model TMF crack-growth parameters more accurately. As a bonus, the 125% constraint TMF tests have the advantage of short lifetimes, i.e., short testing times.

The total strain (directly measured by the extensometer) was controlled in order to realize the abovementioned constraint values. Figure 3 shows the schematic input temperature and total strain profiles for different constraint conditions. Figure 4 presents the controlled temperature and strain profiles of a conducted 75% constraint TMF test, together with resulting cycling out-of-phase stress levels. Figure 5 exhibits typical resulting hysteresis loops, for the lowest (50%) and highest (125%) constraint levels tested, demonstrating that a stable regime sets in already after the first few loops.

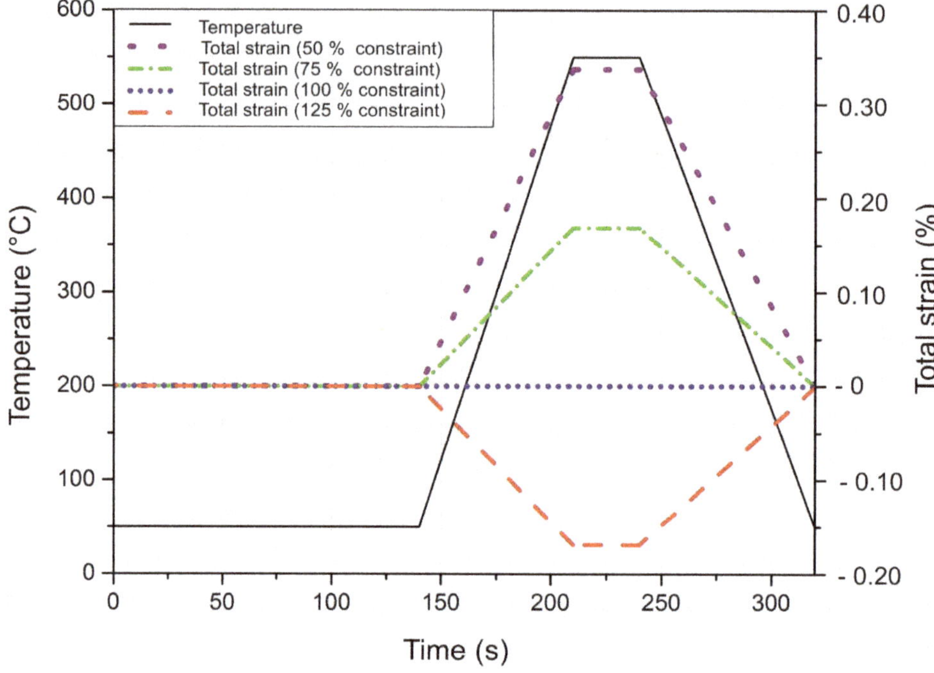

Figure 3. Temperature and total strain profiles for the out-of-phase TMF tests. For all constraint levels, temperature was varied from 50 °C to 550 °C in 70 s, and from 550 °C to 50 °C in 80 s. Holding times of 30 s and 80 s were introduced at 550 °C and 50 °C, respectively.

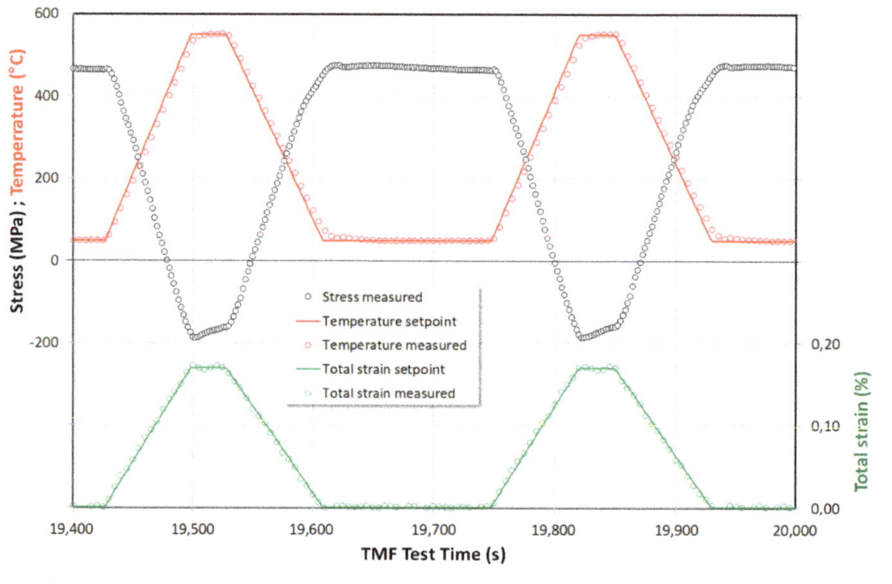

Figure 4. A typical tested TMF test stress–strain response for a 75% TMF test constraint level as an example.

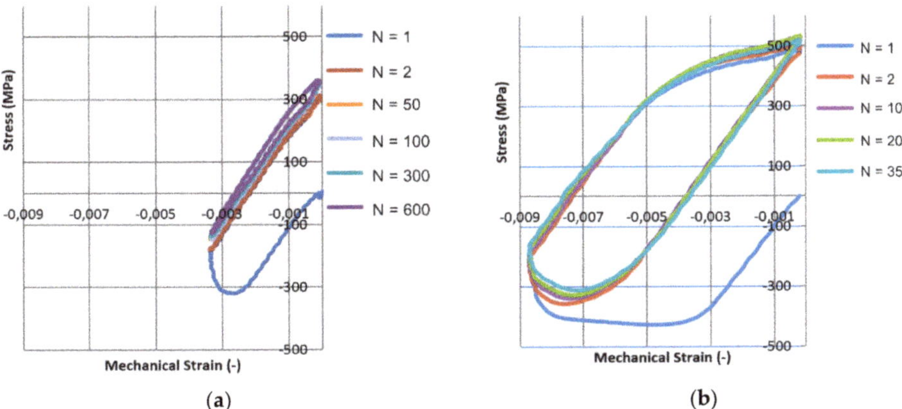

Figure 5. Strain hysteresis loops obtained from TMF tests performed on samples with a notch depth of 0.15 mm (**a**) at 50% partial constraint, and (**b**) at 125% total constraint. Total strain is defined as zero at room temperature. *N* depicts the number of TMF cycles.

2.3. The Necessity and Relevance of Using Notched Specimens

In order to shorten the tests and improve the statistical relevance of the data, the test program necessarily employed specimens with various machined notch depths. In our previous work [3], it was found that the average TMF lifetimes were calculated by taking the notch depth values as initial fatigue crack length [3], which resulted in an accurate match between calculated and experimental TMF lifetimes. Moreover, for unnotched (smooth) specimens, taking the average graphite particle size as initial crack length for the numerical lifetime calculations also produced good results, which indicated the general validity of the initial crack length concept. The quick crack initiation was also microscopically confirmed in recent work on TMF of CGI [8]. The principle of initial crack lengths is adopted in the current research on SGI, the results of which are discussed in Section 4.

The use of various notch depths is essential to determine the crack-growth model parameters. For instance, a model with two unknown model parameters needs at least two distinct boundary conditions to determine the model parameters. Different notch depths (i.e., initial crack lengths) produce distinct TMF lifetimes and can, therefore, provide the necessary boundary conditions to find the crack-growth model parameters. Also, with known model parameters, i.e., from notched specimens, TMF lifetimes of smooth specimens can be calculated by taking the average graphite particle size as initial crack length. This strategy is especially valuable for less severe (i.e., realistic) TMF conditions, since a TMF test for a smooth specimen typically requires a testing time of several weeks.

2.4. TMF Test Constraint Levels and TMF Lifetime

The following constraint test conditions were applied in this work: (i) partial constraint, (ii) total constraint, and (iii) over-constraint. The amount of constraint can be defined as the amount of thermal strain (e_{th}) that is converted into mechanical strain (e_{mech}), according to Equation (1). It is noted that e and S are respectively used for bulk strain and bulk stress levels, while ε and σ are used respectively for local strain and stress values, at the crack-tip level. The mechanical strain is defined as

$$e_{mech} = -\gamma \cdot e_{th},\tag{1}$$

where γ is the amount of constraint. The range of values for γ for the aforementioned constraint conditions are $0 < \gamma < 1$ (partial constraint), $\gamma = 1$ (total constraint), and $\gamma > 1$ (over-constraint) [9].

Thus, the thermal strain is partially, totally, or excessively converted into mechanical strain under partial, total, or over-constraint conditions respectively.

For a uniaxial case, the total strain (e_{total}), as could be measured by an extensometer, can be obtained by making use of the Equations (1)–(3), resulting in Equation (4).

$$e_{total} = e_{th} + e_{mech} \tag{2}$$

where e_{th} can be calculated using the values for the coefficient of thermal expansion (α) and the temperature difference (ΔT) (see Equation (3)).

$$e_{th} = \alpha \cdot \Delta T \tag{3}$$

From Equations (1)–(3), the following relationship is obtained between total strain, amount of constraint, and temperature (Equation (4)):

$$e_{total} = (1 - \gamma)\alpha\Delta T \tag{4}$$

The amount of constraint (γ) is commonly also designated as a percentage, for instance, a situation of total constraint ($\gamma = 1$) is equivalently denoted as 100% constraint.

The experimentally determined TMF lifetimes are denoted as N_{10}, signifying the number of cycles at which σ_{max} drops by 10% relative to the maximum value; the reason for having this criterion was explained in Appendix C of Reference [4].

3. TMF Crack-Growth Models

This work studies the influence of TMF constraint levels on TMF lifetime using two crack-growth models. Both models consider cyclic damage at the crack-tip level to control crack growth. The first model is based on Paris' fatigue crack-growth law and was successfully used in earlier work on CGI [3]. The second model is developed as an extension of the first model by considering the effect of TMF constraint levels on a more fundamental level. As argued subsequently, these models are designated as the local stress (or Paris' law) model and the local strain model, respectively. The term local refers to the crack tip, either sharp or blunt.

3.1. The Local Stress Crack-Growth Model

The Paris law equation establishes a relationship between the crack-growth rate $\frac{da}{dN}$ and a fracture-mechanical parameter, the stress-intensity range ΔK (see Equation (5)).

$$\frac{da}{dN} = C_{Paris}(\Delta K)^m, \tag{5}$$

where a is the crack size, N is the number of load cycles, while coefficient C_{Paris} and exponent m are material-dependent parameters. Since stress intensity characterizes the stress distribution ahead of a crack tip, the Paris model can also be designated as the local stress model.

3.1.1. Calculating TMF Lifetime by Numerical Integration (Local Stress Model)

The cyclic lifetime $N_{\Delta K}$ is obtained from Equation (5) by performing numerical integration by incrementing the crack size with small steps of, e.g., 0.001 mm. It is assumed that a crack initiates immediately from the machined notch or from a graphite particle in the case of an un-notched specimen. Assuming specific values for the Paris parameters C_{Paris} and m, the number of cycles is calculated as needed for the first 0.001 mm of crack growth around the entire circumference of the specimen. This process is repeated for subsequent steps of 0.001 mm, adjusting ΔK in each step in accordance with the increased crack length, until a final crack length value. The chosen final crack length for the iteration process is 2 mm, but this value was not found to be critical for the calculated number of cycles to failure. As shown in Section 4, relatively high TMF crack growth rates are found for crack length values above 1 mm, i.e., the final crack growth stage does not represent a significant portion of the TMF lifetime.

For smooth specimens, the average size of graphite particles (30 µm) is considered as the initial crack size. For notched specimens, the cracks originate from the graphite particle location at or near the notch tip. Thus, the effective initial crack length in the case of notched specimens is constituted by the size of the notch. For calculating ΔK, the K solution for mode I loading in a cylindrical specimen with a circumferential crack, reported in Reference [10], is used (see Equation (6)).

$$K_I = S \sqrt{\pi a} \frac{1}{\left(1 - \frac{a}{r}\right)^{\frac{3}{2}}} \left(1.122 - 1.302\frac{a}{r} + 0.988\left(\frac{a}{r}\right)^2 - 0.308\left(\frac{a}{r}\right)^3\right), \tag{6}$$

where r is the radius of the gauge length of the cylindrical test specimen, and S is the nominal (bulk) stress level. To account for crack closure during compression, ΔK was assumed equal to K_{max} in earlier work [3,4,6]. However, a more detailed analysis of test results shows that most of TMF lifetime is consumed during an initial crack extension of about 0.3–0.5 mm (originating from the machined notch). In this situation, i.e., for short cracks growing from the notch, the effect of crack closure is reported to be limited [11–13]. This indicates that the crack-tip stress intensity range (i.e., $\Delta K = K_{max} - K_{min}$) is more suitable to characterize TMF crack growth of cast iron.

Furthermore, to simplify calculations, the maximum stress range (ΔS) developed during each TMF test was taken to calculate ΔK. The coefficient C_{Paris} and exponent m are considered as model parameters that are fitted to experimental data, whereby only one specific set of values is accepted as best fit for all notch depths employed at a particular constraint level. For a more detailed description regarding the Paris law calculations, the reader is referred to References [3,4,6,10].

3.2. The Local Strain Crack-Growth Model

In Section 4.1, it is discussed that the local stress model, previously used successfully for CGI, also predicts TMF lifetimes for SGI well. However, each TMF constraint level tested for SGI required an adaption of the C_{Paris} parameter value to match calculated and experimental results. The variation in C_{Paris} values discounts the general applicability of the local stress model and inspired the development of a local strain crack-growth model.

To better capture the effect of constraint levels on TMF lifetimes, an improvement of the local stress model is proposed. From the literature, the acknowledged strain-life approach considers cyclic bulk strains to govern low cycle fatigue (LCF) lifetimes, with a combined role for cyclic bulk elastic strains and cyclic bulk plastic strains. For instance, the strain-life approach was used successfully to model the fatigue behavior of three types of metal alloys, with a good match between modeled and experimental results, with cycles to failure ranging from 10^3 to 10^7 cycles [14]. In the present work, the original strain-life approach is the point of departure to develop a crack-growth model based on local strains, i.e., strains acting at the crack-tip level, and assumed to originate from cyclic bulk strains.

A crack-growth model based on both cyclic bulk elasticity and cyclic bulk plasticity suggests a mechanism to be controlled by a combined LEFM and elastic plastic fracture mechanics (EPFM) damage mechanism. During each TMF cycle, the crack tip blunts during tensioning and sharpens again during compressing. Therefore, the crack tip can be considered to experience a distinct sharp stage and blunt stage during each TMF cycle. It is hypothesized that, during the sharp stage, an LEFM damage mechanism is active, while, during the blunt stage, an EPFM damage mechanism takes over. Most notably, the blunting mechanism can be associated with the EPFM concept of crack-tip opening displacement (CTOD). In References [15–17], a model for TMF-lifetime prediction was developed based on a crack-growth law, with the CTOD as main controlling parameter, signifying the relevance of blunting during TMF. However, a sharpening mechanism was not implemented in the referred models. The blunting and sharpening approach proposed here justifies a TMF crack-growth model involving subsequently applying LEFM and EPFM approaches, with a cumulative effect.

The blunting/sharpening concept is employed to translate bulk cyclic strains to local cyclic plastic strains, with separate mechanisms acting during the sharp and blunt stages of the crack tip. As a result, the local strain crack-growth model involves an unconventional combination of LEFM and EPFM.

3.2.1. The Blunting and Sharpening Mechanisms

Figure 6 shows a typical series of measured TMF (100% constraint tested example) hysteresis loops, combined with a sketch of an ideal elastic–plastic TMF hysteresis loop i.e., points A–B–C–D. As an example, a machined notch of 0.15 mm is illustrated, from which a TMF crack extends by about 0.2 mm, creating a crack length of 0.35 mm.

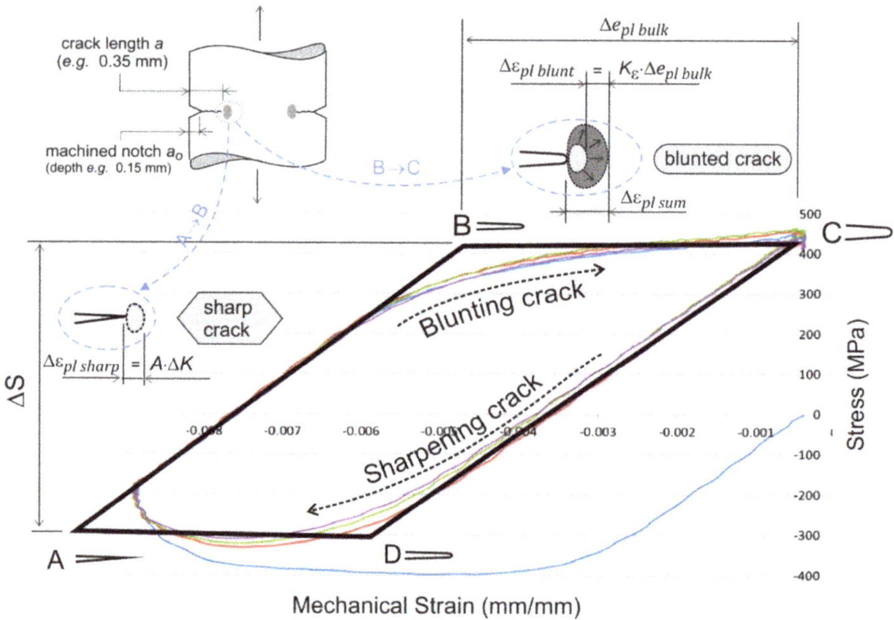

Figure 6. Development of bulk and local plasticity during the different stages of a TMF hysteresis loop. Note: For clarity, strain levels at the crack tip and notch root are illustrated in terms of dimensions of the resulting plastic zones. It is recognized that this representation is valid schematically only.

Starting at point A (550 °C), the tip of a TMF crack is sharp due to high compressive bulk stress levels, and pronounced plastic deformation in the bulk material during compression is present. Along path A → B, i.e., during the cooling phase, the sharp TMF crack tip is loaded toward a high bulk tensile stress level, where crack-tip plasticity is considered to develop according LEFM principles. Subsequently, along path B → C, the plastic deformation of the bulk material surrounding the crack tip blunts the crack tip, transforming the relatively sharp crack tip (point B) into a blunt crack tip (point C). The plastic bulk strain produced along path B → C causes a crack-tip strain development not related to the previously formed LEFM crack-tip plasticity (path A → B). At point C of the hysteresis loop, the total crack-tip strain is considered a superposition of the two independent contributions discussed above.

Following path C → D → A, i.e., heating up to 550 °C, the blunt crack tip is sharpened again due to the development of both high compressive stresses and pronounced compressive bulk plasticity. After reaching point A, the blunt crack transforms into a sharp crack again, as a starting point for the next TMF cycle.

3.2.2. Modeling Local Cyclic Plastic Strains for the Sharp and Blunt Crack Stage

A new model is proposed that incorporates the following elements: (i) applying the renowned strain-life approach from bulk material to the crack tip, (ii) applying the local stress model (Paris), and (iii) the assumed blunting/sharpening concept associated with the LEFM and EPFM mechanisms.

In the strain-life approach [12,14], cyclic bulk strain levels characterize the fatigue lifetime. For this, both the cyclic plastic bulk strain and the cyclic elastic bulk strain are taken into account. This constitutes the total cyclic bulk strain as a characterizing parameter and, therefore, could be referred to as the total strain-life approach. However, in practice, the model is usually concisely referred to as the strain-life approach, i.e., the term "total" is omitted.

The area surrounding the crack tip is not an isolated region, but is an integral part of the adjacent bulk material. A crack tip experiences a level of cyclic plasticity that is related to that in the surrounding bulk material. Therefore, a crack-growth model should involve a parameter related to cyclic bulk plasticity. In addition, the strain-life approach also incorporates cyclic bulk elastic strain as a parameter controlling fatigue lifetime. A crack-growth model should, therefore, also include a parameter reflecting the effect of bulk elasticity.

The local strain crack-growth model is developed respecting the aspects mentioned above. Where the strain-life approach is based on cyclic bulk strains, this crack-growth model is based on cyclic plastic strains at the crack-tip level. A polynomial relation is postulated, which relates the crack-growth rate (da/dN) to the cyclic crack-tip plastic strain ($\Delta\varepsilon_{pl\ sum}$), i.e.,

$$\frac{da}{dN} = B\left(\Delta\varepsilon_{pl\ sum}\right)^{m},$$ (7)

where B and m represent material-related constants.

The crack-tip cyclic plastic strain $\Delta\varepsilon_{pl\ sum}$ is considered to be the superposition of the cyclic crack-tip plastic strain developing during the sharp crack stage of the TMF cycle ($\Delta\varepsilon_{pl\ sharp}$), and that developing subsequently during the blunt crack stage of the TMF cycle ($\Delta\varepsilon_{pl\ blunt}$), i.e.,

$$\Delta\varepsilon_{pl\ sum} = \Delta\varepsilon_{pl\ sharp} + \Delta\varepsilon_{pl\ blunt}.$$ (8)

The local cyclic plastic strain originating from the blunt crack stage, ($\Delta\varepsilon_{pl\ blunt}$), is considered to be related straightforwardly to the cyclic plastic bulk strain, ($\Delta e_{pl\ bulk}$), as mentioned before. A blunt crack can also be perceived as a highly loaded notch, for which a strain concentration factor can be used to define the notch root strain level. Even in the case of full bulk plasticity, the principle of strain concentrations is documented to be still relevant [18]. Therefore, a strain concentration factor (K_ε) is used for estimating the cyclic crack-tip plastic strain, ($\Delta\varepsilon_{pl\ blunt}$), during the blunt crack stage (see Equation (10)). For clarity, it is mentioned that the cyclic plastic bulk strain, ($\Delta e_{pl\ bulk}$), is taken straightforwardly as the width of the stabilized hysteresis loop at a zero-stress level (see Figure 4).

The local cyclic plastic strain contribution for the sharp crack stage ($\Delta\varepsilon_{pl\ sharp}$) is hypothesized to be related to the stress intensity range (ΔK), i.e., the cyclic stress distribution ahead of the sharp crack tip is assumed to also characterize the cyclic strain distribution, within the plastic zone. The value of $\Delta\varepsilon_{pl\ sharp}$ should be considered a characteristic (or average) cyclic plastic strain near the crack tip, affecting TMF crack growth. Its value is assumed to be linearly related to ΔK using a proportionality factor A (see Equation (9)).

To summarize, in the local strain model, crack growth is controlled by cyclic plasticity at the crack-tip level ($\Delta\varepsilon_{pl\ sum}$), with its value being a superposition of the local cyclic plasticities produced during the sequence of the sharp stage and the subsequent blunt stage of each TMF cycle, i.e., $\Delta\varepsilon_{pl\ sharp}$ and $\Delta\varepsilon_{pl\ blunt}$, respectively. The sharp crack-stage contribution is related to cyclic bulk elasticity, while the blunt crack-stage contribution originates from cyclic bulk plasticity. Therefore, the strain-life approach and the local strain model share the same controlling cyclic bulk parameters.

The new local strain model is built-up with the following model equations:

$$\Delta\varepsilon_{pl\ sharp} = A \times \Delta K \text{ , i.e., based on LEFM;} \tag{9}$$

$$\Delta\varepsilon_{pl\ blunt} = K_\varepsilon \times \Delta e_{pl\ bulk} \text{ , i.e., based on EPFM.} \tag{10}$$

Combining Equations (7)–(10) gives the overall representation of the local strain crack-growth model as follows:

$$\frac{da}{dN} = B\left(\Delta\varepsilon_{pl\ sum}\right)^m = B\left(\Delta\varepsilon_{pl\ sharp} + \Delta\varepsilon_{pl\ blunt}\right)^m = B\left(A \times \Delta K + K_\varepsilon \times \Delta e_{pl\ bulk}\right)^m. \tag{11}$$

It can be seen from Equation (11) that the local strain model implicitly incorporates a role for the local stress model (see Equation (5)). For instance, at a 50% constraint level, the amount of cyclic bulk plasticity is negligible ($\Delta e_{pl\ bulk} \approx 0$) and, therefore, the local stress model and the local strain model coincide, i.e.,

$$\frac{da}{dN} = C_{Paris}(\Delta K)^m = B\,(A \times \Delta K)^m. \tag{12}$$

For a 50% constraint level, the local stress model parameter C_{Paris}, and the local strain parameter combination $B\,A^m$ should match, which is covered in Section 4.3.

In Equations (7)–(12), the local cyclic plastic strains produced during the sharp and blunt crack tip stage are represented by $\Delta\varepsilon_{pl\ sharp}$ (m/m) and $\Delta\varepsilon_{pl\ blunt}$ (m/m), respectively. The total local cyclic plastic strain produced during each TMF cycle is $\Delta\varepsilon_{pl\ sum}$ (m/m), the cyclic plastic bulk strain is $\Delta e_{pl\ bulk}$ (m/m), the crack length is a (m), the number of TMF cycles elapsed is N and the stress intensity range is ΔK (MPa $\sqrt{m}$), defined in Section 3.1.1. The units of the local strain model parameters A, B, and m can be deduced from Equations (9) and (11).

3.2.3. Calculating TMF Lifetime by Numerical Integration (Local Strain Model)

For the local strain model, TMF lifetimes are calculated numerically in a similar manner as described for the local stress-based model. Assuming specific values for the model parameters B, A, K_ε, and m, the number of cycles is calculated for the first 0.001 mm of crack extension around the entire circumference of the specimen (see Equations (7) and (11)). This process is repeated for subsequent steps of 0.001 mm, adjusting the values of $\Delta\varepsilon_{pl\ sharp}$ in each step in accordance with the increased crack length (the value of $\Delta\varepsilon_{pl\ blunt}$ is constant). Summing the results of all steps gives the total number of cycles to failure, $N_{\Delta\varepsilon_{pl\ sum}}$ A detailed account of all calculations and model parameters involved is given in Appendix A.

4. Results and Discussion

The results of the TMF tests performed are listed in Tables 2 and 3, with calculated lifetimes using the local stress and local strain crack-growth models, respectively. The tables share the same underlying experimental data, but Table 2 is organized around the constraint level (first column), while Table 3 is structured around the machined notch depth (first column). From the experimental results, a clear decrease in TMF lifetimes is found for higher constraint levels (see Table 2) and larger notch depth values (see Table 3).

In the following sections, it is presented that, by using notched specimens, both local stress and local strain crack-growth models are found to predict TMF lifetimes well for SGI, for all constraint levels tested, within a short testing time and with reduced scatter, while still being representative for TMF behavior of unnotched specimens. The local stress model does not directly address the effect of cyclic bulk plasticity, but accounts for the effect of bulk plasticity by adjusting the values of the (elastic) local stress model parameters. Therefore, the local stress model can be considered useful as a straightforward method to predict TMF lifetimes for a certain TMF constraint level, but does not identify or quantify the underlying contribution of cyclic bulk plasticity as is done in the local strain model.

Table 3. Values for TMF crack-growth parameters for all TMF tests performed, with calculated lifetimes based on the local strain model.

$a_o;K_\varepsilon$		Z [1]	$S \pm$ SD [2]	$\Delta e_{pl\,bulk}$	$N_{10} \pm$ SD [2]	$N_{\Delta\varepsilon_{pl\,sum}} \pm$ SD [2]	Δ [3]	$\Delta\varepsilon_{pl\,sum}$ [4]	R_{EPFM} [5]	ΔK_{a_o}
(mm);(–)	(%)	(–)	MPa	(%)	(–)	(–)	(%)	(%)	(–)	MPa $\sqrt{m}$
	125	-	836	0.37	-	73	-	$0.77\,^{0.27}_{0.49}$	0.64	9.1
0.03;	100	6	$718 \pm 3\%$	0.26	$157 \pm 56\%$	$170 \pm 7\%$	+8	$0.59\,^{0.24}_{0.35}$	0.60	7.9
1.35	75	-	654	0.10	-	636	-	$0.35\,^{0.21}_{0.14}$	0.39	7.1
	50	-	490	0.05	-	2478	-	$0.20\,^{0.16}_{0.07}$	0.30	5.4
	125	3	$836 \pm 3\%$	0.37	$29 \pm 42\%$	$26 \pm 7\%$	−9	$1.28\,^{0.62}_{0.66}$	0.51	21
0.15;	100	3	$772 \pm 3\%$	0.23	$48 \pm 22\%$	$56 \pm 8\%$	+17	$0.99\,^{0.58}_{0.41}$	0.42	19
1.80	75	3	$654 \pm 4\%$	0.10	$168 \pm 19\%$	$183 \pm 13\%$	+9	$0.67\,^{0.49}_{0.18}$	0.27	16
	50	3	$490 \pm 2\%$	0.05	$803 \pm 28\%$	$656 \pm 7\%$	−18	$0.45\,^{0.36}_{0.09}$	0.20	12
	125	2	$808 \pm 3\%$	0.40	$8 \pm 0.5\%$	$7 \pm 6\%$	−18	$2.15\,^{1.03}_{1.13}$	0.52	34
0.40;	100	2	$710 \pm 2\%$	0.25	$13 \pm 11\%$	$16 \pm 5\%$	+26	$1.61\,^{0.90}_{0.71}$	0.44	30
2.90	75	2	$631 \pm 0\%$	0.05	$56 \pm 6\%$	$83 \pm 0\%$	+49	$0.93\,^{0.80}_{0.13}$	0.13	27
	50	4	$480 \pm 7\%$	0.02	$284 \pm 56\%$	$256 \pm 19\%$	−10	$0.68\,^{0.61}_{0.07}$	0.07	20

[1] Number of replicate tests; [2] standard deviation (SD), given as a percentage of the average value; [3] relative difference of the calculate lifetime (local strain model), with the experimental lifetime; [4] superscripts and subscripts are the calculated local cyclic plastic ranges for the sharp and blunt crack stages, i.e., $\Delta\varepsilon_{pl\,sharp}$ and $\Delta\varepsilon_{pl\,blunt}$, respectively. The values represent calculated results for the initial crack length a_o. [5] Example: For $a_o = 0.15$ mm, in 100% constraint, values of $\Delta\varepsilon_{pl\,sharp} = 0.58\%$ and $\Delta\varepsilon_{pl\,blunt} = 0.41\%$ are calculated (see Appendix A). Therefore, $\Delta\varepsilon_{pl\,sum} = (0.58 + 0.41) = 0.99\%$. $R_{EPFM} = 0.41/0.99 = 0.42$. Note: The italic and underlined fonts, for the "smooth" specimens, are based on calculations only, using the values of cyclic bulk stress/strain ranges of the 0.15 notched experiments, in order to estimate TMF test conditions, for tests not actually performed.

4.1. Results of the Local Stress Crack-Growth Model

Results for the local stress crack-growth model (i.e., Paris), are presented in Table 2 and Figure 7. A good match between experimental results (N_{10}) and calculated results ($N_{\Delta K}$) is found. For smooth specimens (i.e., without machined notch), the average graphite particle size of 30 μm was taken as initial crack length.

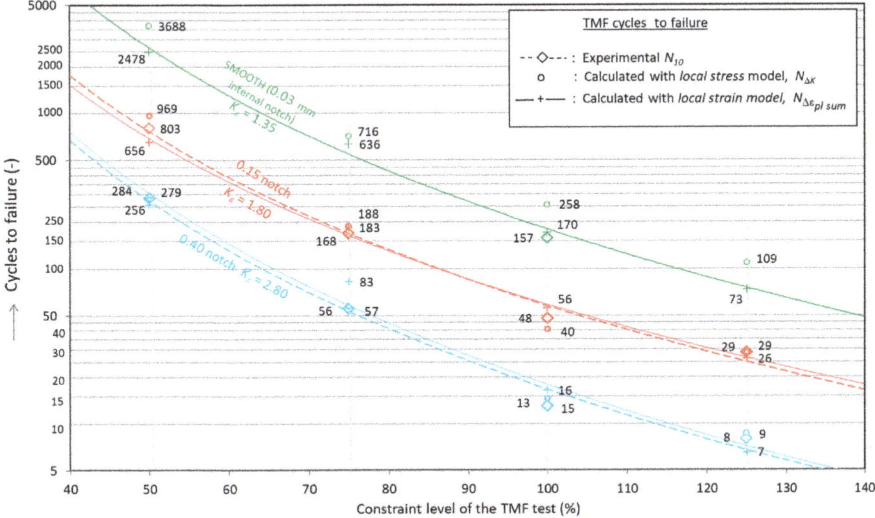

Figure 7. TMF lifetimes calculated according to both the local stress model and local strain model, as a function of constraint level.

As a starting point, the values of the model parameters C_{Paris} and m were roughly estimated, using a straightforward method (see Appendix B). Subsequently, the values of C_{Paris} and m were implemented in the numerical procedure, with values of C_{Paris} and m further optimized, to match calculated and measured TMF lifetimes. A uniform value was found for the local stress parameter $m = 3.58$, for all constraint levels. However, the values of C_{Paris} needed to be adjusted per constraint level. Therefore, this approach does not completely capture the influence of the constraint level on TMF lifetimes. For instance, at a 50% constraint level and a 125% constraint level, the constant C_{Paris} increased by a factor of five, from a value of 1.8×10^{-11} to a value of 9.0×10^{-11}. A clarification for the increase of C_{Paris} at higher constraint levels is given in Section 4.3.

4.2. Results of the Local Strain Crack-Growth Model

Results for the local strain crack-growth model are presented in Table 3 and Figures 7 and 8, revealing a good match between measured and calculated lifetimes for all constraint levels using a fixed set of three model parameters. For smooth specimens (i.e., without machined notch), the average graphite particle size of 30 μm was taken as initial crack length. Model parameter m was copied from the local stress model, signifying the coherence between two approaches.

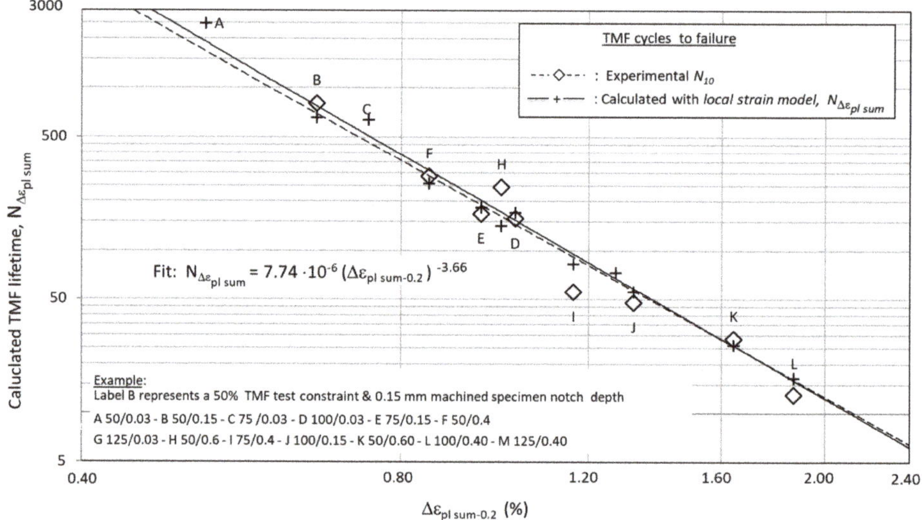

Figure 8. TMF lifetimes plotted as a function of $\Delta\varepsilon_{pl\ sum-0.2}$, with $\Delta\varepsilon_{pl\ sum-0.2}$ being the cumulative local cyclic plastic strain at the crack tip (i.e., $\Delta\varepsilon_{pl\ sum}$), calculated according to the local strain model for a crack length extension of 0.2 mm (ahead of the start crack length, i.e., ahead of the machined notch depth).

The values of model parameters A and B were initially estimated using assumptions explained in Appendix C. Using the estimated values as a starting point, the parameter values of A and B were optimized in the numerical lifetime model to match the experimental results. Values of $A = 3.00 \times 10^{-4}$ and $B = 62.0$ were found to give the best results.

An estimated value for the fourth model parameter, the strain concentration factor K_ε, would ideally be found using Neuber's Equation (12).

$$K_\sigma K_\varepsilon = K_t^2. \tag{13}$$

In Neuber's equation, K_σ and K_ε represent the ratios of the local notch stress/strain levels and the remote nominal stress/strain levels, respectively. The symbol K_t represents the stress concentration factor, for a pure elastic case, having a value depending on notch depth and notch root radius. In view of the extended bulk plasticity, during TMF, the value of K_σ is likely to approach unity ($K_\sigma \approx 1$), and K_ε would equal K_t^2. A problem, however, is that the radius of the blunt crack is not known and K_t, therefore, cannot be determined analytically. Apart from depending on the (unknown) blunt crack-tip radius, the value of K_t also depends on the notch depth and, for small notch depths (i.e., smooth specimens), a value $K_t \approx 1$ is reasonable to assume, i.e., $K_\varepsilon \approx 1$ also. Implementing a value $K_\varepsilon = 1$ in the numerical calculations for smooth specimens already gave a reasonable match between calculated and measured lifetimes. However, a value of $K_\varepsilon = 1.35$ was found to give the best match for smooth specimens. For notched specimens, the value of K_ε can be expected to rise, since K_t increases with notch depth, and, according to Neuber's equation, K_ε would also increase. For notch depths of 0.15 and 0.40 mm, implementing values for K_ε of 1.85 and 2.80, respectively, proved to give the best match between calculated and measured TMF lifetimes.

In the recent work of Besel and Breitbarth [19], plastic zone strain levels were quantified using finite element calculations and digital image correlation techniques. From this work, it was recognized that strain values vary within the plastic zone of a crack tip, with the highest values near the crack tip. In this respect, the value of $\Delta\varepsilon_{pl\ sum}$ (Equation (8)) should be interpreted as an average (or representative) crack-tip plastic strain range, controlling TMF crack growth. However, the levels of local strain at distinct ΔK values (see Table 3) were found to be of the same order of magnitude as reported in the literature for metals [19,20]. It is acknowledged that TMF lifetimes are also successfully determined as a function of plastic CTOD, with TMF crack growth rates found to be approximately linear with plastic CTOD values [15–17]. Both the plastic CTOD parameter and the $\Delta\varepsilon_{pl\ sum}$ parameter (used in Equation (7)) characterize a degree of cyclic plasticity at the crack tip. The plastic CTOD is defined for one specific position, being the exact location of the crack tip (this also holds for its associated plastic strain). However, $\Delta\varepsilon_{pl\ sum}$ (Equations (7) and (8)) constitutes a more averaged level of plasticity, within the crack-tip plastic zone as a whole. Considering the different backgrounds of the plastic CTOD parameter and the $\Delta\varepsilon_{pl\ sum}$ parameters, their role in controlling crack growth cannot be compared directly.

Figure 7 gives a straightforward and useful overview on how TMF lifetimes are affected by constraint levels. However, constraint levels are not the direct physical TMF damaging mechanism, but constitute boundary conditions. Representing TMF lifetimes as a function of the crack-tip cyclic plastic strain ($\Delta\varepsilon_{pl\ sum}$) should give more fundamental information about the TMF crack-growth mechanism. However, $\Delta\varepsilon_{pl\ sum}$ is a crack-tip parameter, increasing in value during crack growth and, therefore, complicating a straightforward characterization. From the numerical results, the development of crack length with the number of elapsed cycles is known, revealing that 80% of TMF lifetime is consumed during a limited crack extension of only 0.4–0.5 mm. Also, the early stage of crack growth should be associated with low crack-growth rates. In other words, the values of $\Delta\varepsilon_{pl.sum}$, present during the early stage of crack growth, dominate the overall TMF lifetimes.

Figure 8 is an alternative of Figure 7, with TMF lifetimes given as a function of a newly defined crack-tip parameter, designated as $\Delta\varepsilon_{pl\ sum-0.2}$. The new crack-tip parameter ($\Delta\varepsilon_{pl\ sum-0.2}$) represents the total local cyclic plastic strain ($\Delta\varepsilon_{pl\ sum}$), as calculated to be present at a crack extension of 0.2 mm (i.e., ahead of the machined notch). According to Figure 8, the TMF lifetimes can be well approximated by a polynomial function of $\Delta\varepsilon_{pl\ sum-0.2}$, qualifying the $\Delta\varepsilon_{pl\ sum-0.2}$ crack-tip parameter as a representative condition during crack growth, apparently able to characterize TMF lifetimes. In the final paragraph of Appendix A, an example for calculating the value of $\Delta\varepsilon_{pl\ sum-0.2}$ is given.

4.3. Comparing the Local Stress and the Local Strain Models

In previous sections, it was observed that for SGI both the local stress and the local strain models predict TMF lifetimes well, for all constraint levels under consideration. The good match for the local strain model can be ascribed to its incorporated $\Delta\varepsilon_{pl\ blunt}$ parameter, which is a function of cyclic bulk plasticity. In contrast, the local stress model (Paris) does not contain a dedicated parameter involving cyclic bulk plasticity; however, paradoxically, it is still capable of predicting satisfactory TMF lifetimes for all constraint levels. The reasoning below can shed some light on this paradox.

In general, fatigue lifetime largely depends on the initial crack-growth rate, since the first phase of crack growth is slow and, thus, consumes most part of fatigue lifetime. In this study, the initial crack-growth rates in the local stress model (Paris) and the local strain model were found to be of the same order of magnitude. Therefore, the two models produced comparable calculated TMF lifetimes. The local stress model adjusts the initial crack-growth rate by choosing the C_{Paris} parameter such that calculated results match experimental results. Therefore, the variation of C_{Paris} for different levels of constraint does not have a physical background, but only expresses the effect of constraint on TMF lifetime. In contrast, the local strain model captures the effect of the constraint level on TMF lifetime with a clear physical parameter, i.e., the local cyclic plastic strain at the crack-tip level ($\Delta\varepsilon_{pl\ sum}$).

The local stress model and the local strain models are, in principle, developed separately. Consequently, the values of the parameters of both models are also determined independently, without obvious interrelation. However, in the case of the 50% constraint TMF tests, the near lack of cyclic bulk plasticity observed enabled a direct comparison between both models, as discussed before in Section 3.2.2, considering Equation (12). Using the values found for the local strain model parameters A, B, and m, the parameter combination ($B\,A^m$) constitutes a value of 1.51×10^{-11}. For the 50% TMF constraint levels, the local stress model parameter C_{Paris} is determined to assume a value of 1.80×10^{-11}. Therefore, for the 50% TMF constraint level, both models predict a TMF crack growth rate of the same order of magnitude, validating the coinciding of both models (for 50% constraint levels).

4.4. The Balance between LEFM and EPFM Mechanisms during TMF (Local Strain Model)

As covered in Section 3.2, the local strain crack-growth model is based on a cyclic blunting/sharpening mechanism of the crack tip during each TMF cycle. As a result, in the local strain approach, crack growth originates from contributions produced separately during the sharp and blunt crack-tip stages. These separate crack-growth contributions are based on distinct principles of LEFM (sharp stage) and EPFM (blunt stage), and originate from cyclic bulk elasticity and cyclic bulk plasticity, respectively.

The separate roles of LEFM and EPFM during TMF can be quantified by considering the ratio of $\Delta\varepsilon_{pl\ blunt}$ and $\Delta\varepsilon_{pl\ sum}$ for each TMF condition tested, reflecting the contribution of the blunt crack stage to the overall TMF lifetime. This strain ratio, $\Delta\varepsilon_{pl\ blunt}/\Delta\varepsilon_{pl\ sum}$, is designated as R_{EPFM}. An R_{EPFM} value of unity indicates the case that TMF is dominated by EPFM crack-growth mechanisms, while a zero value reflects domination by the LEFM mechanism.

Table 3 shows R_{EPFM} for all TMF tests performed. With constraint levels increasing from 50% to 125%, R_{EPFM} was found to consistently increase, ranging from a value of 0.10 to 0.64, respectively. Clearly, an EPFM mechanism gradually takes over TMF crack growth at higher constraint levels due to the associated increased cyclic plastic bulk strain levels. However, it is striking that, even at the maximum 125% constraint level (with only a few cycles to failure), according to the local strain model, TMF is still controlled considerably by an LEFM crack-growth contribution and the associated cyclic bulk elasticity. The considerable role of LEFM, found for pronounced TMF conditions, contradicts the classical Manson–Coffin relationship approach, where LCF/TMF is predominantly attributed to cyclic bulk plasticity [12,18].

Considering Equations (9) and (11), at increased notch depths (i.e., longer initial crack lengths), the related higher initial ΔK values would suggest a transition toward the sharp crack stage mechanism (increase in $\Delta\varepsilon_{pl\ sharp}$). However, on average, per distinct constraint level, similar values for the R_{EPFM} parameter were found for all notch depths. For instance, for the 125% TMF constraint tests, for notch depths of 0.03 mm, 0.15 mm, and 0.40 mm, respective R_{EPFM} values of 0.64, 0.51, and 0.52 were found (see Table 3). The $R_{EPFM} \approx 0.5$ values found for both the 0.15-mm and 0.40-mm notches reflect similar roles for the blunt and sharp crack stages, independent of notch depth. It can be reasoned that, with an increase in notch depth, not only does the value of the initial ΔK level increases (i.e., sharp crack stage), but the strain concentration K_ε also becomes larger (i.e., blunt crack stage, increase in $\Delta\varepsilon_{pl\ blunt}$; see Equation (11)). The increases in both ΔK and K_ε with notch depth are probably in balance, keeping the sharp and blunt crack stage contributions in equilibrium independent of notch depth.

5. Summary and Conclusions

In the present paper, the lifetime was measured and numerically calculated in thermo-mechanical fatigue (TMF) tests under various constraint levels on spheroidal graphite cast iron (SGI) with temperatures cycling between 50 and 550 °C. The tested constraint levels were employed to predict TMF lifetimes more in line with actual service conditions of heavy-duty diesel engines.

For SGI, the fracture mechanical Paris law approach worked well to predict lifetimes for all TMF constraint levels. However, a different C_{Paris} parameter value was found for each TMF constraint level. Therefore, this approach does not completely capture the influence of the constraint level on TMF lifetimes. As this model is based on cyclic crack-tip stress distributions (characterized by ΔK), the Paris model was addressed as the *local stress* approach.

A second crack-growth model was proposed here, based on cyclic plastic strains at the crack-tip level. This model, which was labeled the *local strain* model, was found to predict TMF lifetimes well for all constraint levels, using a fixed set of four model parameters. The local strain model postulates a cyclic blunting and sharpening of the crack tip during each TMF cycle, involving contributions of both linear elastic fracture mechanics (LEFM) and elastic plastic fracture mechanics (EPFM) principles. The LEFM contribution is associated with the stress intensity range (ΔK) and, therefore, is largely controlled by cyclic elastic bulk deformation (i.e., cyclic bulk stress levels). The EPFM contribution is related directly to cyclic plastic bulk strain levels. This means that, in the local strain model, crack growth is induced by both cyclic bulk elasticity (LEFM) and cyclic bulk plasticity (EPFM). Therefore, the local strain crack-growth model and the established strain-life approach share TMF-controlling bulk parameters, demonstrating a coherence between the local strain model and the established strain-life approach. The coherence, however, is weakened for high TMF constraint levels, where the local strain model still involves a considerable role for cyclic bulk stress levels (contradicting the strain-life approach).

Although both the local stress model and the local strain model predict TMF lifetimes satisfactorily, the local strain model can be considered to have a clear physical basis, being the local cyclic plastic strain at the crack tip ($\Delta\varepsilon_{pl\ sum}$). In contrast, the local stress model can be considered a useful fitting method for a distinct TMF constraint level, but it does not physically account for the effect of constraint levels. The local cyclic plastic strain, as calculated to be present 0.2 mm ahead of the initial crack length (i.e., the machined notch depths), was found to be a suitable characterizing parameter to determine TMF lifetimes.

Author Contributions: The work presented in this paper results from a truly shared effort of all authors and originates from the PhD research of S.G. on TMF in Compacted Graphite Iron. In a subsequent MSc graduation project, A.K. validated the work of S.G. for Spheroidal Graphite Iron and additionally designed test methods for different TMF test constraint levels, with interpretation of test results. T.A.C.R. supported the experimental and theoretical work of S.G. and A.K. L.A.I.K. supervised the work of S.G. throughout and critically reviewed the manuscript. More in detail, A.K., S.G. and T.A.C.R. were in charge of the conceptualisation and the methodology. The experimental work was carried out in large part by A.K. The first draft of the paper was written by A.K. and then thoroughly reviewed and extended by S.G. and T.A.C.R. after joint discussion of the experimental results and joint analysis by all co-authors.

Funding: No external funding was provided, however, materials were supplied by DAF Truck N.V.

Acknowledgments: The authors would like to thank DAF Trucks Central Lab., Eindhoven, the Netherlands, for their contributions and provision of materials for this research. Also, the authors are grateful for the valuable contribution of Michael Janssen and Roel Marissen, who provided us with constructive advice during the progress of writing.

Conflicts of Interest: The authors declare no conflict of interest.

Abbreviations and Symbols

a	TMF crack length
A	Proportionality constant, linking ΔK to cyclic (sharp) crack-tip plasticity; see Equation (9)
α	Coefficient of thermal expansion
a_0	Depth of a machined notch, also being the assumed initial crack length
B	Proportionality constant in the local strain crack-growth model; see Equation (7)
C_1	A combination of parameters in the analytical solution for TMF lifetime, for crack growth according to local stress model (i.e., Paris' law)
C_{Paris}	Proportionality constant in the local stress crack-growth law (i.e., Paris' law); see Equation (5)
C_2	see C_1
CGI	Compacted graphite iron
CTE	Coefficient of thermal expansion
da/dN	TMF crack-growth rate
e_{total}	Bulk strain, as measured by the extensometer (i.e., total strain)
e_{mech}	Bulk strain, resulting from stress (i.e., mechanical strain)
e_{th}	Bulk strain, resulting from thermal expansion (i.e., thermal strain)
$\Delta e_{pl\ bulk}$	Cyclic plastic bulk strain, i.e., the width of the hysteresis loop
$\Delta \varepsilon_{pl\ sharp}$	Local cyclic plastic strain at the crack tip, produced during the sharp crack stage of the local strain model; see Equation (9) (mechanical strain)
$\Delta \varepsilon_{pl\ blunt}$	Local cyclic plastic strain at the crack tip, produced during the blunt crack stage of the local strain model; see Equation (10) (mechanical strain)
$\Delta \varepsilon_{pl\ sum}$	The cumulative local cyclic plastic strain at the crack tip; see Equation (8) (mechanical strain)
$\Delta \varepsilon_{pl\ sum(a_0)}$	The value of $\Delta \varepsilon_{pl\ sum}$ at the initial crack length a_0, being the depth of the machined notch (a similar notation is used for initial values of $\Delta \varepsilon_{pl\ sharp}$ and $\Delta \varepsilon_{pl\ blunt}$)
$\Delta \varepsilon_{pl\ sum-0.2}$	The value of $\Delta \varepsilon pl_{sum}$, at a crack length of $(a_0 + 0.2\ \text{mm})$, characterizing TMF lifetime
EPFM	Elastic plastic fracture mechanics
R_{EPFM}	The relative contribution on TMF of the blunt crack stage (local strain crack-growth model)
γ	Relative degree of thermal constraint during a TMF test
K_ε	Strain concentration factor (local strain/nominal strain)
K_t	Geometrical stress concentration factor (defined for elastic strains only)
K_σ	Stress concentration factor (local stress/nominal stress)
K_{min}	Minimum value of the stress concentration factor during TMF
K_{max}	Maximum value of the stress concentration factor during TMF
ΔK	Stress-intensity range $= (K_{max} - K_{min})$
ΔK_{a_0}	Initial stress-intensity range, with the machined notched depth (a_0) taken as initial crack length. For unnotched specimens, the average graphite particle size is taken as initial crack length.
$\Delta K_{\Delta a=0.2}$	The value of ΔK, at a crack length of $(a_0 + 0.2\ \text{mm})$
LCF	Low cycle fatigue
LEFM	Linear elastic fracture mechanics
N_{10}	Number of cycles at a 10% load drop in a TMF test (i.e., experimental cycles to failure)
$N_{\Delta K}$	Number of TMF cycles to failure, calculated by numerical integration of the local stress crack-growth model (i.e., Paris' law)
$N_{\Delta K-an.}$	Number of TMF cycles to failure, given by the analytical solution of the local stress crack-growth model (i.e., Paris' law)
ΔN	Number of TMF cycles elapsed
$N_{\Delta \varepsilon_{pl\ sum}}$	Number of TMF cycles to failure, calculated by numerical integration of the local strain crack-growth model
r	Radius of the cylindrical gauge length of the TMF test specimen
S	Bulk stress
ΔS	Nominal (bulk) stress range
SGI	Spheroidal graphite iron
SiMo	Cast iron with silicon and molybdenum as major alloying elements
ΔT	TMF cycle temperature range
TMF	Thermo-mechanical fatigue
Z	Number of replicate TMF tests

Appendix A. Example Calculations of TMF Lifetimes, Using the Local Strain Model

Example Calculation for 100% Constraint TMF Test Results, for a 0.15-mm Machined Notch

Averaging TMF test results of three replicate tests ($Z = 3$) resulted in a representative (measured) bulk cyclic stress level ΔS of 772 MPa, and a representative measured bulk cyclic plastic strain level ($\Delta e_{pl\ bulk}$) of 0.23%.

At the crack-tip level, the value of the total local cyclic plastic strain ($\Delta\varepsilon_{pl\ sum}$) is calculated below to be 0.99% This result consists of two contributions, being the cyclic crack-tip plastic strain for the sharp crack stage ($\Delta\varepsilon_{pl\ sharp}$) and the (additional) cyclic crack-tip plastic strain developed during the blunt crack stage ($\Delta\varepsilon_{pl\ blunt}$). These two contributions are calculated below under (i) and (ii), respectively.

i. The initial value of ΔK value (ΔK_{a_o}) is calculated according to Equation (6), i.e., $\Delta K_{a_o} = F(a_o/r)\cdot\Delta S\sqrt{\pi a_o}$ considering the notch depth as initial crack length (i.e., 0.15 mm), resulting in $\Delta K_{a_o} = F(0.15/3)\cdot 772\cdot\sqrt{\pi\cdot 0.00015} = 19.17$ MPa $\sqrt{m}$ (see Table 3). Using Equation (9), $\Delta\varepsilon_{pl\ sharp} = A\times\Delta K$, and, using the determined value of constant $A = 3\times 10^{-4}$, $\Delta\varepsilon_{pl\ sharp} = A\times\Delta K = 3\times 10^{-4}\times 19.2\times 100\% = 0.58\%$. It should be noted that the value of A is taken as a constant for all TMF tests performed.

ii. Using Equation (10), $\Delta\varepsilon_{pl\ blunt} = K_\varepsilon\times\Delta e_{pl\ bulk}$, with a value of $K_\varepsilon = 1.80$ and $\Delta e_{bulk} = 0.23\%$ results in $\Delta\varepsilon_{pl\ blunt} = 1.80\times 0.23 = 0.41\%$. It should be noted that the value of $K_\varepsilon = 1.80$ is identical for all TMF tests performed using a 0.15-mm notch depth. Superposition of contributions (i) and (ii), according to Equation (8), gives

$$\Delta\varepsilon_{pl\ sum} = \Delta\varepsilon_{pl\ sharp} + \Delta\varepsilon_{pl\ blunt} = 0.58 + 0.41 = 0.99\%, \tag{A1}$$

which is reported as

$$\Delta\varepsilon_{pl\ sum}{}^{\Delta\varepsilon_{pl\ sharp}}_{\Delta\varepsilon_{pl\ blunt}}\ \text{i.e.}\ 0.99\,{}^{0.58}_{0.41}. \tag{A2}$$

The initial value of $\Delta\varepsilon_{pl\ sum}$ of 0.99%, according to Equation (7) (or Equation (11)), using values of local strain-model constants B and m of 62 and 3.58, respectively, gives an initial crack-growth rate.

$$\frac{da}{dN} = B\left(\Delta\varepsilon_{pl\ sum}\right)^m = 62\times(0.0099)^{3.58} = 4.14\times 10^{-6}\,\frac{m}{\text{cycle}}. \tag{A3}$$

It should be noted that the local strain model constants B and m are identical for all TMF lifetimes calculations, being 62 and 3.58, respectively.

For subsequent discrete steps of 0.001 mm of crack growth, the number of cycles is calculated needed to cover this growth. For instance, for the first iteration step, covering a crack length interval from 0.150 to 0.151 mm, the following is found:

$$\frac{\Delta a}{\Delta N} = 4.13\times 10^{-6}\,\frac{m}{\text{cycle}}\ \text{or}\ \Delta N = \frac{\Delta a}{4.13\times 10^{-6}}\ \text{with}\ \Delta a = 10^{-6}\,m,\ \Delta N = 0.242\ \text{cycle / 0.001 mm}. \tag{A4}$$

Due to the small step size, the results of the second calculated iteration step, being the crack-length increment from 0.151 to 0.152, are almost similar to those of the first step.

Adding the ΔN values of all iteration steps gives a numerically calculated TMF lifetime $N_{\Delta\varepsilon_{pl\ sum}}$ of 56 cycles.

As an example, the situation of the iteration step 0.2 mm ahead of the machined notch tip is considered, i.e., from a crack length of 0.350 mm to 0.351 mm. In this step, due to the longer crack, the value of ΔK raised to a value of 30.31 MPa $\sqrt{m}$ leads to a value of

$$\Delta\varepsilon_{pl\ sharp} = A\times\Delta K = 3\times 10^{-4}\times 30.31\times 100\% = 0.91\%. \tag{A5}$$

The blunt crack-stage contribution ($\Delta\varepsilon_{pl\ blunt}$) is independent of crack length, i.e., 0.41%, as calculated above under (ii). Therefore, at a crack length $a = 0.35$ mm, the local cyclic plastic strain ($\Delta\varepsilon_{pl\ sum}$) can be reported as $1.32\,{}^{0.91}_{0.41}$. The new crack-growth rate is calculated as 1.27×10^{-5} m/cycle, while the increment from 0.350 mm to 0.351 mm consumes a number of cycles $\Delta N = 0.079$ cycle/0.001 mm.

The calculation for a specific crack extension of 0.2 mm was chosen because, in this case, the value of $\Delta\varepsilon_{pl\ sum}$ also constitutes the value of $\Delta\varepsilon_{pl\ sum-0.2}$, as discussed in Section 4.2. The value of $\Delta\varepsilon_{pl\ sum-0.2} = 1.32\%$ can also be observed for label J in Figure 8.

Appendix B. Initial Estimation of the Local Stress Model Parameters

The Local Stress Model Parameters

Considering the local stress model, $\frac{da}{dN} = C_{Paris}(\Delta K)^m$ (Equation (5)), the preliminary values for the parameters C_{Paris} and m are evaluated by considering an estimation of the analytical solution of the local stress crack-growth law. As simplification, the geometrical function $F\left(\frac{a_o}{r}\right)$, as defined in Equation (6), is taken as a constant value, being the initial value for a crack length equal to the machined notch depth. The analytical solution, with the bulk stress range (ΔS) given in MPa, can now be derived from Paris' law, and is also reported in the literature [12].

$$N_{\Delta K-an.} = \frac{1}{C_1 C_2}\left[\left(a_f\right)^{C_2} - \left(a_o\right)^{C_2}\right] \text{ with } C_1 = \left[C_{paris}(\pi)^{\frac{1}{2}m}\right]\left\{F\left(\frac{a_o}{r}\right)\cdot\Delta S\right\}^m ; C_2 = \left(1 - \frac{1}{2}m\right). \tag{A6}$$

Using the 50% constraint test results for 0.15-mm and 0.40-mm notched specimens, experimental results are summarized below. The 50% constraint test results are chosen specifically, because, at 50% constraint, the experiments showed a virtual absence of cyclic bulk plasticity, typical for Paris fatigue crack growth behavior.

0.15-mm machined notch: $(\Delta S)_{0.15} = 490$ MPa, $(N_{10})_{0.15} = 803$ cycles, $F\left(\frac{0.15}{3}\right) = 1.144$.

0.40-mm machined notch: $(\Delta S)_{0.40} = 480$ MPa, $(N_{10})_{0.40} = 284$ cycles, $F\left(\frac{0.40}{3}\right) = 1.196$.

Implementing the experimental lifetimes in the analytical solution (Equation (A6)) is achieved by equating the ratio of experimental lifetimes with the ratio of the analytical lifetimes, as shown below for the 0.15-mm and 0.40-mm machined notches, respectively.

$$\frac{(N_{10})_{0.15}}{(N_{10})_{0.40}} = \frac{(N_{\Delta K-an.})_{0.15}}{(N_{\Delta K-an.})_{0.40}}. \tag{A7}$$

For the analytical solution, the effect of the final crack length (a_f) can be omitted, since the slow crack growth at initial crack length (a_o) dominates the TMF lifetime, which results in the following:

$$\frac{(N_{10})_{0.15}}{(N_{10})_{0.40}} = \frac{\left(F\left(\frac{0.40}{3}\right)(\Delta S)_{0.40}\right)^m}{\left(F\left(\frac{0.15}{3}\right)(\Delta S)_{0.15}\right)^m} \cdot \frac{[-0.15]^{(1-\frac{1}{2}m)}}{[-0.40]^{(1-\frac{1}{2}m)}},$$

$$2.827 = (1.024)^m \times (0.375)^{(1-\frac{1}{2}m)},$$

$$\log(2.827) = \log(1.024)^m + \log(0.375)^{(1-\frac{1}{2}m)},$$

$$0.451 = m \times \log(1.024) + \left(1 - \frac{1}{2}m\right) \times \log(0.375),$$

$$= 0.010m - 0.426 + 0.213m,$$

resulting in the estimation $m = 3.932$.

The estimated value of C_{Paris} is found by substituting the estimated value of $m = 3.932$ in Equation (A6), using the experimental results of the 0.15-mm notch, being $(\Delta S)_{0.15} = 490$ MPa, $(N_{10})_{0.15} = 803$ cycles and $F\left(\frac{0.15}{3}\right) = 1.144$.

$$C_1 = \left[C_{paris}(\pi)^{\frac{1}{2}m}\right]\left\{F\left(\frac{a_o}{r}\right)\cdot\Delta S\right\}^m = C_{paris} \times 9.493 \times \{1.144 \times 490\}^{3.932} = 6.095 \times 10^{11} \times C_{Paris};$$

$$C_2 = \left(1 - \frac{1}{2}m\right) = -0.966;$$

$$N_{\Delta K-an.} = \frac{1}{C_1 C_2}\left[\left(a_f\right)^{C_2} - \left(a_o\right)^{C_2}\right] \approx \frac{1}{C_1 C_2}\left[-\left(a_o\right)^{C_2}\right] \leftrightarrow C_1 = \frac{1}{N_{\Delta K-an.}\cdot C_2}\left[-\left(a_o\right)^{C_2}\right];$$

With $C_1 = 6.095 \times 10^{11} \times C_{Paris}$; $C_2 = -0.966$; $a_o = 0.00015$ m, and $N_{\Delta K-an.} = 803$ cycles yields : $6.095 \times 10^{11} \times C_{Paris} = \frac{-1}{803 - 0.966}[0.00015]^{-0.966} \leftrightarrow C_{Paris} = 1.05 \times 10^{-11}$.

The estimated values of $C_{Paris} = 1.05 \times 10^{-11}$ and $m = 3.932$ were implemented as initial values in the numerical lifetime calculation for the 50% constraint experiments, as discussed in Section 3.1. By adjusting the parameter values incrementally (i.e., by trial and error), the numerical lifetime results were found to match all experimental results optimally for parameter values $C_{Paris} = 1.80 \times 10^{-11}$ and $m = 3.58$.

Appendix C. Initial Estimation of the Local Strain Model Parameters

As discussed in Section 3.2.2, in the local strain model the crack-growth rate is given by

$$\frac{da}{dN} = B\left(A \times \Delta K + K_\varepsilon \times \Delta e_{pl\ bulk}\right)^m. \tag{A8}$$

Preliminary values for the parameters A and B were evaluated by considering the local strain model to be equivalent to the local stress model, for the case of 50% constraint. Also, it was hypothesized that ratios of lifetimes found reflect ratios of initial crack-growth rates, since initial crack-growth rates can be expected to dominate TMF lifetimes.

Considering 50% and 100% constraint tests for 0.15 mm-notched specimens, the following can be found:

50% constraint case (0.15-mm notch): $(N_{10})_{0.15} = 803$ cycles, $(\Delta K)_{50\%} = 12.0$ MPa $\sqrt{m}$ (see Table 3);

100% constraint case (0.15-mm notch): $(N_{10})_{0.15} = 48$ cycles, $(\Delta K)_{100\%} = 19.0$ MPa $\sqrt{m}$ and $\left(\Delta e_{pl\ bulk}\right)_{100\%} = 0.0023$ (m/m) (see Table 3);

$$\frac{\left(\frac{da}{dN}\right)_{50\%}}{\left(\frac{da}{dN}\right)_{100\%}} = \frac{48}{803} = 6.00 \times 10^{-2} = \frac{B\ (A \times \Delta K_{50\%})^m}{B\ (A \times \Delta K_{100\%} + \cdot \Delta e_{pl\ bulk,\ 100\%})^m} = \frac{(A \times \Delta K_{50\%})^m}{(A \times \Delta K_{100\%} + \cdot \Delta e_{pl\ bulk,\ 100\%})^m}. \text{ To summarize, with}$$

$(\Delta K)_{50\%} = 12.0$ MPa $\sqrt{m}$, $(\Delta K)_{100\%} = 19.0$ MPa $\sqrt{m}$, and $\left(\Delta e_{pl\ bulk}\right)_{100\%} = 0.0023$ (m/m) and m = 3.58, a value for parameter A is estimated as 3.16×10^{-4}. The value of parameter B is estimated by considering the local stress and local strain models to coincide for the 50% constraint level, as is discussed in Section 4.3 i.e. $(C_{Paris})_{50\%} = BA^m$. With values of $(C_{Paris})_{50\%} = 1.80 \times 10^{-11}$, parameter A = 3.16×10^{-4} and parameter m = 3.58, the value of parameter B is estimated as 61.2.

The provisionally estimated values of $A = 3.16 \times 10^{-4}$ and $B = 61.2$ were implemented as initial values in the numerical lifetime calculation for the 50% constraint experiments, as discussed in Section 3.2.3. By gradually adjusting these estimated parameter values of A and B (i.e., by trial and error), the numerical lifetime results were found to match all experimental results optimally for parameter values $A = 3.00 \times 10^{-4}$ and $B = 62.0$. Therefore, the estimated and final values found for parameters A and B are in good agreement.

References

1. Gocmez, T.; Awarke, A.; Pischinger, S. A new low cycle fatigue criterion for isothermal and out-of-phase thermomechanical loading. *Int. J. Fatigue* **2010**, *32*, 769–779. [CrossRef]
2. Rémy, L.; Haddar, N.; Alam, A.; Koster, A.; Marchal, N. Growth of small cracks and prediction of lifetime in high-temperature alloys. *Mater. Sci. Eng. A* **2007**, *468–470*, 40–50. [CrossRef]
3. Ghodrat, S.; Riemslag, A.C.; Janssen, M.; Sietsma, J.; Kestens, L.A.I. Measurement and characterization of thermo-mechanical fatigue in compacted graphite iron. *Int. J. Fatigue* **2013**, *48*, 319–329. [CrossRef]
4. Ghodrat, S. Thermo-Mechanical Fatigue of Compacted Graphite Iron in Diesel Engine Components. Ph.D. Thesis, Delft University of Technology, Delft, The Netherlands, 10 June 2013.
5. Ghodrat, S.; Riemslag, A.S.; Kestens, L.A.I.; Petrov, R.H.; Janssen, M.; Sietsma, J. Effects of Holding Time on Thermo-Mechanical Fatigue Properties of Compacted Graphite Iron through Tests with Notched Specimens. *Metall. Mater. Trans. A* **2013**, *44*, 2121–2130. [CrossRef]
6. Ghodrat, S.; Kestens, L. Microstructural dependence of tensile and fatigue properties of compacted graphite iron in diesel engine component. *Steel Res. Int.* **2016**, *87*, 772–779. [CrossRef]
7. Norman, V.; Calmunger, M. On the micro- and macroscopic elastoplastic deformation behaviour of cast iron when subjected to cyclic loading. *Int. J. Plast.* **2019**, *115*, 200–215. [CrossRef]
8. Norman, V.; Skoglund, P.; Leidermark, D.; Moverare, J. The transition from micro- to macrocrack growth in compacted graphite iron subjected to thermo-mechanical fatigue. *Eng. Fract. Mech.* **2017**, *186*, 268–282. [CrossRef]
9. Kalra, A. Thermo-Mechanical Fatigue-Lifetime Determination of Diesel Engine Exhaust Manifolds. Master's Thesis, Delft University of Technology, Delft, The Netherlands, 30 August 2016.
10. Tada, H.; Paris, P.; Irwin, G. *The Stress Analysis of Cracks Handbook*, 2nd ed.; Paris Productions INC: Wilmington, MA, USA, 1985.
11. Vor, K.; Gardin, C.; Sarrazin-Baudoux, C.; Petit, J. Wake length and loading history effects on crack closure of through-thickness long and short cracks in 304L: Part I–Experiments. *Eng. Fract. Mech.* **2013**, *99*, 266–277. [CrossRef]
12. Fuchs, H.O.; Stephens, R.I. *Metal Fatigue in Engineering*, 2nd ed.; John Wiley and Sons: Hoboken, NJ, USA, 2001; Volume 186–234, pp. 103–110, ISBN 978-0-471-51059-8.
13. Ding, F.; Feng, M.; Jiang, Y. Modelling of fatigue crack growth from a notch. *Int. J. Plast.* **2007**, *23*, 1167–1188. [CrossRef]
14. Jiang, Y.; Ott, W.; Baum, C.; Vormwald, M.; Nowack, H. Fatigue life predictions by integrating EVICD fatigue damage model and an advanced cyclic plasticity theory. *Int. J. Plast.* **2009**, *25*, 780–801. [CrossRef]
15. Seifert, T.; Riedel, H. Mechanism-based thermomechanical fatigue life prediction of cast iron, Part I: Models. *Int. J. Fatigue* **2010**, *32*, 1358–1367. [CrossRef]
16. Seifert, T.; Maier, G.; Uihlein, A.; Lang, K.-H.; Riedel, H. Mechanism-based thermomechanical fatigue life prediction of cast iron. Part II: Comparison of model predictions with experiments. *Int. J. Fatigue* **2010**, *32*, 1368–1377. [CrossRef]
17. Antunes, F.V.; Rodrigues, S.M.; Branco, R.; Camas, D. A numerical analysis of CTOD in constant amplitude fatigue crack growth. *Theor. Appl. Fract. Mech.* **2016**, *85*, 45–55. [CrossRef]
18. Manson, S.S.; Halford, G.R. *Fatigue and Durability of Structural Materials*; ASM International: Geauga County, OH, USA, 2006; Volume 179–200, pp. 54–74, ISBN 0-87170-825-6.

19. Besel, M.; Breitbarth, E. Advanced analysis of crack tip plastic zone under cyclic loading. *Int. J. Fatigue* **2016**, *93*, 92–108. [CrossRef]
20. Ghodrat, S.; Riemslag, A.; Kestens, L. Measuring Plasticity with Orientation Contrast Microscopy in Aluminium 6061-T4. *Metals* **2017**, *7*, 108. [CrossRef]

Article

Application of a $\sqrt{\text{area}}$-Approach for Fatigue Assessment of Cast Aluminum Alloys at Elevated Temperature

Roman Aigner [1,*], Christian Garb [2], Martin Leitner [1], Michael Stoschka [1] and Florian Grün [2]

[1] Christian Doppler Laboratory for Manufacturing Process based Component Design, Chair of Mechanical Engineering, Montanuniversität Leoben, 8700 Leoben, Austria; martin.leitner@unileoben.ac.at (M.L.); michael.stoschka@unileoben.ac.at (M.S.)

[2] Chair of Mechanical Engineering, Montanuniversität Leoben, 8700 Leoben, Austria; christian.garb@alumni.unileoben.ac.at (C.G.); florian.gruen@unileoben.ac.at (F.G.)

* Correspondence: roman.aigner@unileoben.ac.at; Tel.: +43-3842-402-1454

Received: 5 November 2018; Accepted: 4 December 2018; Published: 6 December 2018

Abstract: This paper contributes to the effect of elevated temperature on the fatigue strength of common aluminum cast alloys EN AC-46200 and EN AC-45500. The examination covers both static as well as cyclic fatigue investigations to study the damage mechanism of the as-cast and post-heat-treated alloys. The investigated fracture surfaces suggest a change in crack origin at elevated temperature of 150 °C. At room temperature, most fatigue tests reveal shrinkage-based micro pores as their crack initiation, whereas large slipping areas occur at elevated temperature. Finally, a modified $\sqrt{\text{area}}$-based fatigue strength model for elevated temperatures is proposed. The original $\sqrt{\text{area}}$ model was developed by Murakami and uses the square root of the projected area of fatigue fracture-initiating defects to correlate with the fatigue strength at room temperature. The adopted concept reveals a proper fit for the fatigue assessment of cast Al-Si materials at elevated temperatures; in detail, the slope of the original model according to Murakami should be decreased at higher temperatures as the spatial extent of casting imperfections becomes less dominant at elevated temperatures. This goes along with the increased long crack threshold at higher operating temperature conditions.

Keywords: aluminum cast; fatigue strength; defects; hardness; tensile tests; elevated temperature

---

1. Introduction

Aluminum cast parts enable the manufacturing of quite complex geometries, due to their sound castability [1]. Furthermore, the investigated alloys also feature a proper fatigue strength; hence the achievement of lightweight design goals is additionally supported [2]. Therefore, Al-Si cast alloys are commonly used materials for automotive engine components, as reported in [3–6]. However, the manufacturing process, as well as the subsequent heat treatments, must be considered as they possess distinctive impacts on the resulting mechanical properties [7–9]. As such cast aluminum components are also exposed to elevated temperatures during service, the characterization of the material properties under these conditions is inevitable. Hence, the material properties of Al-Si cast alloys are investigated under elevated temperatures in a preliminary study [10]. This work extends the investigations from Garb et al. [10] by means of sampling positions, the statistical evaluation of fatigue fracture-initiating inhomogeneities and a defect-based fatigue strength model at elevated temperature of 150 °C, which was chosen in this study to reflect typical service conditions [11]. Another study [12] proposed the estimation of fatigue life under enhanced temperatures from tensile test results. Yet, increased temperatures significantly influence the microstructure as well as the fatigue lifetime [13,14]. The corresponding investigated sampling position parts can inherit different local microstructures due to the varying local cooling conditions. Therefore, the inhomogeneities and their statistically distribution regarding shape and spatial extent differ from each sample position to another. Tiryakioğlu proposed in [15] that fatigue-initiating defect sizes in aluminum cast material can be properly described by Gumbel [16], see Equation (1) or the Generalized Extreme Value distribution [17], see Equation (2).

$$G_{(d_{eq})} = exp\left[-exp\left(-\frac{d_{eq} - \lambda}{\delta}\right)\right]$$ (1)

With $G_{(d_{eq})}$ being the cumulative distribution function of the Gumbel distribution, its course is characterized by the distribution parameters λ and δ, also referred to as location and scale parameter.

$$P_{(d_{eq})} = exp\left\{-\left[1 + \xi\left(\frac{d_{eq} - \mu}{\sigma}\right)^{-\frac{1}{\xi}}\right]\right\}$$ (2)

Furthermore, $P_{(d_{eq})}$ equals the cumulative distribution function of the Generalized Extreme Value distribution depending on the equivalent circle diameter d_{eq}, the location μ, the scale σ and the shape parameter ξ. Furthermore, it was shown that the fatigue strength in the finite life region can be assessed by linking the cumulative probability density of fracture-initiating defect size distribution with the Paris-Erdoğan equation for stable crack growth [18]. Due to the fact that the focus of this work is the fatigue strength assessment of Al-Si cast materials in the long-life fatigue region, Murakami's $\sqrt{area}$ concept is used, see [19]. Hereby, the author in [19] stated that in the presence of extrinsic heterogeneities, such as flaws, the fatigue strength correlates well with the size of those defects. The spatial extent of the fatigue fracture-initiating defects is characterized by he square root of defect projection area $\sqrt{area}$, whereby the area is projected onto the plane perpendicular to the direction of maximum stress. A summary of this procedure is additionally given in [20]. The $\sqrt{area}$ is used as the size parameter of fatigue fracture-initiating flaws, due to the sound correlation with the maximal stress intensity factor at the crack tip K_{max}, see [20]. Murakami's approach is based on the material and defect location dependent coefficient C_1, the material constants C_2 and m, as well as on the Vickers hardness HV, see Equation (3).

$$\sigma_{f,1E7} = C_1 \cdot \frac{(HV + C_2)}{\sqrt{area}^{1/(2 \cdot m)}}$$ (3)

Murakami proposed the exponent m of the defect size to possess a constant value of 3, which revealed sound results for preliminary studies [21]. Furthermore, Murakami estimated the material and defect location coefficient C_1 as 1.43 for surface and 1.56 for subsurface cracks and the material dependent parameter C_2 to possess a constant value of 120 by applying the least squares method [19].

To assess the fatigue strength for operating conditions, this paper focuses on the influence of elevated temperature on the mechanical properties of two Al-Si cast materials, namely EN AC-46200 and EN AC-45500. Both tensile tests to assess the quasi-static properties as well as fatigue tests are performed at room and at elevated temperature to assess an advanced fatigue strength model based on the $\sqrt{area}$-approach by Murakami [19,20]. In summary, this paper scientifically contributes with the following working packages:

- Investigation of the fatigue strength of Al-Si cast materials at an elevated temperature of 150 °C
- Assessment of statistically defect distribution regarding spatial extent and shape
- Investigation of damage mechanisms at enhanced operating temperatures
- Evaluation of the material constants C_1, C_2 and m of Murakami's $\sqrt{area}$ concept for elevated temperatures

2. Materials and Methods

2.1. Materials

The examined materials are two Al-Si cast alloys with Strontium (Sr) as eutectic modifier and post-heat treatments to obtain T5 and T6 condition, see previously performed studies [10,22,23]. The T5 heat treatment consists of a quenching process and subsequent artificial aging whereas the T6 heat treatment usually involves three stages; the solution treatment at high temperatures, the quenching, and the age hardening process [24]. A summary of the T6 heat-treatment procedure and its impact on the resulting mechanical properties, depending on the chemical composition of the alloy and the exposure temperature and time, is given in [25]. The specimens are taken from gravity casted crankcases (CC) and cylinder-heads (CH) from varying positions (denoted as Pos #1 to Pos #3) exhibiting different local cooling conditions and therefore a variation in microstructure such as secondary dendrite arm spacing (sDAS) and micro pore size distribution. The nominal chemical composition of the investigated alloys is given in Table 1. The eutectic modifier Sr acts as micro alloying element and is measured in the ppm-range in the final cast material condition.

Table 1. Nominal chemical composition of the investigated cast alloys in weight percent.

Alloy	Si [%]	Cu [%]	Fe [%]	Mn [%]	Mg [%]	Ti [%]	Al [-]
EN AC-46200	7.5–8.5	2.0–3.5	0.8	0.15–0.65	0.05–0.55	0.25	balance
EN AC-45500	6.5–7.5	0.2-0.7	0.25	0.15 5	0.45	0.20	balance

In addition, an overview of the component's material specifications is listed in Table 2.

Table 2. Material specifications.

Part	Position	Alloy	Modifier	Heat Treatment	sDAS [μm]
CH	Pos #1	EN AC-46200	Sr	T5	24 ± 4.8
CC	Pos #2	EN AC-46200	Sr	T6	30 ± 7.3
CC	Pos #3	EN AC-46200	Sr	T6	72 ± 24.9
CH	Pos #1	EN AC-45500	Sr	T6	27 ± 6.6

2.2. sDAS and Metallographic Analysis

The sDAS is evaluated by means of a procedure proposed by [26], referred to as measurement method D. Hereby, metallographic specimens are taken out at the very sample positions and prepared by polishing. Afterwards, the specimens are investigated by digital optical microscopy. The secondary dendrite arm spacing is then calculated invoking the number of secondary arms along one side of a primary arm, such that dendrite asymmetry does not affect this method. This procedure is used, as it provides the most accurate estimation of secondary dendrite arm spacing, according to [26]. The investigation of metallographic sections in the testing region proposes a similar secondary dendrite arm spacing of Pos #1 for both alloys. The investigation of metallographic parameters in Pos #2 proposes a slightly enhanced local sDAS of 30 μm, regarding the values of Pos #1. The evaluated sDAS in sampling position Pos #3 is significantly higher though, due to lower cooling conditions compared to Pos #1 and #2, see [27]. The hardness measurements are conducted by means of a system of the type Zwick ZHU 2.5 TS1S (Ulm, Germany) and the metallographic samples are prepared by means of a Buehler BETA and a Struers CitPress system.

2.3. Quasi-Static and Fatigue Testing

The evaluation of the quasi-static material properties is carried out for room and elevated temperature, see [28] and [29] respectively. Each test series contains a minimum number of three specimens to statistically assess quasi-static properties. All tensile specimens and are conducted strain-controlled at a strain rate of 3.6×10^{-3} 1/s with a gauge length of $l_0 = 25$ mm, using an extensometer. Both the tensile tests and the fatigue tests are conducted at a hydraulic Instron Schenk testing system. To additionally evaluate both the fatigue strength and yield strength of the investigated materials not only at room but also at the ET of 150 °C, an Instron SFL 3119-400 Series temperature controlled chamber (Darmstadt, Germany) is used. The tensile test specimen geometry is displayed in Figure 1.

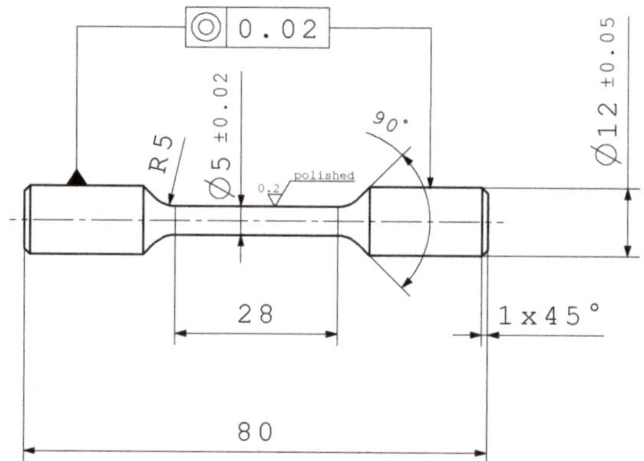

Figure 1. Tensile test specimen geometry in units of millimetre.

2.4. X-ray Computed Tomography

The investigated samples possess a low density and a small specimen diameter. Thus, the non-destructive investigation of selected specimens is carried out by a Phoenix/X-ray Nanotom 180, enabling a resolution of just 5 µm voxel-size, such that extrinsic flaws with a spatial extent of about 15 µm can be evaluated properly.

3. Results

3.1. Tensile Tests

Preliminary studies [30] stated that the fatigue resistance correlates linearly with the quasi-static material properties as the ultimate tensile strength (UTS) or the yield stress (YS) at corresponding temperatures. Furthermore, it is well known that exposure of Al-Si cast material at enhanced temperature leads to a significant decrease of the fatigue resistance [14]. Therefore, quasi-static tensile tests are not only conducted at room temperature, but also at an elevated temperature of 150 °C, using a heat chamber. An overview of the quasi-static test results is provided in Table 3.

Table 3. Results of quasi-static tensile tests.

Abbreviation	Temperature	UTS [MPa]	YS [MPa]	A [%]
EN AC-46200 Pos #2	RT	326	277	1.58
EN AC-46200 Pos #2	ET	265	245	2.96
EN AC-46200 Pos #3	RT	208	207	0.18
EN AC-46200 Pos #3	ET	187	187	0.19
EN AC-46200 Pos #1	RT	287	198	2.31
EN AC-46200 Pos #1	ET	234	184	5.25
EN AC-45500 Pos #1	RT	334	273	6.78
EN AC-45500 Pos #1	ET	259	233	9.55

EN AC-45500 Pos #1 reveals the highest UTS with 334 MPa at room temperature. In addition, EN AC-45500 Pos #1 also possesses the highest fracture elongation values *A* at room and elevated temperature. As listed in Table 3, the fracture elongation increases if tested at elevated compared to room temperature.

The results confirm preliminary studies [31], which proposed a function of the ascending elongation with increasing testing temperature. Yet, the authors in [31] do not register a significant increase of *A* below 270 °C. Another study [32] shows a strong increase of the elongation at fracture only for testing temperatures above 300 °C testing temperature. However, the test results observed in this study propose an increased fracture elongation already at 150 °C for the examined alloys, see Figure 2. Figure 2 displays representative tensile test results for the varying sampling positions at room and elevated temperature. Furthermore, representative microstructures at the very specimen locations are illustrated in detail in Figure 3.

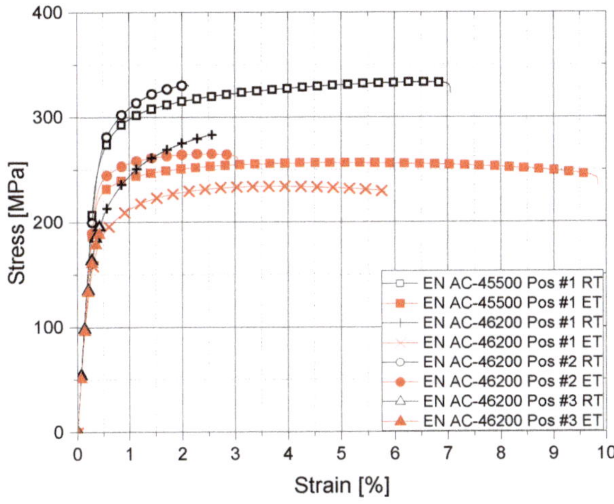

Figure 2. Representative tensile test curves at RT and ET.

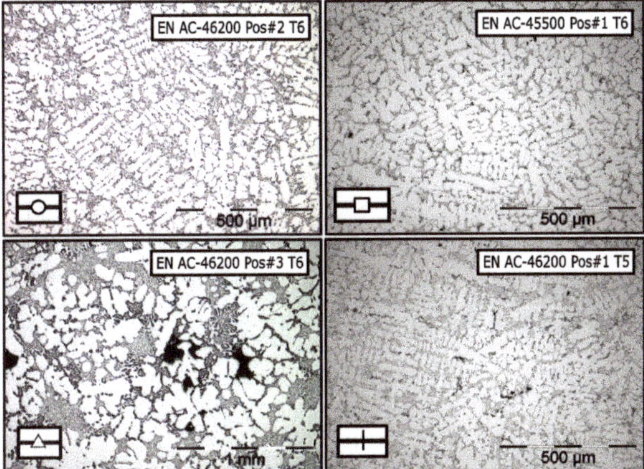

Figure 3. Microstructure at the investigated sample positions.

On the other hand, the yield strength in general decreases at elevated temperature, as listed in Table 3. The mean reduction of the UTS from room to elevated temperature is approximately 17%. This goes along with preliminary studies in [31–33].

3.2. Hardness Measurements

As stated in [34,35] the quasi-static material properties such as the yield strength exhibit a basic coherence with the macrohardness. Also, the investigation of the position depending hardness is from significance to set up Murakami's $\sqrt{area}$ model [19] in a uniform way for the investigated specimen series. Therefore, Vickers hardness measurements are executed at room temperature in line with [36] applying a testing force of 3 kp. A preliminary study [37] describes a linear relationship of Al-Zn-Mg alloys between the Vickers hardness HV and the flow stress. In [38] it is also stated that the Brinell hardness BHN corresponds well to the YS for both A356 and A357 alloys. In [39] an evaluation of models for hardness-yield strength relationships is presented. Drouzy and Richard [40] presented a quite simple linear ratio between the YS and the Brinell hardness BHN of underaged Al-Si alloys. However, it is also shown that the YS-BHN relationship of Al-Si alloys can be rather described by a hysteresis, which shows a significantly higher slope in the underaged region than the proposed relationship from [40]. Murakami's $\sqrt{area}$ model takes the local Vickers hardness HV into account, as it relates well to the fatigue strength [20]. Therefore, the local hardness for the sample positions must be investigated not only at room but also at elevated temperatures.

To estimate the hardness at an elevated temperature, a linear fit, using the least square method, between the measured hardness and calculated yield strength, is carried out, see Figure 4. The evaluated slope of the fit is almost two times the proposed value of approximately 2.85 from [40]. Each test series contains a minimum number of three tensile specimens and three hardness tests at the corresponding position to tone down outlier values. Subsequently, the yield strength of tensile tests at elevated temperatures is re-inserted in the estimated YS-HV fit, to estimate the Vickers hardness at elevated temperatures.

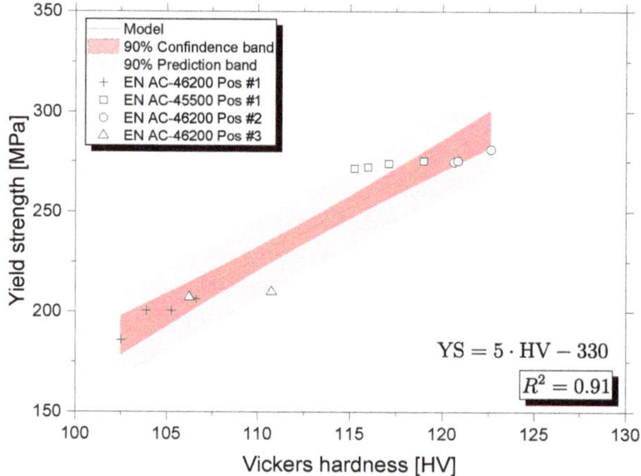

Figure 4. Evaluated correlation between Vickers hardness and Yield strength for Al-Si castings.

3.3. Fatigue Tests

The fatigue tests are conducted under alternating tension-compression testing. This deduces a load stress ratio, lower load level to upper load level, of R $= -1$. Therefore, the mean stress is equal to zero. All fatigue tests are executed at a hydraulic testing machine with a testing frequency of 30 Hz until either specimen burst failure or run out at 1×10^7 load cycles. In line with the quasi-static tests, the fatigue tests at elevated temperature are performed using a heat chamber. The fatigue resistance in the finite life region is statistically evaluated by means of the ASTM E 739 standard [41]. This methodology assumes a constant variance in the finite life region. Furthermore, the fatigue data in the run out region is statistically investigated based on the $arcsin\sqrt{p}$ methodology, see [42]. This procedure approximates the probabilities of survival in the long-life region by a $arcsin\sqrt{p}$ function and is a proven methodology to statistically estimate the mean value and standard deviation of the fatigue strength and likewise minimum life, see [43]. As proposed in [44], the slope k_2 of the run out region is assumed to be five times the slope k_1 in the finite life region. This assumption is also verified in [45–47]. All fatigue data has been normalized by the UTS at room temperature of EN AC-46200 -Pos #2, evaluated to 326 MPa, see Table 3.

The tests are performed until a total number of 1×10^7 load cycles, because preliminary studies showed defect correlated specimen failure in the HCF region between 1×10^6 and 1×10^7 load cycles, see [3,48,49]. All evaluated S/N-curves are displayed including the 10 and 90 % probability of survival scatter band. The statistical evaluation of scatter bands in the run out region is conducted by means of the *arcsin*$\sqrt{\bar{p}}$ methodology [42,43]. The evaluated fatigue strength data for each sample position as well as the scatter band in the run out region is listed in Table 4. In Figure 5 the S/N-curve of EN AC-45500 with T6 heat treatment at Pos #1 at room and at elevated temperature is displayed. While the slope k_1 is almost identical for RT and ET, the number of cycles N_D for the transition region rises with increased testing temperature from approximately 1×10^6 to about 4.27×10^6 load cycles. The investigated fatigue strength of EN AC-46200 with T5 heat treatment at Pos #1 states also a similar k_1 at room temperature, see Figure 6. On the other hand, the high-cycle fatigue strength at 1×10^7 load cycles $\sigma_{f,1 \times 10^7}$ significantly decreases by about 21 % at 150 °C. The investigation of the fatigue strength from EN AC-46200 with T6 heat treatment at Pos #2 assumes a slightly shallower S/N-curve at elevated temperature, represented by the slope in the finite life region k_1. Furthermore, the transition point N_D rises to 1.75×10^6 load cycles, with a decrease of about 7% in fatigue strength $\sigma_{f,1 \times 10^7}$, as seen in Figure 7. Finally, EN AC-46200 with T6 heat treatment at Pos #3 shows a significant shallower slope in the finite life region k_1, while the evaluated fatigue resistance $\sigma_{f,1E7}$ decreases only by 2% at 150 °C testing temperature, see Figure 8. It must be pointed out that the depicted minor reduction of the fatigue strength in Pos #3 is within the scatter band of the S/N-curves.

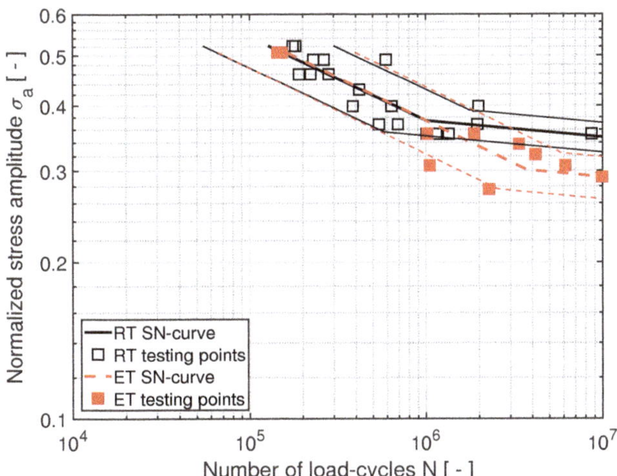

Figure 5. S/N-curves of EN AC-45500 T6 Pos #1 at RT and ET.

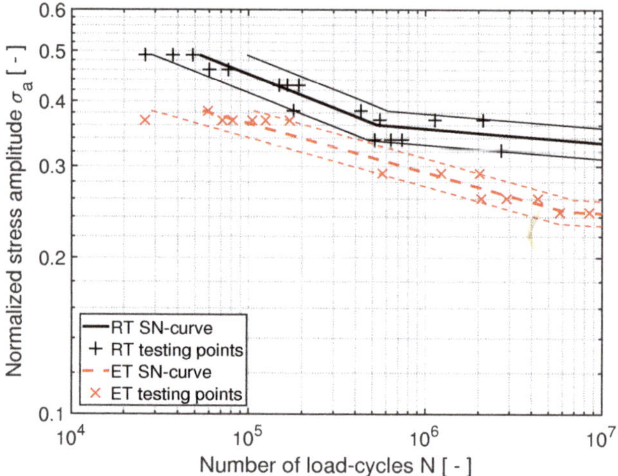

Figure 6. S/N-curves of EN AC-46200 T5 Pos #1 at RT and ET.

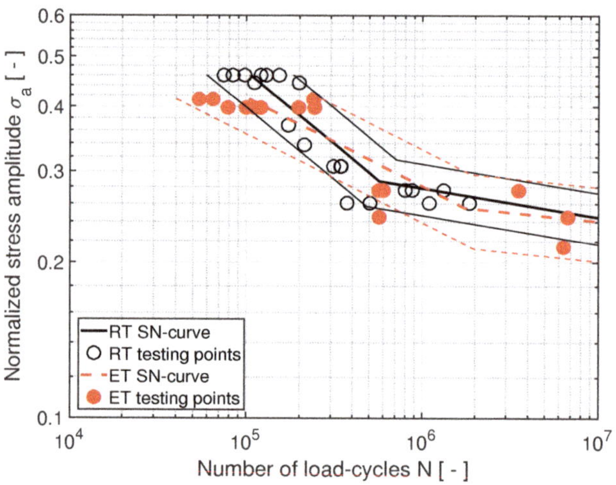

Figure 7. S/N-curves of EN AC-46200 T5 Pos #2 at RT and ET.

Table 4. Fatigue test results at room and elevated temperature.

Abbreviation	Temperature	$\sigma_{1\times10^7,norm}$	N_D	k_1	$\frac{1}{T_s}$
EN AC-46200 Pos#2	RT	0.245	5.71×10^5	3.58	1.256
EN AC-46200 Pos#2	ET	0.241	1.75×10^6	5.85	1.391
EN AC-46200 Pos#3	RT	0.176	3.92×10^6	5.13	1.278
EN AC-46200 Pos#3	ET	0.172	2.68×10^6	6.41	1.210
EN AC-46200 Pos#1	RT	0.332	5.29×10^5	7.40	1.147
EN AC-46200 Pos#1	ET	0.244	6.16×10^6	10.65	1.115
EN AC-45500 Pos#1	RT	0.348	9.99×10^5	6.22	1.140
EN AC-45500 Pos#1	ET	0.291	4.27×10^6	7.48	1.206

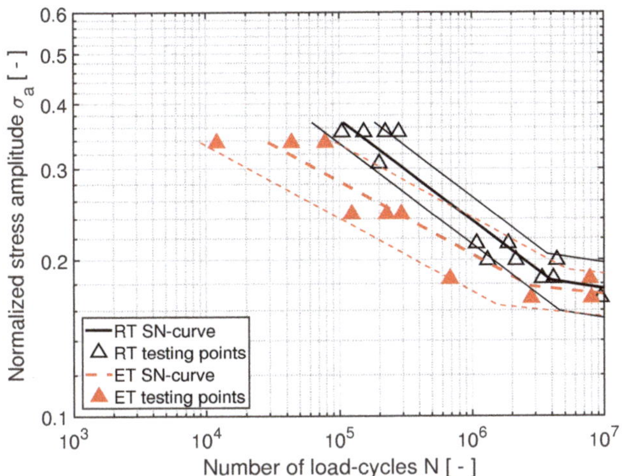

Figure 8. S/N-curves of EN AC-46200 T5 Pos #3 at RT and ET.

To investigate the fracture-initiating defects, it is from utmost importance to analyze all tested specimens either by means of digital microscope as well as by SEM. The detected crack-initiating flaws are characterized by means of geometrical parameters such as the square root of the effective defect area. In line with the procedure proposed in [20], the size of fatigue fracture-initiating defects is characterized by the square root of the projected area of the flaw, perpendicular to the load direction. This methodology is displayed in Figure 9 using the example of a fracture-initiating heterogeneity at Pos #3. Furthermore, to characterize not only the fracture-initiating defects by means of fractography, selected specimens are investigated non-destructively with X-ray computed tomography (XCT). This methodology supports the holistic characterization of the defect population respectively its spatial extent and is further described in [22,48].

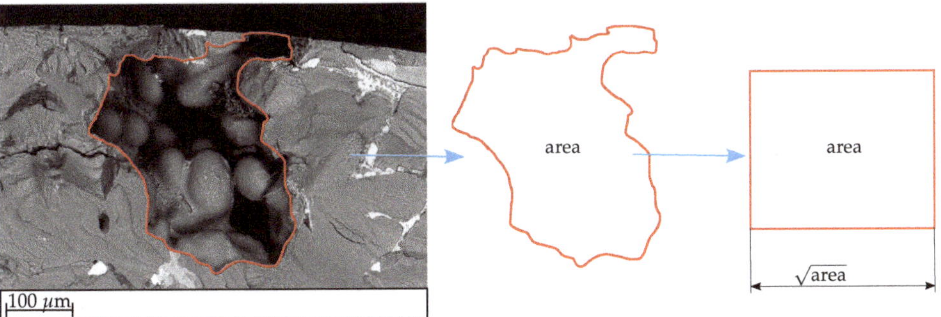

Figure 9. Defect size measurement methodology and representative fracture-initiating micropore at Pos#3.

Figures 9 and 10 show a fracture surface with a crack origin at a micro pore. At room temperature, all tested EN AC-46200 specimens initiate from a such micro pores. On the other hand, Figure 11 displays a different cause of failure. At ET, the stress intensity is enhanced in the defect-near area while the activation energy of slip-planes decreases due to the increased operating temperature. Therefore, specimens tested at a higher temperature activate a different failure mechanism.

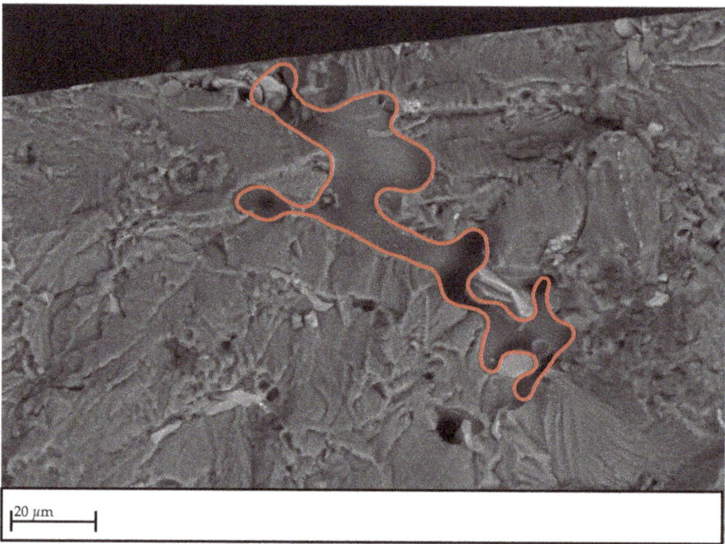

Figure 10. Representative fracture-initiating micropore at Pos#2.

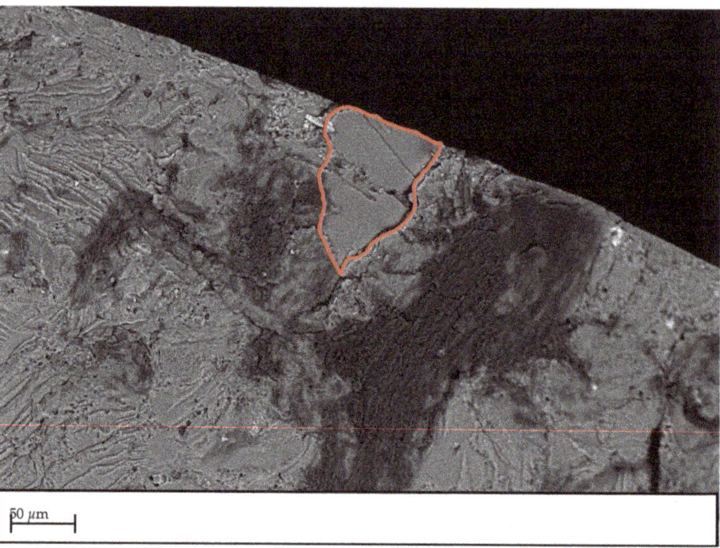

Figure 11. Representative fracture-initiating slipplane at Pos#1.

Some specimens would reveal a crack initiation right at the surface along with crack propagation at large slip bands. Nevertheless, some investigated specimens revealed mixed defect mechanisms of micro pores and large slipping areas. At the latter ones, the crack initiates at an intrinsic inhomogeneity and possesses a stable crack growth as with increasing crack length, the stress intensity rises near the crack tip. If the stress intensity factor and the activation energy based on the thermal energy reach a certain threshold, the crack starts to slip over a slip-band area during one single load-cycle and therefore significantly increases the crack growth. As a result, the remaining fatigue strength is reduced compared to an arbitrary defect with a similar $\sqrt{area}$.

As the fractography results show, this failure mechanism occurs especially at EN AC-46200 T5 and EN AC-45500 T6, where a major part of the specimens at Pos #1 inherit a slip-band-induced failure. It must be noted that EN AC-45500 T6 Pos #1 even shows a slip-band-like failure mechanism also at room temperature. In Figure 12, the different damage mechanism fractions of the corresponding positions and alloys are displayed.

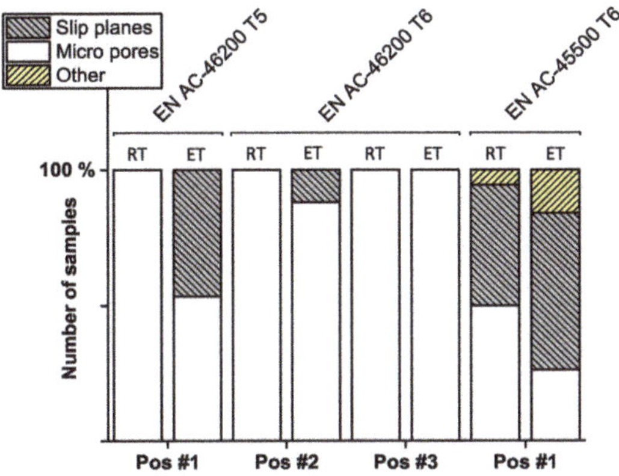

Figure 12. Comparison of the crack-initiating failures at RT and ET.

As proposed by [50] either Gumbel or GEV distributions are applicable for fatigue-initiating defects. The distribution parameters are evaluated by means of a maximum likelihood method, as presented in [51].

$$\phi = \frac{1}{n} \sum_{i=1}^{n} \frac{d_{eq,i}}{d_{max,i}} \tag{4}$$

The probability of occurrence of an arbitrary d_{eq}, based on the cumulative density function of the GEV-fit for each sampling position is displayed in Figure 13. The probability of occurrence of a d_{eq} of 200 µm is less than 10% for sampling positions #1 and #2, whereas at Pos #3 the probability of occurrence for the same equivalent diameter possesses a value of just above 97%.

The cumulative density function of the distribution is computed by means of Equation 2, using the equivalent circle diameter (d_{eq}) from the most critical defects, whereas ξ is denoted as the shape σ the scale and μ the location parameter. The equivalent circle diameter d_{eq} of one flaw can easily be derived by multiplying $\sqrt{area}$ with the factor $\frac{2}{\sqrt{\pi}}$. The ratio of the equivalent circle diameter d_{eq} to the maximum diameter d_{max} describes the shape ϕ of the crack-initiating pores [52], see Equation (4). The shape factor ϕ therefore ranges between zero and one, indicating the roundness of a crack-initiating pore. The lower ϕ, the more complexly shaped is the defect. Thus, a circle-shaped flaw would possess a shape factor $\phi = 1$. The shape of representative defects with varying sphericity is presented in detail in [53].

Both the evaluated distribution parameters as well as the mean shape of the defects are listed in Table 5. The investigation of the different fracture-initiating defect sizes proposes a significant different micro pore size distribution at Pos #3, in line with the increased local sDAS. The mean shape ϕ of the flaws in Pos #3 also reckon them to possess a more spherical shape.

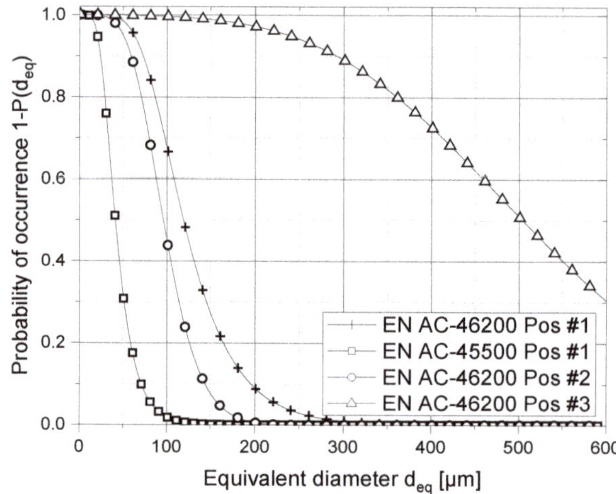

Figure 13. Probability of occurrence of fracture-initiating defects.

Table 5. GEV-fit parameters based on evaluated d_{eq} of fracture-initiating defects.

Abbreviation	ξ	σ	μ	ϕ
EN AC-46200 Pos #1	0.03	39.03	104.60	0.52
EN AC-46200 Pos #2	−0.12	29.83	85.10	0.48
EN AC-46200 Pos #3	−0.19	168.57	444.88	0.60
EN AC-45500 Pos #1	0.06	14.50	36.08	0.49

3.4. Fatigue Assessment Model

To assess the fatigue strength of Al-Si alloys incorporating manufacturing process-based inhomogeneities, a defect size-based material model is set up. The main causes of failure, evaluated in both EN AC-46200 Pos #1 an EN AC-45500 Pos #1, are basically not based on micro pores. Therefore, the defect-based material model from Murakami [19] is mainly set up for EN AC-46200 Pos #2 and Pos #3 as herein the main failure cause can be assigned to micro porosity. However, specimens with defect correlated crack initiation from Pos #1 of both alloys are also displayed in the adopted material model, see Figure 14. A parameter set, containing the stress amplitude at 1×10^7 load cycles $\sigma_{f,1\times10^7}$ of run-outs, the stress amplitude σ_a and number of load cycles N of tests in the finite life region, as well as the corresponding defect size $\sqrt{area}$ are required. To increase the applicable data a power-function like projection method for specimens failed in the finite life region is executed. This method is suitable to increase the data points in the run out region of $N = 1\times10^7$ without significant falsification of the scatter band in the HCF region $1/T_s$, see [21].

The original data from Murakami [19] proposes a coefficient of $m = 3$ for the exponent of the defect size. The parameters C_1 and C_2 are material dependent constants. This approach provides reasonable defect-based material models for Al-Si alloys at room temperature. However, the exponent of the defect size can vary at elevated temperatures, as it represents the slope of the material model. Therefore, the coefficient m is not further considered to be a constant, see Equation (5).

$$\sigma_{1E7} = C_{1,T} \cdot \frac{(HV + C_{2,T})}{\sqrt{area}^{1/(2 \cdot m_T)}} \tag{5}$$

This model is supplied with all parameter sets depending on their alloy, position, and testing temperature. Furthermore, the hardness of the corresponding positions at elevated temperature is estimated based on the experimentally evaluated yield strength, as discussed in Section 3.2. Next, the parameters $C_{1,T}$ and $C_{2,T}$ as well as the slope m_T are estimated applying a non-linear solver, using the least square method. The evaluated parameters maintain a slope of $m_T = 3.02$ for specimens tested at room temperature, which agrees with the proposed constant of $m = 3$ in the original model, see Equation (3). However, the data at elevated temperature leads to a change of the slope value. The least square method proposes a coefficient of $m_T = 4.05$ for specimens tested at an elevated temperature of 150 °C. The defect-based material model for elevated temperatures with the 90 and 10% probability scatter bands is displayed in Figure 14. The defect correlated fatigue strength is restricted by two major limits. On the one hand, the upper boundary is set by the fatigue strength of near defect-free material where the area of the crack-initiating inhomogeneities tends to zero. On the other hand, the lower boundary is determined by huge defects, such that the stress intensity factor along the internal crack meets the long crack threshold $\Delta K_{th,lc}$, see Equation (6).

$$\Delta K_{th,lc} = Y \Delta \sigma_{1E7} \sqrt{\pi \sqrt{area}} \tag{6}$$

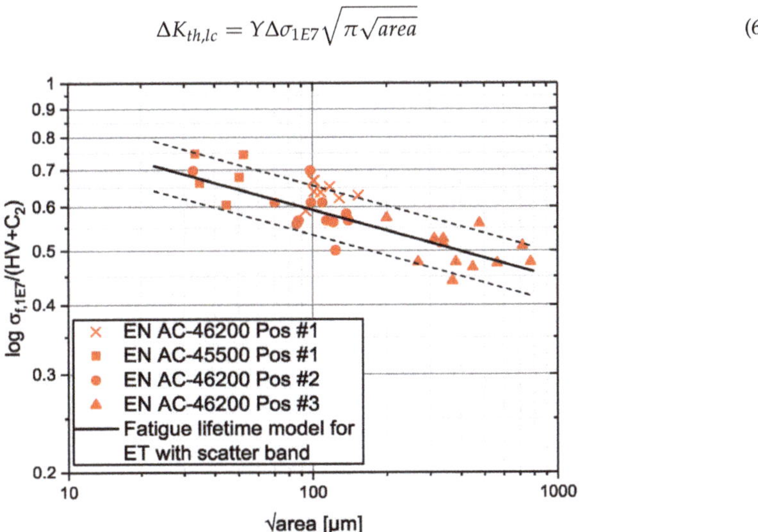

Figure 14. Defect correlated fatigue lifetime model with evaluated coefficient m at elevated temperature.

Preliminary studies [54–56] revealed that $\Delta K_{th,lc}$ rises in line with increasing testing temperature, until a critical temperature is met. This results from an increased plastic zone in front of the crack tip. The size of the monotonic plastic zone can be estimated by Irwin's estimation, based on the YS of the material, see Equation (7) [57].

$$r_{pl} = \frac{K^2}{\pi YS^2} \tag{7}$$

The relationship proposed by Irwin applies for a monotonic plastic zone with no crack closure occurring [58]. To evaluate the cyclic plastic zone, incorporating crack closure effects, K is superimposed by ΔK_{eff}, such that:

$$\Delta r_{pl} = \frac{\Delta K_{eff}^2}{\pi 4YS^2} \tag{8}$$

As the YS decreases with elevated testing temperature at an average of 10.7%, the size of the cyclic plastic zone Δr_{pl} therefore rises by 25.4%, see Equation (8). Hence, the plastic-induced crack closure effects significantly increase, in line with the extended plastic zone [59–61], resulting in an elevated long crack threshold $\Delta K_{th,lc}$ at higher temperatures. As a result, defects with large spatial extent do not affect the fatigue strength at elevated temperatures as significantly as for room temperatures, see Equation (6). The increased slope of the defect-based material model, represented by the coefficient m_T, thus is deduced by increased $\Delta K_{th,lc}$ at elevated temperatures.

4. Discussion

In this paper, fatigue strength, hardness and quasi-static tests are executed at both RT and an ET of 150 °C. The investigated materials are different Al-Si alloys with subsequent T5 or T6 heat treatment and varying local solidification times, leading to locally adjusted microstructural features, such as secondary dendrite arm spacing and micropore distribution. Furthermore, fractographic analysis is conducted to evaluate the fracture-initiating defects and their spatial extent. To consider the decreasing hardness at elevated temperature, a linear relation between the yield strength and Vickers hardness is evaluated to establish a link between room- and elevated temperature data. The hardness at elevated temperatures is subsequently estimated based on quasi-static test results at 150 °C testing temperature, using the evaluated YS-HV relationship displayed in Figure 4. The investigation of the fatigue strength at elevated temperature reveals a significant decrease of 21% for specimen Pos #1 of both observed alloys in respect to the fatigue strength at room temperature. The evaluation of fatigue strength at elevated temperature at specimen Pos #2 and Pos #3 revealed a minor decrease of 2%, though. The tensile test data at elevated temperature reveals an overall increase of elongation at fracture of about 65%, as well as a decrease of Young's modulus of about 4%, due to elevated dislocation mobility. The specimen positions inherit major differences in fatigue fracture-initiating defects, as the statistical evaluation of flaw sizes states. This is mainly caused by the local significantly varying cooling rates between the sample positions and results in a broad spectrum of fatigue fracture-initiating material heterogeneities. Hence, the commonly used $\sqrt{area}$ fatigue strength assessment model proposed by Murakami can be invoked as basic fatigue assessment strategy. The extended model, introduced in this paper, takes a modified and temperature dependent slope value of $m_T \approx 4$ into account to properly assess the fatigue strength in the long-life region at elevated temperatures, see Equation (5).

As shown in Figure 15, the proposed fatigue assessment model is compared with the normalized experimental results. The model for elevated temperatures properly meets the experimental fatigue lifetime data. The scatter band and the mean value are statistically estimated, approaching the methodology from [41]. The statistically evaluation of the fatigue assessment model reveals a comparably minor scatter band $1/T_m$ of 1.18.

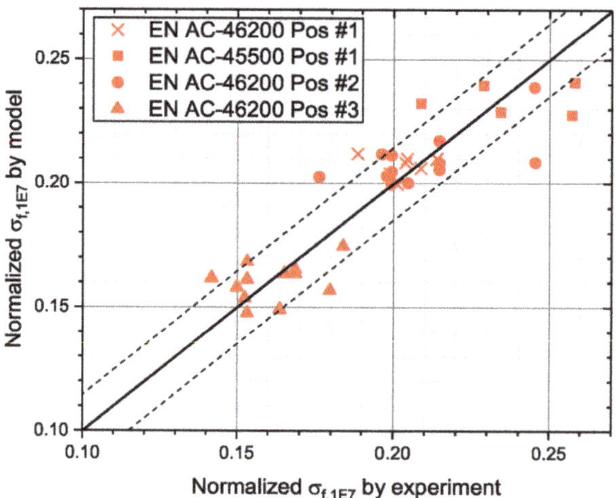

Figure 15. Validation of the $\sqrt{area}$ model for elevated temperatures.

5. Conclusions

Based on the conducted research work and assessed results, the following conclusions can be drawn:

- The statistically evaluated fatigue strength of all tested alloys drops when being exposed at elevated temperatures compared to fatigue lifetime at room temperature.
- A significant change in damage mechanism at elevated temperatures is observed. A major part of the specimens taken from both EN AC-46200 Pos #1, as well as EN AC-45500 Pos #1, maintain slipping areas as crack origins.
- While the original $\sqrt{area}$ model proposed by Murakami provides a sound fit for room temperature with a slope of $m = 3$, the slope changes at elevated temperature. The estimated slope $m_T \approx 4$ suggests an increased long crack threshold at elevated temperature, caused by more pronounced plasticity-induced crack closure effects. Therefore, the impact of increasing defect sizes on the anticipated fatigue resistance generally declines at elevated temperatures.
- Comparing the fatigue strength of the introduced extension of Murakami's model for higher operating temperatures, the experiments reveal a proper relationship. The mean value of the suggested model turns out to be slightly conservative.

Subsequent work focuses on the validation of the proposed model for further elevated temperature values incorporating additional fatigue tests. Moreover, the yield strength versus hardness relationship may be investigated in more detail at higher temperatures by additional testing series. Finally, near defect-free material must be experimentally analyzed in terms of fatigue strength at elevated temperatures to define the very upper boundary of the defect-based material model most accurately.

Author Contributions: Conceptualization, R.A., C.G. and M.L.; methodology, R.A., C.G. and M.L.; software, R.A. and C.G.; validation, R.A. and M.L.; formal analysis, R.A. and M.L.; investigation, R.A. and C.G.; resources, M.S. and F.G.; data curation, R.A. and C.G.; writing—original draft preparation, R.A.; writing—review and editing, M.L. and M.S.; visualization, R.A.; supervision, M.S. and M.L.; project administration, M.S. and F.G.; funding acquisition, M.S. and F.G.

Acknowledgments: The financial support by the Austrian Federal Ministry for Digital and Economic Affairs and the National Foundation for Research, Technology and Development is gratefully acknowledged. Furthermore, I would like to thank the industrial partners BMW München Group and Nemak Dillingen GmbH for the excellent mutual scientific cooperation within the CD-laboratory framework. The authors would like to thank Nemak Linz and Nemak Dillingen for providing the cast components. Further on, execution of the CT scans at the Materials Center Leoben GmbH is highly appreciated. Finally, financial project support was gladly provided by MAGMA Gießereitechnologie GmbH and AVL List GmbH. Financial support by the Austrian Federal Government (in particular from Bundesministerium für Verkehr, Innovation und Technologie and Bundesministerium für Wissenschaft, Forschung und Wirtschaft) represented by Österreichische Forschungsförderungsgesellschaft mbH and the Styrian and the Tyrolean Provincial Government, represented by Steirische Wirtschaftsförderungsgesellschaft mbH and Standortagentur Tirol, within the framework of the COMET Funding Programme is gratefully acknowledged.

Conflicts of Interest: The authors declare no conflict of interest. The founding sponsors had no role in the design of the study; in the collection, analyses, or interpretation of data; in the writing of the manuscript, or in the decision to publish the results.

Abbreviations

The following abbreviations are used in this manuscript:

RT	Room temperature
ET	Elevated temperature
λ	Location parameter of the Gumbel distribution
δ	Scale parameter of the Gumbel distribution
GEV	Generalized Extreme Value distribution
ξ, σ, μ	Shape, scale, and location parameter of the GEV

$\sigma_{f,1\times10^7}$	Long-life fatigue strength amplitude at 1×10^7 load cycles
$\Delta\sigma_{1\times10^7}$	Long-life fatigue strength range at 1×10^7 load cycles
σ_a	Fatigue strength amplitude
HV	Vickers hardness
C_1, C_2,m	Constants of the $\sqrt{area}$ approach by Murakami
sDAS	Secondary dendrite arm spacing
YS	Yield strength
UTS	Ultimate tensile strength
HCF	High-cycle fatigue
E	Young's modulus
A	Fracture elongation
N_D	Number of load cycles at transition knee point of S/N-curve
N	Number of load cycles until failure
k_1	Slope in the finite life region of S/N-curve
T_S	Scatter band of S/N-curve in the HCF region
BHN	Brinell hardness number
SEM	Scanning-electron-microscopy
XCT	X-ray computed tomography
d_{eq}	Equivalent circle diameter
d_{max}	Maximal spatial extent of an inhomogeneity
ϕ	Shape factor
n	Number of defects
$C_{1,T}$, $C_{2,T}$,m_T	Parameters of the extended $\sqrt{area}$ approach for elevated temperatures
K	Stress intensity factor
K_{max}	Maximal stress intensity factor
$\Delta K_{th,lc}$	Long crack threshold
ΔK_{eff}	Effective crack threshold
Y	Geometry factor
Δr_{pl}	Cyclic plastic zone
r_{pl}	Monotonic plastic zone
T_m	Scatter band of the validation of the fatigue assessment model

References

1. Di Sabatino, M.; Arnberg, L. Castability of aluminium alloys. *Trans. Indian Inst. Met.* **2009**, *62*, 321–325. [CrossRef]
2. Canales, A.A.; Carrera, E.; Silva, J.T.; Valtierra, S.; Colás, R. Mechanical properties in as-cast and heat treated Al-Si-Cu alloys. *Int. J. Microstruct. Mater. Prop.* **2012**, *7*, 281. [CrossRef]
3. González, R.; González, A.; Talamantes-Silva, J.; Valtierra, S.; Mercado-Solís, R.D.; Garza-Montes-de Oca, N.F.; Colás, R. Fatigue of an aluminium cast alloy used in the manufacture of automotive engine blocks. *Int. J. Fatigue* **2013**, *54*, 118–126. [CrossRef]
4. Mattos, J.; Uehara, A.Y.; Sato, M.; Ferreira, I. Fatigue properties and micromechanism of fracture of an AlSiMg0.6 cast alloy used in diesel engine cylinder head. *Procedia Eng.* **2010**, *2*, 759–765. [CrossRef]
5. Mbuya, T.O.; Sinclair, I.; Moffat, A.J.; Reed, P. Micromechanisms of fatigue crack growth in cast aluminium piston alloys. *Int. J. Fatigue* **2012**, *42*, 227–237. [CrossRef]
6. Javidani, M.; Larouche, D. Application of cast Al–Si alloys in internal combustion engine components. *Int. Mater. Rev.* **2014**, *59*, 132–158. [CrossRef]
7. Tabibian, S.; Charkaluk, E.; Constantinescu, A.; Guillemot, G.; Szmytka, F. Influence of process-induced microstructure on hardness of two Al–Si alloys. *Mater. Sci. Eng. A* **2015**, *646*, 190–200. [CrossRef]
8. Lu, L.; Nogita, K.; Dahle, A.K. Combining Sr and Na additions in hypoeutectic Al–Si foundry alloys. *Mater. Sci. Eng. A* **2005**, *399*, 244–253. [CrossRef]
9. Yang, H.; Ji, S.; Fan, Z. Effect of heat treatment and Fe content on the microstructure and mechanical properties of die-cast Al–Si–Cu alloys. *Mater. Des.* **2015**, *85*, 823–832. [CrossRef]

10. Garb, C.; Leitner, M.; Grün, F. Effect of elevated temperature on the fatigue strength of casted AlSi8Cu3 aluminium alloys. *Procedia Struct. Integr.* **2017**, *7*, 497–504. [CrossRef]

11. Molina, R.; Amalberto, P.; Rosso, M. Mechanical characterization of aluminium alloys for high temperature applications Part1: Al-Si-Cu alloys. *Metall. Sci. Tecnol.* **2011**, *29*, 5–15.

12. Özdeş, H.; Tiryakioğlu, M. On estimating high-cycle fatigue life of cast Al-Si-Mg-(Cu) alloys from tensile test results. *Mater. Sci. Eng. A* **2017**, *688*, 9–15. [CrossRef]

13. Ceschini, L.; Morri, A.; Toschi, S.; Seifeddine, S. Room and high temperature fatigue behaviour of the A354 and C355 (Al–Si–Cu–Mg) alloys: Role of microstructure and heat treatment. *Mater. Sci. Eng. A* **2016**, *653*, 129–138. [CrossRef]

14. Konečná, R.; Nicoletto, G.; Kunz, L.; Riva, E. The role of elevated temperature exposure on structural evolution and fatigue strength of eutectic AlSi12 alloys. *Int. J. Fatigue* **2016**, *83*, 24–35. [CrossRef]

15. Tiryakioğlu, M. Statistical distributions for the size of fatigue-initiating defects in Al–7%Si–0.3%Mg alloy castings: A comparative study. *Mater. Sci. Eng. A* **2008**, *497*, 119–125. [CrossRef]

16. Gumbel, E.J. *Statistics of Extremes*; Columbia University Press: New York, NY, USA, 1958.

17. Jenkinson, A.F. The frequency distribution of the annual maximum (or minimum) values of meteorological elements. *Q. J. R. Meteorol. Soc.* **1955**, *87*, 145–158. [CrossRef]

18. Tiryakioğlu, M. Relationship between Defect Size and Fatigue Life Distributions in Al-7 Pct Si-Mg Alloy Castings. *Metall. Mater. Trans. A* **2009**, *40*, 1623–1630. [CrossRef]

19. Murakami, Y. *Metal Fatigue: Effects of Small Defects and Nonmetallic Inclusions*; Elsevier: Amsterdam, The Netherlands, 2002.

20. Murakami, Y. Material defects as the basis of fatigue design. *Int. J. Fatigue* **2012**, *41*, 2–10. [CrossRef]

21. Garb, C.; Leitner, M.; Grün, F. Application of $\sqrt{area}$-concept to assess fatigue strength of AlSi7Cu0.5Mg casted components. *Eng. Fract. Mech.* **2017**, *185*, 61–71. [CrossRef]

22. Leitner, M.; Garb, C.; Remes, H.; Stoschka, M. Microporosity and statistical size effect on the fatigue strength of cast aluminium alloys EN AC-45500 and 46200. *Mater. Sci. Eng. A* **2017**, *707*, 567–575. [CrossRef]

23. Garb, C.; Leitner, M.; Grün, F. Fatigue Strength Assessment of AlSi7Cu0.5Mg T6W Castings Supported by Computed Tomography Microporosity Analysis. *Procedia Eng.* **2016**, *160*, 53–60. [CrossRef]

24. DIN EN 515. *Aluminium and Aluminium Alloys-Wrought Products-Temper Designations*; Beuth Verlag: Berlin, Germany, 2017.

25. Sjölander, E.; Seifeddine, S. The heat treatment of Al–Si–Cu–Mg casting alloys. *J. Mater. Process. Technol.* **2010**, *210*, 1249–1259. [CrossRef]

26. Vandersluis, E.; Ravindran, C. Comparison of Measurement Methods for Secondary Dendrite Arm Spacing. *Metallogr. Microstruct. Anal.* **2017**, *6*, 89–94. [CrossRef]

27. Zhang, L.Y.; Jiang, Y.H.; Ma, Z.; Shan, S.F.; Jia, Y.Z.; Fan, C.Z.; Wang, W.K. Effect of cooling rate on solidified microstructure and mechanical properties of aluminium-A356 alloy. *J. Mater. Process. Technol.* **2008**, *207*, 107–111. [CrossRef]

28. DIN EN ISO 6892-1. *Metallic Materials-Tensile Testing-Part 1: Method of Test at Room Temperature*; Beuth Verlag: Berlin, Germany, 2016.

29. DIN EN ISO 6892-2. *Metallic Materials-Tensile Testing-Part 2: Method of Test At Elevated Temperature*; Beuth Verlag: Berlin, Germany, 2011.

30. Zhu, X.; Shyam, A.; Jones, J.; Mayer, H.; Lasecki, J.V.; Allison, J.E. Effects of microstructure and temperature on fatigue behavior of E319-T7 cast aluminum alloy in very long life cycles. *Int. J. Fatigue* **2006**, *28*, 1566–1571. [CrossRef]

31. Rincón, E.; López, H.F.; Cisneros, M.M.; Mancha, H.; Cisneros, M.A. Effect of temperature on the tensile properties of an as-cast aluminum alloy A319. *Mater. Sci. Eng. A* **2007**, *452-453*, 682–687. [CrossRef]

32. Zamani, M.; Seifeddine, S.; Jarfors, A.E. High temperature tensile deformation behavior and failure mechanisms of an Al–Si–Cu–Mg cast alloy—The microstructural scale effect. *Mater. Des.* **2015**, *86*, 361–370. [CrossRef]

33. Mohamed, A.; Samuel, F.H.; kahtani, S.A. Microstructure, tensile properties and fracture behavior of high temperature Al–Si–Mg–Cu cast alloys. *Mater. Sci. Eng. A* **2013**, *577*, 64–72. [CrossRef]

34. Khan, S.; Ourdjini, A.; Named, Q.S.; Alam Najafabadi, M.A.; Elliott, R. Hardness and mechanical property relationships in directionally solidified aluminium-silicon eutectic alloys with different silicon morphologies. *J. Mater. Sci.* **1993**, *28*, 5957–5962. [CrossRef]
35. Cahoon, J.R.; Broughton, W.H.; Kutzak, A.R. The determination of yield strength from hardness measurements. *Metall. Trans.* **1971**, *2*, 1979–1983. [CrossRef]
36. DIN EN ISO 6507. *Metallic Materials-Vickers Hardness Test-Part 1: Test Method*; Beuth Verlag: Berlin, Germany, 2018.
37. Tiryakioğlu, M.; Robinson, J.S.; Salazar-Guapuriche, M.A.; Zhao, Y.Y.; Eason, P.D. Hardness–strength relationships in the aluminum alloy 7010. *Mater. Sci. Eng. A* **2015**, *631*, 196–200. [CrossRef]
38. Rometsch, P.; Schaffer, G. An age hardening model for Al–7Si–Mg casting alloys. *Mater. Sci. Eng. A* **2002**, *325*, 424–434. [CrossRef]
39. Tiryakioğlu, M.; Campbell, J.; Staley, J.T. On macrohardness testing of Al–7 wt.% Si–Mg alloys: II. An evaluation of models for hardness-yield strength relationships. *Mater. Sci. Eng. A* **2003**, *361*, 240–248. [CrossRef]
40. Drouzy, M.; Jacob, S.; Richard, M. Interpretation of tensile results by means of quality index and probable Yield Strength-application to Al-Si7 Mg foundry alloys. *Int. Cast Met. J.* **1980**, *5*, 43–50.
41. ASTM International E 739. *Standard Practice for Statistical Analysis of Linear or Linearized Stress-Life (S-N) and Strain Life (E-N) Fatigue Data*; ASTM International: West Conshohocken, PA, USA, 1998.
42. Dengel, D.; Harig, H. Estimation of the fatigue limit by progressively-increasing load tests. *Fatigue Fract. Eng. Mater. Struct.* **1980**, *3*, 113–128. [CrossRef]
43. Dengel, D. Die arc sin $\sqrt{P}$-Transformation—Ein einfaches Verfahren zur grafischen und rechnerischen Auswertung geplanter Wöhlerversuche. *Materialwissenschaft und Werkstofftechnik* **1975**, *6*, 253–261. [CrossRef]
44. Leitner, H. Simulation des Ermüdungsverhaltens von Aluminiumguslegierungen. Ph.D. Thesis, Montanuniversität Leoben, Leoben, Austria, 2001.
45. McKelvey, S.A.; Lee, Y.L.; Barkey, M.E. Stress-based uniaxial fatigue analysis using methods described in FKM-guideline. *J. Failure Anal. Prev.* **2012**, *12*, 445–484. [CrossRef]
46. Sonsino, C.M.; Dieterich, K. Einfluss der Porosität auf das Schwingfestigkeitsverhalten von Aluminium-Gusswerkstoffen. II. *Giessereiforschung* **1991**, *43*, 131–140.
47. Sonsino, C.M.; Dieterich, K. Einfluss der Porosität auf das Schwingfestigkeitsverhalten von Aluminium-Gusswerkstoffen. I. *Giessereiforschung* **1991**, *43*, 119–130.
48. Aigner, R.; Leitner, M.; Stoschka, M. Fatigue strength characterization of Al-Si cast material incorporating statistical size effect. *MATEC Web Conf.* **2018**, *165*, 14002. [CrossRef]
49. González, R.; Martínez, D.I.; González, J.A.; Talamantes, J.; Valtierra, S.; Colás, R. Experimental investigation for fatigue strength of a cast aluminium alloy. *Int. J. Fatigue* **2011**, *33*, 273–278. [CrossRef]
50. Tiryakioğlu, M. On the size distribution of fracture-initiating defects in Al- and Mg-alloy castings. *Mater. Sci. Eng. A* **2008**, *476*, 174–177. [CrossRef]
51. Beretta, S.; Murakami, Y. Statistical analysis of defects for fatigue strength prediction and quality control of materials. *Fatigue Fract. Eng. Mater. Struct.* **1998**, *21*, 1049–1065. [CrossRef]
52. Souza, D.; Menegalli, F.C. Image analysis: Statistical study of particle size distribution and shape characterization. *Powder Technol.* **2011**, *214*, 57–63. [CrossRef]
53. Garb, C.; Leitner, M.; Tauscher, M.; Weidt, M.; Brunner, R. Statistical analysis of micropore size distributions in Al-Si castings evaluated by X-ray computed tomography. *Int. J. Mater. Res.* **2018**, *109*, 889–899. [CrossRef]
54. Makhlouf, K.; Jones, J. Near-threshold fatigue crack growth behaviour of a ferritic stainless steel at elevated temperatures. *Int. J. Fatigue* **1992**, *14*, 97–104. [CrossRef]
55. Alvarez Holston, A.M.; Stjärnsäter, J. On the effect of temperature on the threshold stress intensity factor of delayed hydride cracking in light water reactor fuel cladding. *Nuclear Eng. Technol.* **2017**, *49*, 663–667. [CrossRef]
56. Ritchie, R.O. Near-threshold fatigue-crack propagation in steels. *Int. Met. Res.* **2013**, *24*, 205–230. [CrossRef]
57. Irwin, G.R. Analysis of stresses and strains near the end of a crack traversing a plate. *J. Appl. Mech.* **1957**, *24*, 361–364.
58. Pippan, R.; Hohenwarter, A. Fatigue crack closure: A review of the physical phenomena. *Fatigue Fract. Eng. Mater. Struct.* **2017**, *40*, 471–495. [CrossRef]

59. McClung, R.C. Crack closure and plasitc zone sizes in fatigue. *Fatigue Fract. Eng. Mater. Struct.* **1991**, *14*, 455–468. [CrossRef]
60. Hou, C.; Charng, J. Estimation of plasticity-induced crack closure in a pre-existing plastic zone. *Int. J. Fatigue* **1996**, *18*, 463–474. [CrossRef]
61. Zhang, W.; Liu, Y. Plastic zone size estimation under cyclic loadings using in situ optical microscopy fatigue testing. *Fatigue Fract. Eng. Mater. Struct.* **2011**, *34*, 717–727. [CrossRef]

Article

Influence of Strain Rate and Waveshape on Environmentally-Assisted Cracking during Low-Cycle Fatigue of a 304L Austenitic Stainless Steel in a PWR Water Environment

Thibault Poulain [1,2], Laurent de Baglion [2], Jose Mendez [1] and Gilbert Hénaff [1,*]

[1]   Pprime Institute, UPR 3346 CNRS-ENSMA-Université de Poitiers, ISAE-ENSMA, Téléport 2,
     1 avenue Clément Ader, BP 40 109-86961 Futuroscope Chasseneuil CEDEX, France;
     thipoulain@gmail.com (T.P.); jose.mendez@ensma.fr (J.M.)
[2]   Framatome, Tour Areva - 1 place Jean Millier, 92 084 La Défense CEDEX, France;
     laurent.debaglion@framatome.com
*   Correspondence: gilbert.henaff@isae-ensma.fr; Tel.: +33-054-949-8233

Received: 18 December 2018; Accepted: 6 February 2019; Published: 8 February 2019

Abstract: In this paper, the low cycle fatigue resistance of a 304L austenitic stainless steel in a simulated pressurized water reactor (PWR) primary water environment has been investigated by paying a special attention to the interplay between environmentally-assisted cracking mechanisms, strain rate, and loading waveshape. More precisely, one of the prime interests of this research work is related to the consideration of complex waveshape signals that are more representative of solicitations encountered by real components. A detailed analysis of stress-strain relation, surface damage, and crack growth provides a preliminary ranking of the severity of complex, variable strain rate signals with respect to triangular, constant strain-rate signals associated with environmental effects in air or in PWR water. Furthermore, as the fatigue lives in PWR water environment are mainly controlled by crack propagation, the crack growth rates derived from striation spacing measurement and estimated from interrupted tests have been carefully examined and analyzed using the strain intensity factor range ΔK_ε. It is confirmed that the most severe signal with regards to fatigue life also induces the highest crack growth enhancement. Additionally two characteristic parameters, namely a threshold strain $\varepsilon_{th}{}^*$ and a time T^*, corresponding to the duration of the effective exposure of the open cracks to PWR environment have been introduced. It is shown that the T^* parameter properly accounts for the differences in environmentally-assisted growth rates as a function of waveshape.

Keywords: stainless steel; environmentally-assisted cracking

1. Introduction

The aging management of nuclear power plants is one of the main challenges for the energy stakeholders worldwide in the coming years. In particular, the potential damage induced by corrosion and environmentally-assisted cracking needs to be more thoroughly investigated in order to assess the residual structural integrity of the components. In the case of pressurized water reactors (PWR), water is used both as a nuclear reaction moderator and as a heat carrier medium. In order to maintain the water in a liquid state throughout the entire primary circuit, a pressure of about 140 bars is applied while its temperature varies between 290 and 350 °C. This water, with a definite and controlled chemistry, represents the PWR medium. Additionally, the primary circuit pipings, commonly made of austenitic stainless steel, are subjected to complex thermomechanical loadings in the low-cycle fatigue domain, due to numerous operating transients, in conjunction with this exposure to the PWR environment.

The design against fatigue of components in nuclear power plants is fundamentally based on the fatigue life established in air at room temperature. Different transposition factors are thereafter applied to these data so as to account for the respective influence of different parameters which mainly are the scale effects, the material variability, the load history, the surface finish, etc. In recent years, nuclear safety and inspection authorities have paid a special attention to environmental effects, and more precisely to the influence of the PWR environment on fatigue damage. In this context, among the various factors controlling the fatigue life in PWR medium, strain rate is extremely important. Indeed, most of the environmentally-assisted cracking mechanisms are time-dependent and, therefore, depend to some extent on the exposure duration [1–4]. Nevertheless, one cannot exclude complex interactions between the environmental exposure duration and the load signal waveform to account for damage mechanisms at the crack tip.

Additionally, in conventional fatigue testing, simple input signal shapes, such as triangular or saw-tooth signals, are used. In such cases, the strain rate is easily controlled and kept constant during a large part of the load period. Typically, strain-control triangular signals are commonly applied to establish the data required in design codes. However, especially when investigating the very low strain rate domain, saw-tooth signals with a slow loading rate and a fast unloading rate are also considered in order to reduce the test duration, by assessing that the fatigue lives are only dependent of the rising load part duration [2–4]. However, this type of loading does not fully account for the actual loading conditions in industrial components. For instance, in PWR, the Safety Injection System (SIS) is used in case of incident to introduce boron-containing cold water under high pressure to moderate the nuclear reaction and ensure a proper cooling of the reactor core.

The signal shown in Figure 1 corresponds to the resulting mechanical loading imposed on the pipes of primary circuit when using the SIS circuit. It can be seen that the strain rate continuously varies during the fatigue cycle. Thus, the first part of the signal corresponds to the injection of cold water (20 °C) into the primary circuit. The component is then rapidly stressed in tension because the material shrinks. Thereafter, the hot water (300 °C) normally circulating in the primary circuit causes a further expansion which results into a compressive deformation. Finally, the temperature becomes homogeneous when recovering equilibrium conditions. The evolution of the mechanical deformation resulting from such thermal loading can be used as a loading signal in isothermal tests in simulated PWR environment. This signal will be denoted in the following as the SIS A signal.

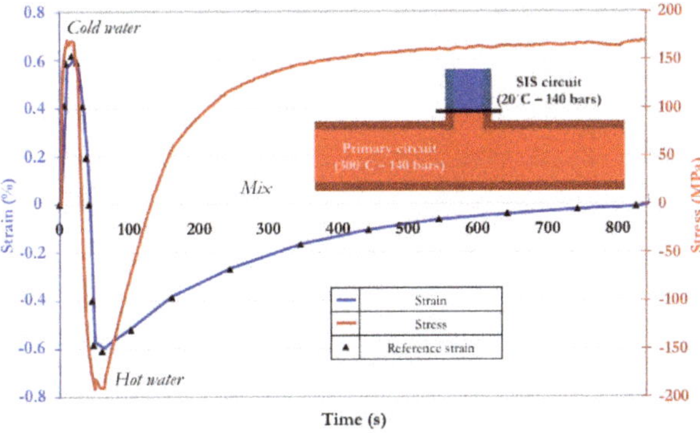

Figure 1. Variation of strain and stress during one cycle of the SIS A signal.

Therefore, it seems important to investigate the effect of such a more complex and representative signal shape on the fatigue resistance in PWR medium, in order to assess the conservatism of the design rules. In particular, in comparison with triangular signals, it can be expected that such complex loadings, with a varying strain rate during one loading period, will not only affect the cyclic stress-strain behavior, in relation with dynamic strain aging (DSA), but also the exposure of the crack tip to the medium and subsequently influence the fatigue damage and the crack growth rates [5], through interactions between elementary processes of deformation and damage. Typically, DSA can introduce strain localization [6–8], which, in turn, can modify the action of the corrosive medium.

With this aim, the influence of strain rate and SIS complex loading signal shape on the LCF resistance in PWR environment of a 304L austenitic stainless steel is addressed in the present paper. More precisely, the material cyclic stress-strain (CSS) behavior and fatigue lives with complex signals were first determined in PWR medium and then compared to data obtained using triangular signals in the same environment, but also to data obtained in air in similar loading conditions at total strain amplitude of 0.6%. Furthermore, previous studies have shown that in the PWR environment the initiation stage is very short [9]. This is the reason why a special attention has been paid here to a fine analysis of the propagation stage, by combining measurements of striation spacing with crack growth data derived from interrupted tests, using a methodology explained in [9]. The objective is to identify a parameter that can account for the crack growth behavior under different strain rates and loading signals. While this work is a part of a more comprehensive study, encompassing the examination of the effect of surface finishing, only results obtained on polished samples are considered here for the sake of clarity. The aspects peculiar to the ground surface finishing, which is also representative of some actual surface conditions in a component, were addressed in another paper [9].

2. Experimental Conditions

2.1. Material and Specimens

The material used in this study is a 304L austenitic stainless steel from a 30 mm thick plate elaborated by Creusot Loire Industries (CLI, Creusot, France). It was obtained by rolling and subjected to a solution-treatment at 1100 °C before being water-quenched. Its chemical composition is (in weight %): C-0.029, Si-0.37, Mn-1.86, P-0.029, S-0.004, Ni-10.0, Cr-18.00, Mo-0.04, N-0.056, and Fe balance. The grain size is about 80 μm.

The tensile properties of this specific cast have been determined in a previous study [10]. The results are given in Table 1. Additional information about the cyclic stress-strain behavior and low-cycle fatigue resistance as a function of temperature and environment can be found in [11–14].

Table 1. Tensile properties of the CLI cast 304L stainless steel.

	L Direction		T Direction	
	20 °C	**300 °C**	**20 °C**	**300 °C**
Yield strength $\sigma_{y\,0.2}$ (MPa)	214	130	224	133
Ultimate Tensile strength (MPa)	592	414	593	409
Tensile Elongation A%	57	39	57	39
Reduction in section area Z%	84	78	81	75

Fatigue specimens are machined with the loading axis parallel to the rolling direction of the plate. The gauge length is 13.5 mm and the diameter is 9 mm. The surface of the fatigue samples were thereafter mirror-polished by mechanical polishing down to 1 μm grade diamond paste. The specimen geometry and tensile mechanical properties were described in more details elsewhere [9]. In addition, flanges were machined on both specimen shoulders in order to fix the strain measurement system described in the next section, which requires a calibration to relate the elongation applied between flanges to the strain amplitude targeted in the gauge length.

2.2. Test Procedures

LCF testing were performed in air at 300 °C using an electromechanical INSTRON 1362 machine (Instron, Norwood, MA, USA). For tests in simulated PWR primary water environment, a MTS (MTS,

Eden Prairie, MN, USA) servohydraulic loadframe equipped with a pressure vessel which enables a maximum pressure of 140 bars and a temperature of 300 °C was used (Figure 2a). The simulated PWR water environment is characterized by:

- Dissolved oxygen content: lower than 0.01 ppm;
- Hydrogen: 25–35 cc (STP)/kg PWR water;
- Cl and F: lower than 0.05 ppm;
- B: ~1000 ppm (adjusted by boric acid additions);
- Li: quantity needed for adjustment of pH (~2 ppm);
- PH: [7–10]; and
- Conductivity: 2–40 µS/cm.

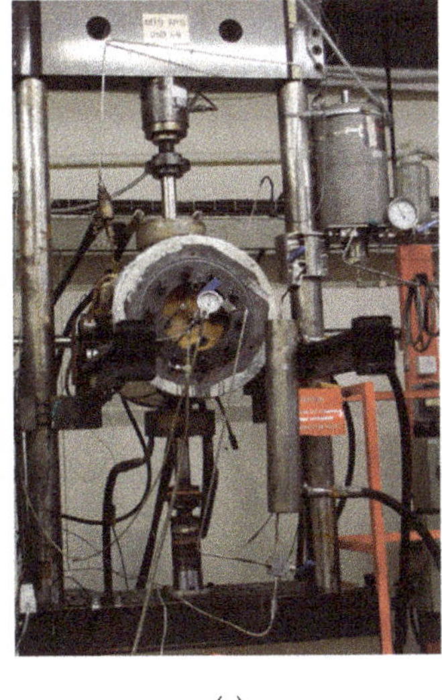

(a)

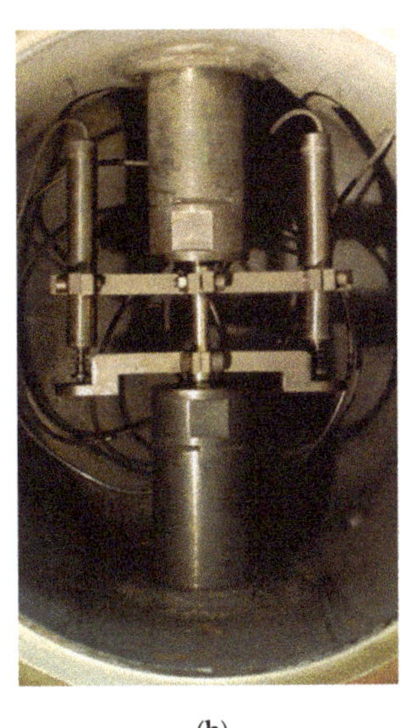

(b)

Figure 2. PWR test rig (a) general view of the autoclave and loadframe, and (b) close view of the specimen equipped with LVDT sensors for strain control.

These conditions are controlled in a loop and the PWR water is introduced in the pressure vessel mounted on the loadframe.

Push-pull tests ($R_\varepsilon = -1$, where R_ε denotes the ratio between the minimum and the maximum strain) were performed under total axial strain control with a total strain amplitude of ±0.6%. Strain was controlled using two linear variable differential transformers (LVDT, RDP LIN 56, DP Electronics Ltd., Wolverhampton, UK) measuring the displacement between flanges placed on specimen shoulders as shown in Figure 2b. This remote control of deformation requires a prior calibration to relate the displacement imposed at the level of the flanges via the LVDT sensor and the deformation on the gauge length of the testpiece.

Reference tests were conducted with a triangular waveshape at different strain rates. Additional tests were also performed using complex signals noted SIS A and SIS B.

The SIS A signal is representative of the loading conditions induced by water mixture in a tee junction as presented in the introduction. As shown in Figure 3a, this SIS A signal, with a total period of 840 s, can be divided in five elementary segments (Figure 3a) characterized by a duration and a mean strain rate as listed in Table 2. Four segments are characterized by a positive strain rate and a strain range corresponding to one quarter of the total strain range while the last segment corresponds to the unloading part of the fatigue cycle, associated with a negative strain rate. Since the strain range is identical but the strain rate is varying, the duration of the segments varies too.

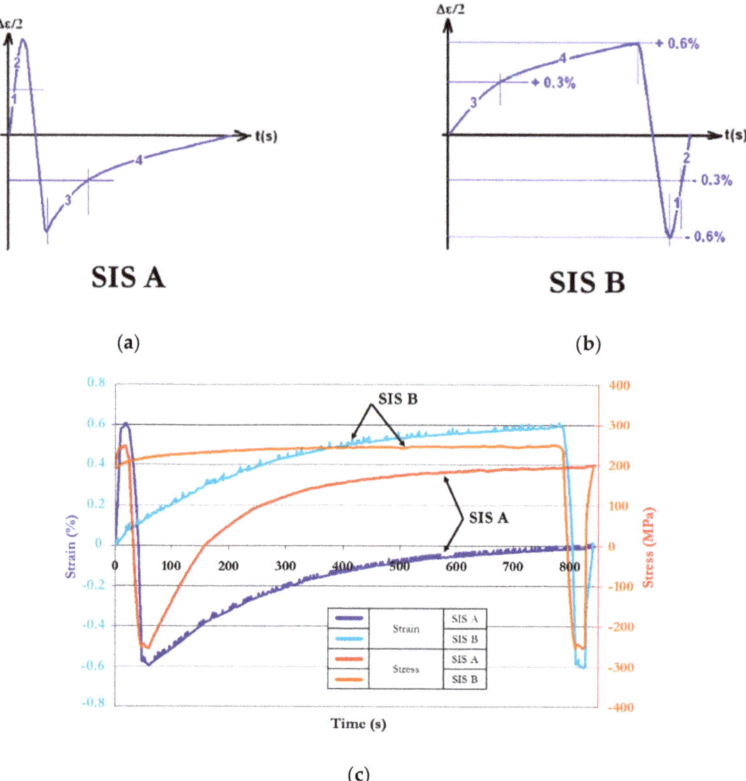

Figure 3. Presentation of the complex signals (**a**) SIS A, (**b**) SIS B, and (**c**) comparison of the two signals with the stress response.

Table 2. Description of the five elementary segments composing the SIS A signal.

	Segment 1	Segment 2	Negative Strain Rate	Segment 3	Segment 4
Duration (s)	10	10	40	140	640
Mean strain rate (s^{-1})	3.00×10^{-4}	3.00×10^{-4}	3.00×10^{-4}	1.88×10^{-5}	4.84×10^{-6}

In order to examine in more details, the influence of this loading signal and in particular of the slowest part (segment 4), different combinations of these elementary segments have been considered in previous studies [15–17]. The results indicated that the signals where the longest segment 4 is applied in tension, that means with a positive deformation, are more damaging than the signals where this slow strain rate segment is applied in the compressive part of the cycle, that means with a negative deformation [12]. In the present study, only the so-called SIS B signal, presented in Figure 3b, will be considered in addition to the SIS A signal. The comparison presented in Figure 3c with the associated stress response indicates that one of the major differences between the two signals resides in the duration of the cycle part with a positive strain, which is much larger in the case of the SIS B signal.

As it has been previously pointed out [9], the passage of the moving rod within the autoclave along with an efficient sealing are delicate to ensure without friction along the load axis. Since the load cell is located outside the autoclave, such frictions may slightly disrupt the measurement of the load.

This can explain some discrepancies in the stress amplitudes between tests performed in air and in PWR simulated environment. However, it must be kept in mind that the value of the applied strain amplitude is not affected and is the same in air and in PWR environment. To analyze the effect of the loading signal on the stress response, the results from tests conducted in the air environment can be considered as the more reliable.

Tests were conducted to failure or intentionally interrupted after selected number of cycles in order to assess the damage in terms of crack density, crack depth and crack front shape. For the first type of tests, the fatigue life, noted N_5, corresponds here to the number of cycles required to detect a decrease of 5% in the quasi-stabilized cyclic stress response. This criterion was selected principally in order to be consistent with existing internal databases. However, it must be noted that, given the 9 mm diameter of the specimen, the number of cycles to conduct the specimen to failure can be slightly superior of 100 to 1000 cycles, depending on the loading conditions. For the second type of tests, namely interrupted tests, after the achievement of requested LCF cycles at 300 °C, the specimens were fatigued to failure in air at room temperature under stress-control (R_σ = 0.2, where R_σ denotes the ratio between minimum and maximum stress) at high frequency (10 Hz). The main objective of this last step is twofold: the first one is to achieve failure as rapidly as possible, the second one is to prevent any damage on fracture surfaces prior to observations by applying a positive stress ratio. A stress amplitude of $\Delta\sigma/2$ = 100 MPa was first applied until failure or 6×10^6 cycles at most. When necessary, this amplitude was then increased to $\Delta\sigma/2$ = 120 MPa. SEM analyses of the fracture surfaces were finally conducted. Differences in fracture surfaces morphology in LCF and HCF periods allows then to provide data of both the crack depth and the crack front shape of the largest cracks as a function of the number of cycles applied during the LCF nominal solicitations.

3. Results

3.1. Influence of Strain Rate and/or Complex Loading on Cyclic Behavior and Fatigue Life in Air

Figure 4 presents the variation of the cyclic peak stress as a function of the number of cycles for the loading conditions considered in air. It is worth noticing that the estimated duration of tests to failure using complex signals with a long period was so high in this environmental condition that it was decided to interrupt the tests after 1350 cycles. Moreover, the presence of a Portevin-Le Chatelier effect was noticed during the slowest strain rate part (segment 4) of both SIS signals. Concerning the tests using standard triangular waveshape, those performed at the intermediate and the highest strain rates were conducted until failure while the one carried out at the slowest strain rate was stopped before failure, after 2500 cycles, because of the very long duration of the test in such condition. For triangular waveshape tested specimens, the analysis of the peak stress response indicates a shift of the peak hardening towards higher number of cycles when the strain rate is reduced and more generally an increase in the cyclic stress response associated with DSA. The intrinsic effect of the strain rate on fatigue life and behavior has been previously shown by de Baglion [11] on the basis of experiments performed in vacuum on the same material. Indeed, the results obtained in air are already influenced by interactions between fatigue and environmental effects [18].

It can be observed from Figure 4 that the stresses associated with the SIS A signal are close to the ones recorded using a triangular signal with a strain rate of 1×10^{-4} s^{-1}, while the SIS B signal induces a cyclic stress response similar to the one observed with a triangular signal at 1×10^{-5} s^{-1}. The DSA phenomenon, in term of enhanced cyclic hardening, is thus particularly effective when the positive slow strain rate is applied in tension, i.e., with a positive deformation. This result is, however, not in agreement with the findings reported in [3] which indicate no difference in peak hardening in light water reactor (LWR) environment.

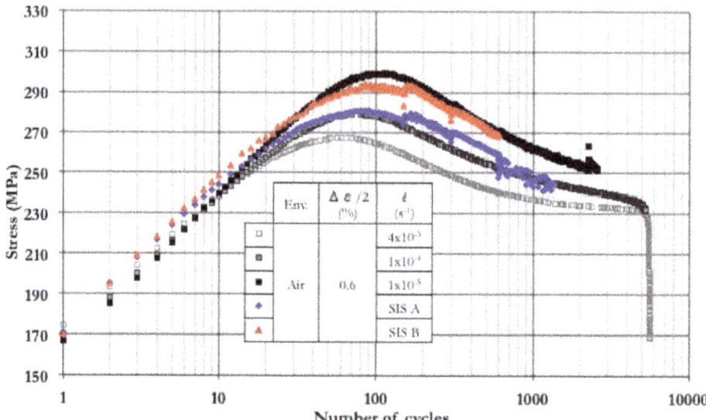

Figure 4. Peak stress as a function of the number of cycles in air with triangular signals at different strain rates and with SIS A and SIS B signals.

Samples from interrupted tests were fatigued to failure under load control at high frequency. The residual number of cycles to failure, N_R, can be considered as a damage indicator, using the assumption that, for a same stress amplitude during HCF loading, the lower the N_r value, the higher the damage extend after the 1350 initial LCF cycles. The values reported in Table 3 indicate a higher damage induced by the SIS B signal. The fracture surfaces were then carefully examined, with a special attention to the initiation area as exemplified in Figure 5. The damage has been quantitatively assessed by determining the maximum crack depth, the surface crack length of this main crack and the total cracked area on the fracture surface. These results, reported in Table 3, confirm the enhanced damaging effect of the SIS B signal, regardless of the parameter under consideration. Additionally, it should be mentioned that the crack shape features generated under these complex waveshapes are not much different from those of cracks produced using triangular signals with a constant strain rate of 1×10^{-4} and 1×10^{-5} s^{-1}. On that basis and due to the limited number of data obtained under complex signals, it will be assumed in the following that the variation of the crack aspect ratio during the propagation in the case of complex signals is the same than in the case of these two triangular signals [9].

Table 3. Quantitative damage measurements for tests in air interrupted after 1350 cycles with SIS A and SIS B signals.

Signal	N_R (Cycles)	Maximum Crack Depth (μm)	Surface Crack Length (μm)	Cracked Area (μm²)
SIS A	2,163,000	92	305	0.20
SIS B	654,600	129	580	0.26

Since no data from tests interrupted at 1350 cycles were available for triangular signals, the maximum crack depths measured after 1350 cycles for SIS signals are compared in Figure 6 to macroscopic crack growth laws established for triangular signals [9] on the basis of interrupted tests. The curves presented are obtained by fitting all the data from tests to failure and interrupted test by integrating a Paris law-type crack growth equation as explained in [9]. It is thus confirmed that while the crack depth by applying the SIS A signal is nearly similar to that obtained with crack growth laws associated to various strain rates, the crack depth for the SIS B signal is significantly higher.

Additional views at higher magnification of the fracture surfaces are presented in Figure 7 for SIS A and in Figure 8 for SIS B. It can be seen that the general aspect of these surfaces is similar, with the formation of large planar facets reminiscent of those observed under a triangular signal with a strain rate of 1×10^{-5} s^{-1}. In addition, striations can be distinguished at small crack depth. However, it has to be noticed that since the variation in striation spacing covered during 1350 cycles is too narrow, these data will not be considered in the following.

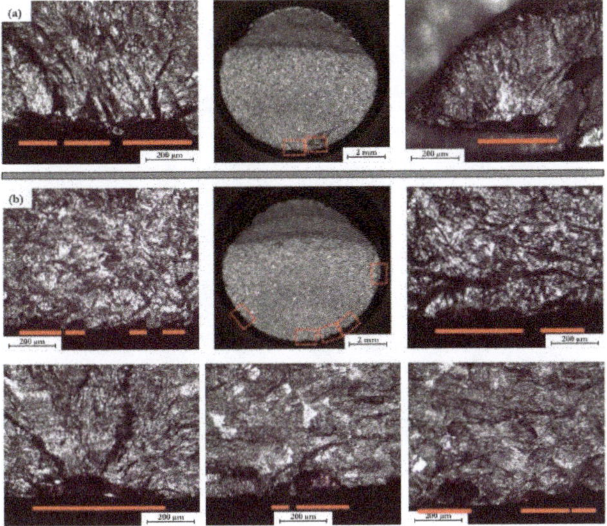

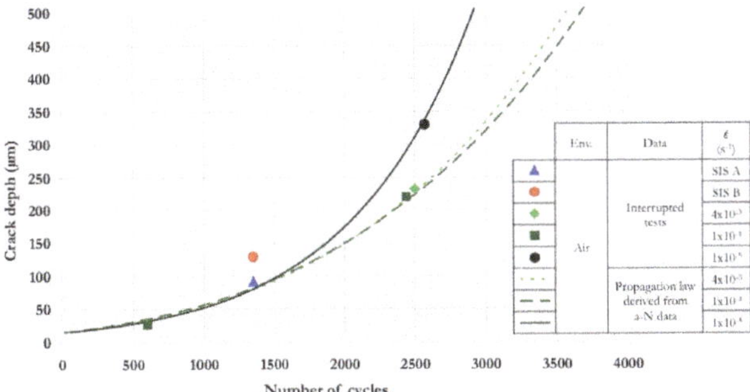

Figure 5. Fracture surfaces with details of the initiation area in test in air at $\Delta\varepsilon_t/2 = 0.6\%$ interrupted after 1350 cycles using the (**a**) SIS A signal and (**b**) SIS B signal.

Figure 6. Crack depth as a function of the number of cycles derived from interrupted tests for different loading conditions, namely, triangular signals with different strain rates, SIS A, and SIS B signals in air.

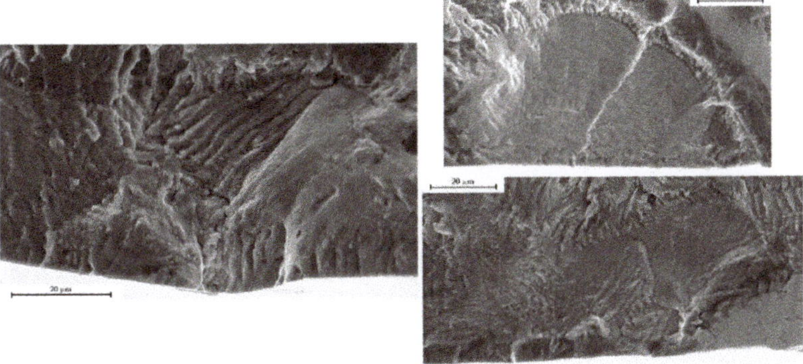

Figure 7. SEM observations of the fracture surfaces produced in air using the SIS A signal.

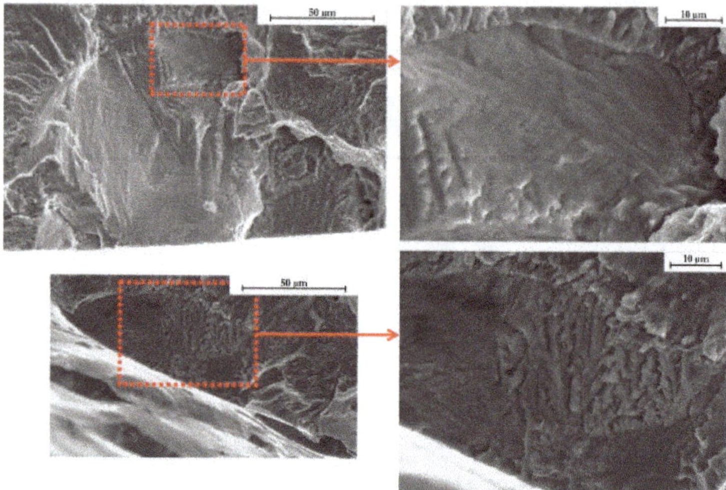

Figure 8. SEM observations of the fracture surfaces produced in air using the SIS B signal.

3.2. Influence of Strain Rate and/or Complex Loading on Cyclic Behavior and Fatigue Life in the PWR Environment

Figure 9 presents the peak stress versus the number of cycles for different loading conditions (triangular signals with different strain rates and complex signals) in the PWR environment. By comparing with the data presented in Figure 4, it can be seen that for a given triangular signal, the peak stress is nearly the same while the fatigue life is significantly lower in PWR. As regards complex signals, in the case of the SIS B signal, the stress response is extremely similar to the one observed in air while with SIS A signal a significant discrepancy is noticed. However, as pointed out before, the load measurement via a load cell which is placed outside the pressure vessel during the tests in PWR medium can be affected by the frictions along the loading rod, in particular at slow strain rate. Nevertheless, it is worth noticing that the applied strain amplitude is not affected.

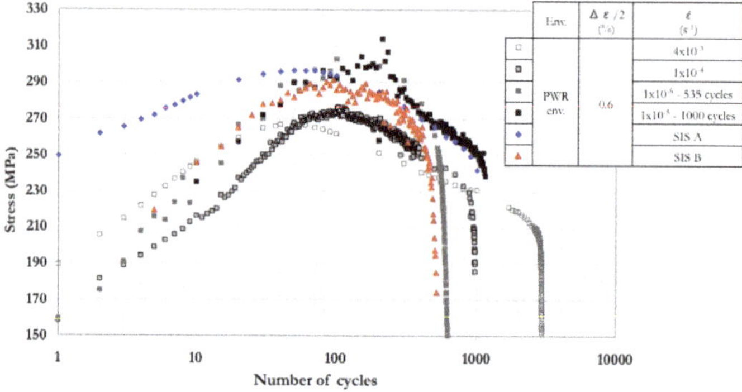

Env.	$\Delta \varepsilon /2$ (%)	$\dot{\varepsilon}$ (s^{-1})
□		4×10^{-3}
▫		1×10^{-4}
▪	PWR env.	1×10^{-5} - 535 cycles
▪		1×10^{-5} - 1000 cycles
◆	0.6	SIS A
▲		SIS B

Figure 9. Peak stress as a function of the number of cycles with triangular signals at different strain rates and with SIS A and SIS B signals in the PWR environment.

Apart from this discrepancy, the cyclic stress response generated by SIS signals is intermediate between those obtained using triangular signals at 1×10^{-4} and 1×10^{-5} s^{-1}, confirming the observations in air.

The fatigue lives determined in PWR environment using complex signals are $N_5 = 1200$ cycles for SIS A and $N_5 = 440$ cycles for SIS B. Therefore, the longer fatigue life obtained with the SIS A signal seems consistent when compared with the results in air presented in the previous section and indicating a less extend of damage with the SIS A signal. These trends are furthermore in agreement with similar data previously obtained on a different grade of 304L [12,17], even though the fatigue resistance of these two alloys are different.

In order to grain a deeper insight into the influence of the strain rate and load signal on the development of fatigue damage, additional tests interrupted in the early stage of the fatigue life have been carried out in this environment. The conditions and the corresponding results are given in Table 4.

Table 4. Quantitative damage measurements from interrupted tests with SIS A and SIS B signals in the PWR environment.

Signal	N_{LCF} (Cycles)	N_R (Cycles)	Maximum Crack Depth (µm)	Surface Crack Length (µm)	Cracked Area (µm²)
SIS A	300	6,858,800 (+1,045,800 *)	69	0.10	0.02
	600	472,400	360	0.75	0.23
SIS B	300	483,200	362	0.59	0.18

For the test interrupted after 300 cycles with the SIS A signal, since no failure was achieved after 6,858,800 cycles under a load amplitude of 100 MPa, the stress amplitude has been raised to a value of 120 MPa (indicated by *); 1,045,800 additional cycles were still required at this higher amplitude to achieve failure, which is indicative of a lower extend of damage introduced during the initial 300 cycles in this loading condition. General views of the corresponding fracture surfaces are presented in Figure 10. It is first noticed that the number of initiation sites is lower using complex signals than with triangular signals. Damage has been quantified further by using the same parameters than in air. The comparison after 300 cycles once again indicates a higher damage induced by SIS B with respect to SIS A, regardless of the characteristic damage parameter under consideration. Apart from a lower value of the cracked area, the damage after 300 cycles using SIS B signal is actually close to the one obtained after 600 cycles with SIS A.

When comparing the results on crack depth as a function of the number of cycles as reported in Figure 11, it can be seen that the damage generated by SIS A signal is in between the damage corresponding to a triangular signal with a strain rate of 4×10^{-3} and 1×10^{-5} s^{-1} while the SIS B generated damage is similar to that of the triangular signal with a strain rate of 1×10^{-4} s^{-1}. A similar ranking of the severity of the different signals is obtained by considering the cracked area. These observations suggest the following ranking in terms of severity:

$$D_{SISB} \sim D_{10^{-4}} > D_{10^{-5}} > D_{SISA} > D_{4 \times 10^{-3}}$$

where D represents the damage as characterized by the crack depth or the cracked area and the subscript the type of signal or the strain rate in the case of triangular signals.

This analysis confirms that in PWR environment the fatigue damage induced by the SIS B signal is more important than with SIS A signal. It is thus determined that, more than the average strain rate, the location in the signal of slow strain rate segments is a key parameter.

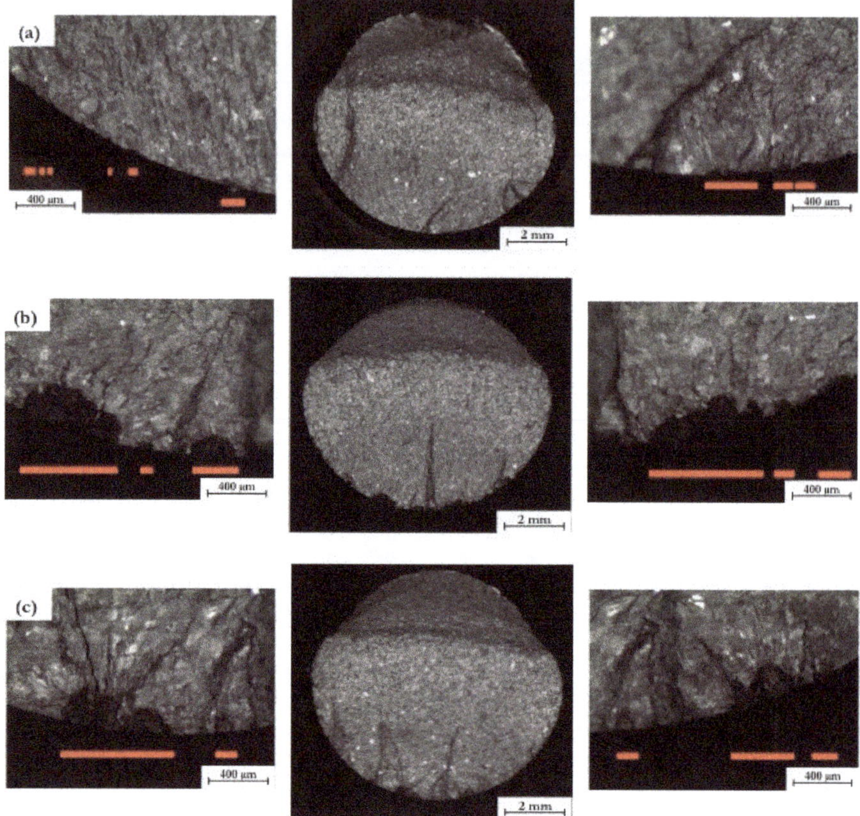

Figure 10. Fracture surfaces and details of the initiation zone obtained from interrupted tests in PWR environment (**a**) SIS A, 300 cycles, (**b**) SIS A, 600 cycles, and (**c**) SIS B, 300 cycles.

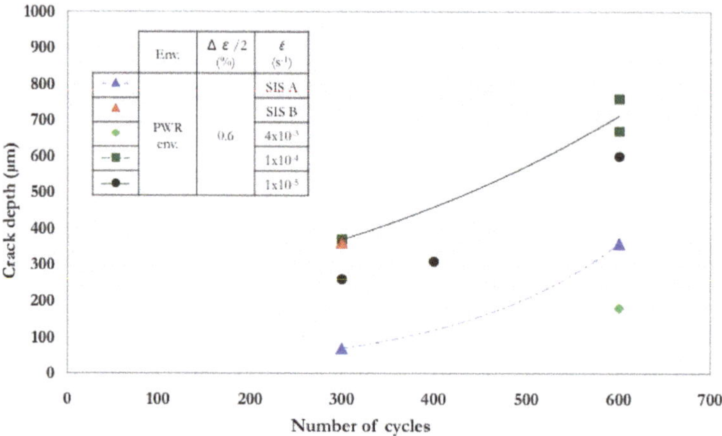

Figure 11. Comparison of the measured crack depths during interrupted tests for different loading conditions.

4. Discussion and Analysis

4.1. Analysis of the Propagation Stage

The general aspect of the fracture surfaces of specimens using SIS A and SIS B signals in the PWR environment is presented in Figure 12. Similarly to that observed in the air environment, the surfaces produced by SIS A and SIS B signals in PWR medium are extremely close to each other and not much different from the surfaces obtained by using a triangular signal at a strain rate of $1 \times 10^{-5}\,\mathrm{s}^{-1}$, all exhibiting a planar aspect. With such complex signals, striations can, however, be observed at smaller crack depths and the corresponding cracked areas are larger. Accordingly, striation spacing measurements can, therefore, be plotted in Figure 13 as a function of the strain intensity factor ΔK_ε [9,19,20]: $\Delta K_\varepsilon = F(a,b) \times \Delta \varepsilon_t \times (\pi \times a)^{\frac{1}{2}}$, where a is the crack depth, b the surface length, and $F(a,b)$ is a geometrical correction factor as defined in linear fracture mechanics. Indeed, as detailed in a previous paper [9], it was shown that the growth of the main crack during LCF in air as in simulated PWR conditions is properly accounted for by the use of the strain intensity factor range ΔK_ε as a crack driving force parameter, according to the original proposal of Kamaya [20]. Moreover, in PWR water, the striation spacing can be equated to the average crack growth rate per cycle, which is not the case in air at 300 °C [9]. One can notice that the spacing with SIS B is larger than with SIS A signal, indicating a higher growth rate. In fact, the spacings measured for SIS A signal are similar to those obtained with a triangular signal at a strain rate of $1 \times 10^{-4}\,\mathrm{s}^{-1}$ or $1 \times 10^{-5}\,\mathrm{s}^{-1}$ while the spacings associated with SIS B signal are the largest. This suggests that the differences in damage or fatigue life might be attributed to the initiation and micro-propagation stages. In addition, the two slopes characteristics of crack growth curves established for triangular signals cannot be clearly identified in the case of SIS A and SIS B signals. The fatigue lives of specimen tested in simulated PWR environment can be estimated on the basis of these data as it has been previously done for tests carried out using triangular signals [9], by integrating the corresponding $\mathrm{da/dN} = C \times \Delta K_\varepsilon{}^m$ power-law equation between an initial crack length of 20 μm and a final crack length of 3 mm. The results are given in Table 5. It can be noticed that the agreement between calculation and experimental data is fair. This supports the fact that striation spacing actually corresponds to the crack advance during one cycle for these complex signals also in the PWR environment.

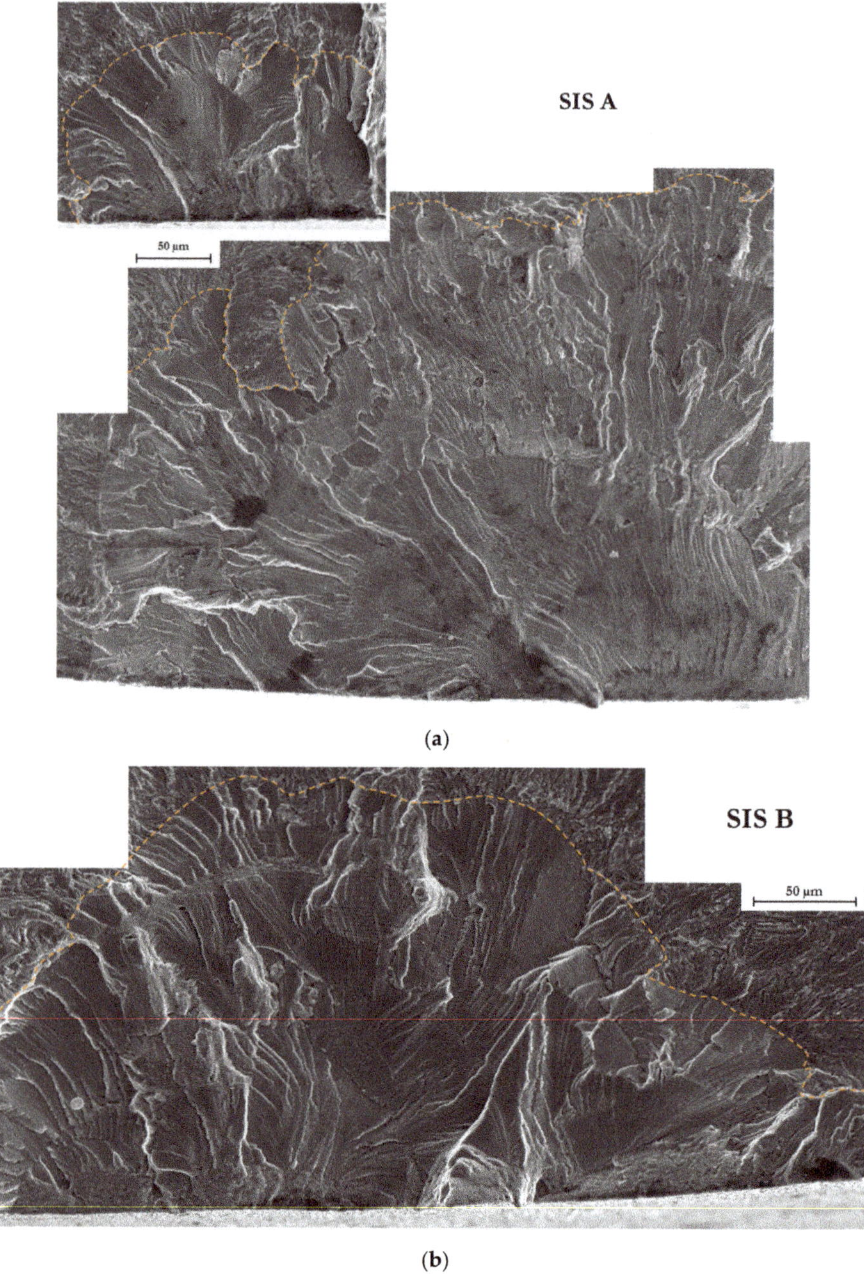

Figure 12. Fracture surfaces from interrupted tests in PWR environment (**a**) SIS A and (**b**) SIS B. The dotted lines correspond with the crack front at the transition between LCF and HCF loadings.

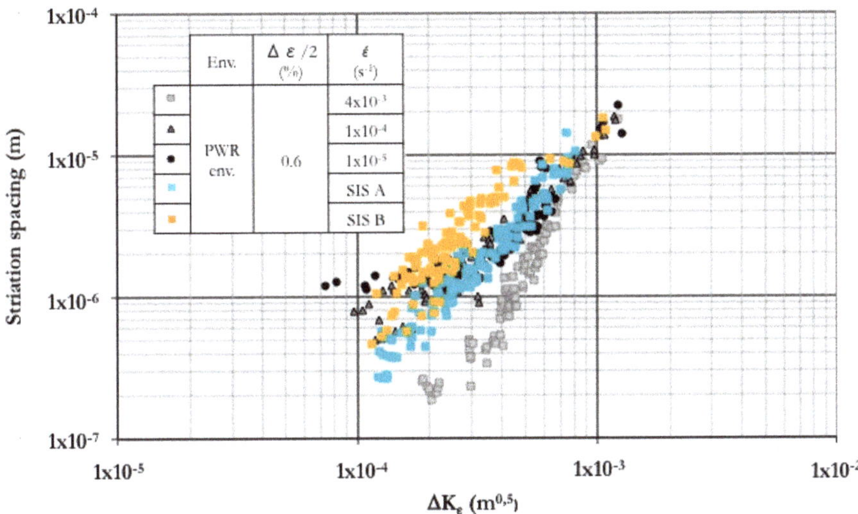

Figure 13. Striation spacing as a function of the strain intensity factor range ΔK_ε for different signals in PWR environment.

Table 5. Estimated fatigue lives as compared to experimental data for fatigue testing with SIS A and SIS B signals in PWR environment (The value of C is determined for da/dN expressed in m/cycle and ΔK_ε in m$^{1/2}$).

	Fatigue Crack Growth Law Parameters		Np (20 μm to 3 mm) (Cycles)	N5 (Cycles)
	C	**M**		
SIS A	12	1.96	952	1200
SIS B	0.84	1.54	452	440

4.2. Analysis of the Influence of the Signal Shape on Fatigue Crack Growth in a PWR Environment

The results obtained in vacuum have evidenced an intrinsic effect of strain rate which induces higher crack growth rates and a reduction in fatigue lives when the strain rate decreases [12]. In a PWR environment, the growth rates and consequently the fatigue lives are additionally conditioned by the action of environmentally-assisted fatigue crack growth mechanisms. Thus, decreasing the strain rate from 4×10^{-3} s^{-1} to 1×10^{-4} s^{-1} in vacuum induces a life reduction by a ratio equal to 1.3, while this ratio is equal to 3.2 in the PWR environment. Such an enhanced environmental effect at slow strain rate can be at least partly attributed to a longer exposure of the crack tip to PWR water [21]. However, isolating the respective influence of the numerous factors controlling the growth rate is made particularly difficult by the fact that the variation of certain parameters, such as strain amplitude or strain rate not only affects the exposure duration, but also the intrinsic response of the material as previously discussed. The situation is even more complicated when considering complex signals since the strain rate varies during one cycle. The next section constitutes a first attempt to tackle this issue.

In order to evidence a possible time-controlled crack growth regime, different authors [22–24] have proposed to relate the crack advance during one cycle to the rise time T_R, i.e., the duration of the part of the loading cycle characterized by a positive strain rate. This type of approach was originally applied by Shack and Kassner [25] to account for fatigue behavior of stainless steels in LWR environments. The use of the T_R parameter was justified by considering that environment affects crack propagation only when the crack is opening, that means during the rising strain part ($\dot{\varepsilon} > 0$) of the fatigue cycle. This was supported by the results of tests performed under saw-tooth signals in

BWR environment. In Figure 14, the growth rates estimated from striation spacings, are plotted as a function of the T_R parameter for selected values of the strain intensity factor range ΔK_ε calculated as indicated in [24], from data obtained with triangular signals at different strain rates (4×10^{-3}, 1×10^{-4}, 1×10^{-5} s^{-1}) and total strain amplitudes (0.3 and 0.6%). It can be seen that at low ΔK_ε value, the crack growth rates increase as the T_R value is increased up to T_R~175 s. Beyond this value, a saturation in the crack growth enhancement is noticed. Such a saturation is almost immediately reached at high ΔK_ε value. It is noteworthy that, when considering triangular signals, the different times are always proportional. As a consequence the ranking of the different signals in terms of crack growth enhancement will be independent of the part of the signal considered for the definition of T^*. Additionally, the T_R parameter cannot account for the differences observed in crack growth rates between SIS A and SIS B signals since its value, calculated on the same straining segments, is identical in both cases.

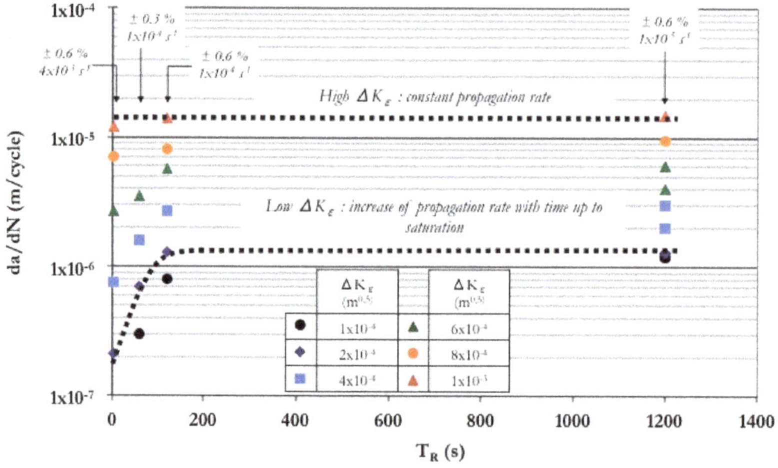

Figure 14. Fatigue crack growth rates as a function of the T_R parameter for selected ΔK_ε values.

Therefore, in the present study, a different approach in identifying a characteristic time, noted T^*, was developed in order to account for the influence of PWR environment under complex signals. The T^* parameter is here considered to represent the time during one fatigue cycle where environment is actually affecting the crack growth process. In a first step, it is assumed that the effective environment action takes place only during the rising straining part of the cycle. However, from the work by Tsutumi et al. [2] this time would depend on a threshold strain ε_{th} that would control the interactions between crack tip deformation and environmental effects. Indeed, they have performed tests in PWR environment on a 316NG stainless steel by using complex signals consisting in different combinations of fast and slow loading stages during one fatigue cycle. Their results [2] indicate that under a certain threshold in terms of strain noted ε_{th}, deforming the material with a slow or a fast strain rate has no effect on growth rates as well as on fatigue lives. For a value of the total strain amplitude of 0.6%, this ε_{th} threshold value is equal to -0.2% while it is equal to 0% when the strain amplitude is reduced to 0.3%. Thus, the environment would enhance the crack growth rate only when $\varepsilon > \varepsilon_{th}$.

However, applying this value to the data collected in the present study does not provide a proper description of the crack growth rates as a function of strain rate and load signal. This is the reason why it was decided to apply an inverse method in order to identify the value of the strain threshold ε_{th}^* and the characteristic time T^* pertaining to the experimental conditions (material, load signal, etc.). Indeed, as shown in Figure 15, a small variation in the ε_{th}^* value close to 0 has a very large impact in terms of relative values on the corresponding T^* in the case of the SIS A signal which is in any case small, while the much larger value associated with SIS B is almost unaffected by a such a small variation in the ε_{th}^* value. This value was tuned by a "trail-and-error" procedure and finally identified as: $\varepsilon_{th}^* = -0.0065\%$. In the following, crack growth rate and fatigue life data will be analyzed along this ε_{th}^* value and associated T^* values. In Figure 16 the crack growth rates measured at selected values of ΔK_ε are plotted as a function of the T^* parameter. It can be seen that while the tendency already noticed with the T_R parameter for triangular signals is still observed, the data corresponding to SIS A and SIS B signals become consistent with this type of analysis using T^*, which means higher crack growth rates with SIS B than with SIS A signal for a given value of ΔK_ε. More importantly, this diagram suggests that the environmentally-induced fatigue crack growth enhancement is controlled by the T^* parameter, in relation with the exposure of the crack tip to the PWR environment, until this enhancement achieves a kind of saturation with a plateau for $T^* > 50$ s.

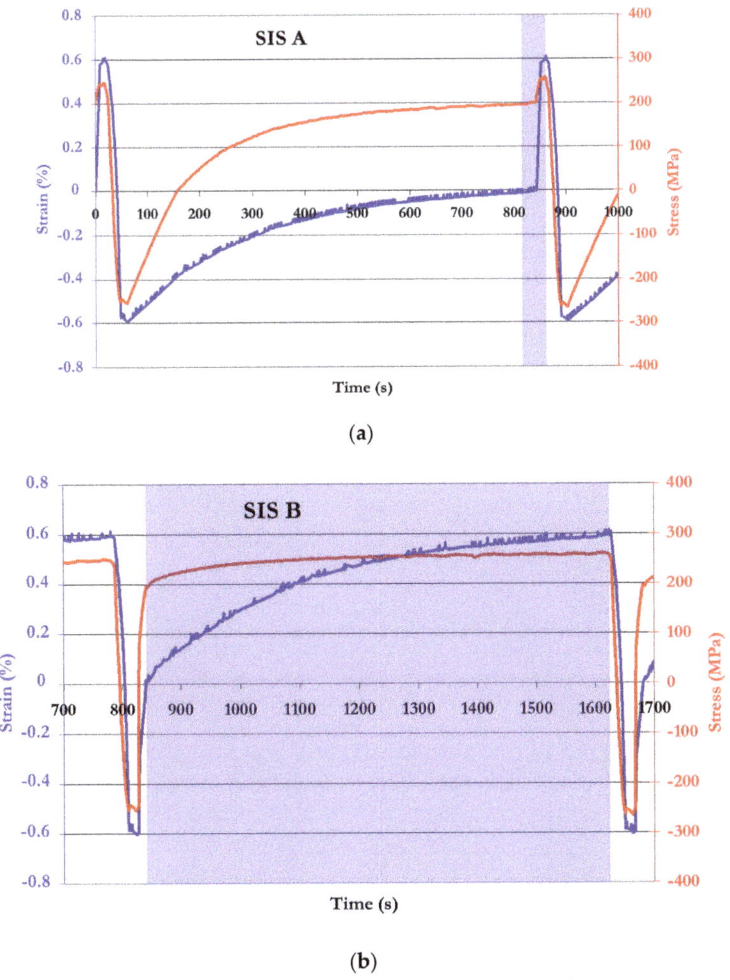

Figure 15. Representation of the T^* time in the SIS A and in SIS B signals.

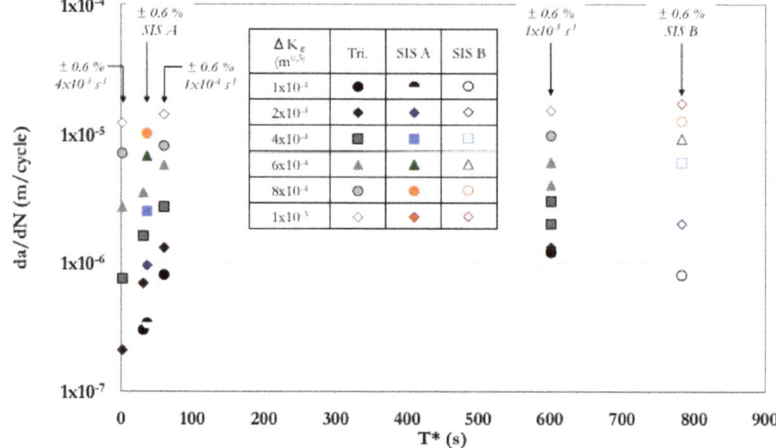

Figure 16. Fatigue crack growth rates plotted as a function of the T^* parameter for selected ΔK_ε values.

This characteristic time defined by the T^* parameter can furthermore be considered to analyze the fatigue lives in PWR environment under different loading conditions. Such an analysis is performed in Figure 17 where the fatigue lives obtained for triangular and SIS signals appear to be well ordered with the variation in T^* values, i.e., the longer T^*, the lower the life. Moreover, consistently to what was noticed on crack growth rate data, a significant decay in fatigue resistance is noticed for $T^* < 50$ s while the fatigue life remains almost unaffected at higher values.

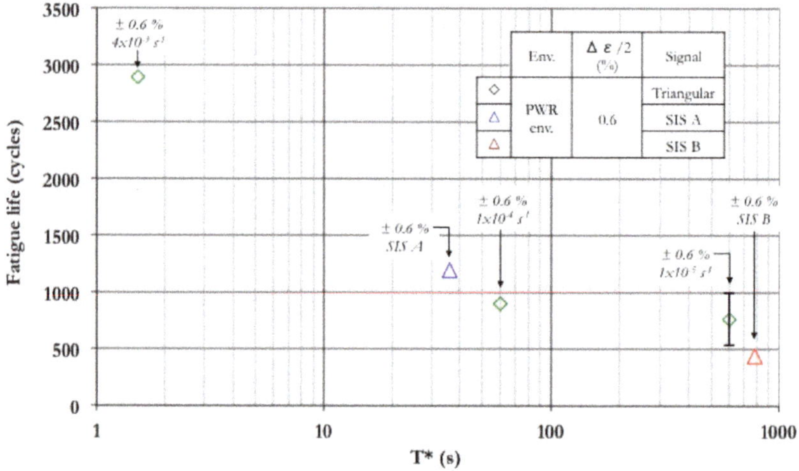

Figure 17. Fatigue lives in the PWR environment as a function of the T^* parameter.

Hence it has been shown that the use of the characteristic time T^* accounts for the differences noticed in fatigue lives and crack growth rates between both SIS A and SIS B signals, which is not the case when considering other characteristic times, like the cycle period or the total time spent in tension loading T_R, which are identical for both SIS signals. By its present definition, the characteristic time T^* is similar to the one proposed by Tsutsumi et al. [2] in the case of slow-fast-fast (SFF) and fast-slow-fast (FSF) signals tested on 316NG steel. This differs, however, by the value of the deformation threshold ε_{th}^*, which could be due to differences in the stress-strain behaviors of these alloys and/or in their susceptibility to the PWR environment effects. Thus, the effective time T^* during which the material is exposed to environmental effects has been defined from a deformation threshold ε_{th}^* which has been identified at -0.0065% of strain. This finding is consistent with observations by Vormwald and Seeger [26] who have shown that under large-scale yielding conditions, the crack opening can actually be connected to a deformation threshold located in the negative strain part of the cycle. Thus, experimental measurements of this opening deformation would be necessary to further assess the signification of this ε_{th}^* parameter which has been fitted to experimental data so far. Furthermore, it is here assumed in a first analysis that this ε_{th}^* threshold does not depend on the strain amplitude because of the limited data at different amplitudes.

Finally, a saturation in environmental effects has been noticed on fatigue lives as on growth rates for $T^* > 50$ s (Figures 16 and 17). This response in terms of crack growth rates and fatigue lives might be related to a saturation of one of the sequential elementary steps involved in environmentally-assisted crack growth [27], namely transport to the crack tip, adsorption, surface reactions and repassivation, absorption of atoms produced by surface reactions, and diffusion. Critical experiments are still required to gain further insights into the kinetics of these different processes.

5. Summary and Conclusions

In this study, the influence of complex load signals (SIS A and SIS B) on the LCF resistance of a 304L austenitic stainless steel has been investigated in relation with the occurrence of the strain rate variations during the loading cycle, with a special attention paid to environmental effects by comparing damage observed in air and in a simulated PWR water environment.

In particular, the analysis of the LCF tests using both SIS signals has evidenced differences in the stress-strain behavior engendered by the two signals. Indeed, as evidenced under triangular signals, dynamic strain ageing intervenes at low strain rates, and as a consequence influences the stress-strain response depending on the signal considered. A classification of the severity of the different signals has furthermore been established on the basis of the fatigue life results and detailed observations of the cracking features on the samples. More particularly, it is shown that the SIS B signal is more damaging than SIS A in air and in PWR environment. Additionally, striation spacing measurements indicate that this is mainly due to higher fatigue crack growth rates.

As what more specifically concerns the interplay between intrinsic strain-rate effects and time-dependent mechanisms during environmentally-assisted cracking in PWR environment, a threshold strain ($\varepsilon_{th}^* = -0.0065\%$) has been identified. Below this threshold, it is assumed that cracks are not fully opened, so that the PWR environment cannot affect the growth process. Reciprocally, a characteristic time T^*, corresponding to the duration where environment effectively assists the growth of open cracks has been introduced. It is then shown that the large difference in the T^* values between SIS A and SIS B signals indeed accounts for the difference noticed in growth rates and consequently in fatigue lives in accordance with the effect of strain rate. Furthermore, it is shown that for T^* values lower than 50 s, the crack growth enhancement is proportional to the T^* value, while a saturation in environmental assistance is noticed above 50 s.

Further studies will aim in elucidating the environmentally-assisted cracking mechanisms controlling these effects and precise the signification of the T^* characteristic time.

Author Contributions: Conceptualization: J.M., G.H. and L.B.; Methodology: J.M., G.H., T.P. and L.B.; Investigation: T.P.; Data curation: T.P., J.M., G.H., and L.B.; Writing—original draft preparation: G.H. and T.P.; Writing—review and editing: J.M., G.H., T.P. and L.B.; Project administration: J.M., G.H. and L.B.

Funding: This research received no external funding.

Acknowledgments: The authors thank Framatome for the financial support and the technical center of Framatome in Le Creusot for tests in PWR water environment.

Conflicts of Interest: The authors declare no conflict of interest.

References

1. Chopra, O.K.; Shack, W.J. The effect of lwr coolant environments on the fatigue life of reactor materials. In Proceedings of the ASME 2006 Pressure Vessels and Piping/ICPVT-11 Conference, Vancouver, BC, Canada, 23–27 July 2006; American Society of Mechanical Engineers: New York, NY, USA, 2006; pp. 191–204.
2. Tsutsumi, K.; Dodo, T.; Kanasaki, H.; Nomoto, S.; Minami, Y.; Nakamura, T. Fatigue behavior of stainless steel under conditions of changing strain rate in PWR primary water. In *Pressure Vessel and Piping Codes and Standards*; ASME: New York, NY, USA, 2001; pp. 135–141.
3. Higuchi, M.; Sakaguchi, K.; Nomura, Y. Effects of strain holding and continuously changing strain rate on fatigue life reduction of structural materials in simulated lwr water. In Proceedings of the ASME 2007 Pressure Vessels and Piping Conference, San Antonio, TX, USA, 22–26 July 2007; American Society of Mechanical Engineers: New York, NY, USA, 2007; pp. 123–131.
4. Huin, N. Environmental Effect on Cracking of an 304L Austenitic Stainless Steels in PWR Primary Environtment under Cyclic Loading. Ph.D. Thesis, ISAE-ENSMA National School of Mechanics and Aerotechics-Poitiers, Poitiers, France, 2013.
5. Atkinson, J.D.; Forrest, J.E. Factors influencing the rate of growth of fatigue cracks in rpv steels exposed to a simulated PWR primary water environment. *Corros. Sci.* **1985**, *25*, 607–631. [CrossRef]
6. Alain, R.; Violan, P.; Mendez, J. Low cycle fatigue behavior in vacuum of a 316L type austenitic stainless steel between 20 and 600 °C: 1. Fatigue resistance and cyclic behavior. *Mater. Sci. Eng. A* **1997**, *229*, 87–94. [CrossRef]
7. Gerland, M.; Alain, R.; Saadi, B.A.; Mendez, J. Low cycle fatigue behaviour in vacuum of a 316L-type austenitic stainless steel between 20 and 600 °C: 2. Dislocation structure evolution and correlation with cyclic behaviour. *Mater. Sci. Eng. A* **1997**, *229*, 68–86. [CrossRef]
8. Mannan, S.L. Role of dynamic strain-aging in low-cycle fatigue. *Bull. Mater. Sci.* **1993**, *16*, 561–582. [CrossRef]
9. Poulain, T.; Mendez, J.; Henaff, G.; de Baglion, L. Analysis of the ground surface finish effect on the lcf life of a 304L austenitic stainless steel in air and in PWR environment. *Eng. Fract. Mech.* **2017**, *185*, 258–270. [CrossRef]
10. Petitjean, S. Influence de l'état de surface sur le comportement en fatigue à grand nombre de cycles de l'acier inoxydable austénitique 304L. Ph.D. Thesis, Université de Poitiers, Poitiers, France, 2003.
11. De Baglion, L.; Mendez, J. Low cycle fatigue behavior of a type 304L austenitic stainless steel in air or in vacuum, at 20 °C or at 300 °C: Relative effect of strain rate and environment. *Procedia Eng.* **2010**, *2*, 2171–2179. [CrossRef]
12. De Baglion, L.; Mendez, J.; Le Duff, J.-A.; Lefrancois, A. Influence of PWR primary water on lcf behavior of type 304L austenitic stainless steel at 300 °C—comparison with results obtained in vacuum or in air. In Proceedings of the ASME 2012 Pressure Vessels and Piping Conference, Toronto, ON, Canada, 15–19 July 2012; American Society of Mechanical Engineers: New York, NY, USA, 2012; pp. 463–472.
13. Huin, N.; Tsutsumi, K.; Couvant, T.; Henaff, G.; Mendez, J. Fatigue life of the strain hardened austenitic stainless steel in simulated pressurized water reactor primary water. *J. Press. Vessel Technol.* **2014**, *136*, 031405. [CrossRef]
14. Tsutsumi, K.; Huin, N.; Couvant, T.; Henaff, G.; Mendez, J.; Chollet, D. Fatigue life of the strain hardened austenitic stainless steel in simulated PWR primary water. In Proceedings of the ASME 2012 Pressure Vessels and Piping Conference, Toronto, ON, Canada, 15–19 July 2012; American Society of Mechanical Engineers: New York, NY, USA, 2012; pp. 155–164.
15. Le Duff, J.A.; Lefrancois, A.; Vernot, J.P. Effects of surface finish and loading conditions on the low cycle fatigue behavior of austenitic stainless steel in PWR environment comparison of LCF test results with NUREG/CR-6909 life estimations. In *Proceedings of the ASME 2008 Pressure Vessels and Piping Conference*; Chicago, IL, USA, 27–31 July 2008, American Society of Mechanical Engineers: New York, NY, USA, 2008; pp. 453–462.
16. Le Duff, J.A.; Lefrancois, A.; Vernot, J.P. Effects of surface finish and loading conditions on the low cycle fatigue behavior of austenitic stainless steel in PWR environment for various strain amplitude levels. In Proceedings of the ASME 2009 Pressure Vessels and Piping Conference, Prague, Czech Republic, 26–30 July 2009; American Society of Mechanical Engineers: New York, NY, USA, 2009; pp. 603–610.

17. Le Duff, J.A.; Lefrancois, A.; Vernot, J.P.; Bossu, D. Effect of loading signal shape and of surface finish on the low cycle fatigue behavior of 304L stainless steel in PWR environment. In Proceedings of the ASME 2013 Pressure Vessels and Piping Conference, Paris, France, 14–18 July 2013; American Society of Mechanical Engineers: New York, NY, USA, 2013; pp. 233–241.

18. Mendez, J. On the effects of temperature and environment on fatigue damage processes in ti alloys and in stainless steel. *Mater. Sci. Eng. A* **1999**, *263*, 187–192. [CrossRef]

19. Boettner, R.C.; Laird, C.; McEvily, A.J. Crack nucleation and growth in high strain-low cycle fatigue. *Trans. AIME* **1965**, *233*, 379–387.

20. Kamaya, M. Low-cycle fatigue crack growth prediction by strain intensity factor. *Int. J. Fatigue* **2015**, *72*, 80–89. [CrossRef]

21. Sakaguchi, K.; Nomura, Y.; Suzuki, S.; Tsutsumi, K.; Kanasaki, H.; Higuchi, M. Effect of factors on fatigue life in PWR water environment. In Proceedings of the ASME 2006 Pressure Vessels and Piping/ICPVT-11 Conference, Vancouver, BC, Canada, 23–27 July 2006; American Society of Mechanical Engineers: New York, NY, USA, 2006; pp. 103–111.

22. Shoji, T.; Takahashi, H.; Suzuki, M.; Kondo, T. A new parameter for characterizing corrosion fatigue crack growth. *J. Eng. Mater. Technol.* **1981**, *103*, 298–304. [CrossRef]

23. Tice, D.R. Assessment of environmentally assisted cracking in PWR pressure vessel steels. *Int. J. Press. Vessels Pip.* **1991**, *47*, 113–126. [CrossRef]

24. Gilman, J.D. Further development of a model for predicting corrosion fatigue crack growth in reactor pressure vessel steels. *J. Press. Vessel Technol.* **1987**, *109*, 340–346. [CrossRef]

25. Schack, W.J.; Kassner, T.F. *Review of Environmental Effects on Fatigue Crack Growth of Austenitic Stainless Steel*; Argonne National Laboratory: Argonne, IL, USA, 1994.

26. Vormwald, M.; Seeger, T. The consequences of short crack closure on fatigue crack growth under variable amplitude loading. *Fatigue Fract. Eng. Mater. Struct.* **1991**, *14*, 205–225. [CrossRef]

27. Wei, R.P.; Simmons, G.W. Recent progress in understanding environment-assisted fatigue crack growth. *Int. J. Fract.* **1981**, *17*, 235–247. [CrossRef]

Article

An Isotropic Model for Cyclic Plasticity Calibrated on the Whole Shape of Hardening/Softening Evolution Curve

Jelena Srnec Novak [1,*], Francesco De Bona [1] and Denis Benasciutti [2]

[1] Politechnic Department of Engineering and Architecture (DPIA), University of Udine, via delle Scienze 208, 33100 Udine, Italy
[2] Department of Engineering, University of Ferrara, via Saragat 1, 44122 Ferrara, Italy
* Correspondence: jelena.srnec@uniud.it; Tel.: +39-0432-558-297

Received: 14 June 2019; Accepted: 28 August 2019; Published: 30 August 2019

Abstract: This work presents a new isotropic model to describe the cyclic hardening/softening plasticity behavior of metals. The model requires three parameters to be evaluated experimentally. The physical behavior of each parameter is explained by sensitivity analysis. Compared to the Voce model, the proposed isotropic model has one more parameter, which may provide a better fit to the experimental data. For the new model, the incremental plasticity equation is also derived; this allows the model to be implemented in finite element codes, and in combination with kinematic models (Armstrong and Frederick, Chaboche), if the material cyclic hardening/softening evolution needs to be described numerically. As an example, the proposed model is applied to the case of a cyclically loaded copper alloy. An error analysis confirms a significant improvement with respect to the usual Voce formulation. Finally, a numerical algorithm is developed to implement the proposed isotropic model, currently not available in finite element codes, and to make a comparison with other cyclic plasticity models in the case of uniaxial stress and strain-controlled loading.

Keywords: cyclic plasticity; kinematic model; isotropic model; hardening/softening

---

1. Introduction

Thanks to their favorable combination of mechanical and thermal properties, metals are widely employed in industrial applications in which components are subjected to high thermo-mechanical loadings. During the component's service life, high temperatures combined with high mechanical stresses may induce in the component a plastic deformation, even only locally. If thermo-mechanical loadings also vary cyclically, the resulting cyclic elasto-plastic response may lead to some kind of fatigue damage. To perform a durability assessment, it is often advantageous to use a numerical approach based on the finite element (FE) method. The accuracy of results depends significantly on the capability of the material model, in the numerical code, to describe the cyclic plasticity behavior of the material, as it is observed experimentally.

A noteworthy example—considered in this work—is the case of copper alloys used in components of steel making plants (e.g., mold for continuous steel casting, anode for electric arc furnace, etc.) [1,2]. The material model in FE model needs to be calibrated on experimental cyclic plasticity data, which for copper alloys are rarely available in the literature. In [3], parameters for nonlinear kinematic and isotropic plasticity models were obtained for pure copper and CuCrZr alloys at a temperature range of between 20 °C and 550 °C. Cyclic hardening behavior of pure copper was studied in [4] but only limited to low-cycle fatigue curves. In a recent experimental study on CuAg0.1 [5], the identification of material parameters was performed with specific attention to nonlinear kinematic and isotropic models (Armstrong and Frederick, Chaboche and Voce), as generally they are already available in most common commercial finite element codes. While the kinematic model provided a quite precise fitting, the isotropic model (Voce) seemed less capable of following the softening trend of the CuAg0.1 alloy [5]. In fact, as it will be shown, the "S-shaped" curves characterizing the exponential law governing the Voce model hardly fitted the trend of experimental data, making the determination of the speed of stabilization fairly inaccurate. This aspect was already observed by [6,7] in the case of different types of steels. Neither the method—proposed in [8]—for changing the hardening modulus in the kinematic part, nor that proposed in [9] where the hardening model is obtained by superimposing several parts of kinematic and isotropic hardening, seem to solve the matter, as both appear only suitable for materials (like stainless steel) for which the hardening evolution depends on strain range. A possible way to capture more precisely a cyclic hardening/softening trend is that of superimposing two or more Voce models, as proposed in [10].

As the ability of plasticity models to accurately represent the material behavior plays an important role in numerical simulations and could have a direct influence on the accuracy of results, the aim of this work is to develop a new isotropic model that is able to replicate better the softening evolution of the copper alloy here investigated.

2. Materials and Methods

2.1. Experimental Testing

The experimental results of low-cycle fatigue (LCF) tests described in [5] are considered in this work. To characterize the cyclic stress-strain behavior and the fatigue strength of CuAg0.1 alloy (classified in ASTM B 124 standard [11]) at room and high temperature, strain-controlled tests with fully reversed ($R_\varepsilon = -1$) triangular waveform at a strain rate of 0.01 s^{-1} were performed at three temperature levels (20 °C, 250 °C, 300 °C). However, only data at room temperature will be considered from now on.

2.2. Kinematic and Isotropic Plasticity Models: Theoretical Background

Plasticity is characterized by the irreversible straining that occurs once a certain level of stress is reached. If plastic strains are assumed to develop independently of time, several theories are available to characterize the response of materials. Plasticity theory distinguishes three main aspects: yield criterion, flow rule and material (hardening) models [12–17].

The yield criterion determines the stress level at which yielding occurs. In this work, the von Mises yield criterion is considered and the following expression of the plastic potential f is introduced [12–14]:

$$f = \sqrt{\frac{3}{2}(\boldsymbol{\sigma}' - \boldsymbol{X}') : (\boldsymbol{\sigma}' - \boldsymbol{X}')} - R - \sigma_0 = 0 \tag{1}$$

where $\boldsymbol{\sigma}'$ is the deviatoric stress tensor, $\boldsymbol{X}'$ is the deviatoric part of the back stress, R is the drag stress and σ_0 is the initial yield stress (bold symbols indicate tensors).

Once yielding occurs, a flow rule is needed to relate stresses and plastic strains. In the literature, several flow rules are available, nevertheless in the following the associated flow is adopted, in which the plastic flow is connected or associated with the yield criterion [12–17]:

$$d\varepsilon_{\mathrm{pl}} = d\lambda \frac{\partial f}{\partial \boldsymbol{\sigma}} \tag{2}$$

The direction of the plastic strain increment $d\varepsilon_{\mathrm{pl}}$ depends on both the plastic multiplier $d\lambda$ (non-negative scalar) and the plastic potential, which determine the amount and the direction of plastic straining, respectively.

Finally, hardening models (kinematic and isotropic) describe the change of yield surface as a function of plastic strain.

2.2.1. Kinematic Material Model

It is experimentally observed in metals under cyclic loading that the center of the yield surface moves in the direction of the plastic flow (the Bauschinger effect) [17]. A kinematic model captures the aforementioned effect, since it assumes that under a progressive yielding, the yield surface translates in the stress space, maintaining a constant size [12,15–17]. The translation of the yield surface center is described by the back stress X.

Different kinematic models have been developed to relate the increment of the back stress dX with the plastic strain ε_{pl} and usually also with the accumulated plastic strain $d\varepsilon_{pl,acc} = (2/3d\varepsilon_{pl}{:}d\varepsilon_{pl})^{1/2}$. For uniaxial loading $d\varepsilon_{pl,acc} = d\varepsilon_{pl}$ [15].

The Prager model assumes that X is proportional to the plastic strain increment [12–17]:

$$dX = \frac{2}{3}Cd\varepsilon_{pl} \tag{3}$$

where C is the hardening modulus. Armstrong and Frederick's (AF) nonlinear model adds a recall term to Equation (3):

$$dX = \frac{2}{3}Cd\varepsilon_{pl} - \gamma X d\varepsilon_{pl,acc} \tag{4}$$

where γ is the non-linear recovery parameter that defines the rate at which the hardening modulus starts to decrease as the plastic strain develops. Finally, the Chaboche model is obtained by superimposing several AF models [8]:

$$X = \sum_{i=1}^{3} X_i dX_i = \frac{2}{3}C_i d\varepsilon_{pl} - \gamma_i X_i d\varepsilon_{pl,acc} \tag{5}$$

2.2.2. Isotropic Material Model

The isotropic model assumes that, at any stage of loading, the center of the yield surface remains at the origin and the surface expands homotetically in size as plastic strain develops. Very often, a nonlinear isotropic model (known also as the Voce model [18]) is adopted. Expansion of the yield surface is described with the change of drag stress R [12–15]:

$$R = R_\infty\left[1 - \exp\left(-b\varepsilon_{pl,acc}\right)\right] \tag{6}$$

where b is the speed of stabilization and R_∞ is the stabilized stress. Parameter R_∞ can be positive or negative representing either cyclic hardening ($R_\infty > 0$) or softening ($R_\infty < 0$), respectively. Differentiation of Equation (6) gives [3–15]:

$$dR = b(R_\infty - R)d\varepsilon_{pl,acc} \tag{7}$$

Furthermore, the evolution of R can also be interpreted as the relative change of the maximum stress $\sigma_{max,i}$ in the Nth cycle with respect to the maximum stress in the first ($\sigma_{max,1}$) and in the stabilized ($\sigma_{max,s}$) cycle [12–14]:

$$\frac{\sigma_{max,i} - \sigma_{max,1}}{\sigma_{max,s} - \sigma_{max,1}} \approx \frac{R}{R_\infty} = 1 - \exp\left(-b\varepsilon_{pl,acc}\right) \tag{8}$$

In case of strain-controlled loading, the plastic strain range per cycle $\Delta\varepsilon_{pl}$ is approximately constant and the plastic strain accumulated after N cycles becomes [12,14]:

$$\varepsilon_{pl,acc} \approx 2\Delta\varepsilon_{pl}N \tag{9}$$

Figure 1 plots Equation (8) for different values of b and versus the accumulated plastic strain on a log scale. As can be noticed, by increasing the speed of stabilization the curve shifts to the left, while its

"S-shape" remains essentially unaffected. In other words, a material with a higher value of b reaches its stabilized condition at a lower value of $\varepsilon_{pl,acc}$ (i.e., in a smaller number of cycles).

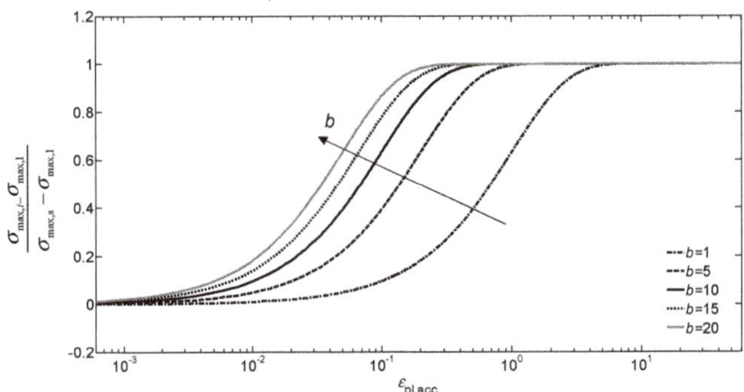

Figure 1. Plot of Equation (8) for different values of b parameter.

As with the kinematic model, also two or more Voce models can be superimposed, as proposed in [10]. Such a procedure requires a higher number of parameters to be estimated (minimum four: two parameters, b_1 and b_2, controlling the speed of stabilization and two others, $R_{\infty,1}$ and $R_{\infty,2}$, defining the stress saturation). While the sum $R_{\infty} = R_{\infty,1} + R_{\infty,2}$ clearly identifies the total saturated stress, the two components $R_{\infty,1}$ and $R_{\infty,2}$, taken individually, seem not to have a so evident physical meaning. The same observation holds true for parameter b in the original Voce model. If the model is described by more than one b, its physical meaning (i.e., speed of stabilization) seems to be lost, especially when one parameter is much bigger than the other (i.e., $b_1 >> b_2$). This observation is even more evident once the accelerated technique is adopted [19], in which the speed of stabilization is fictitiously increased. Generally, this approach is particularly suitable when dealing with computationally demanding simulations, i.e., in circumstances when small plastic strains occur during loading and a material model needs a huge number of cycles to reach a complete stabilization.

Finally, kinematic and isotropic models need to be jointly used if both the Bauschinger effect and the cyclic hardening/softening are to be simulated numerically.

2.2.3. Material Model Calibration

After having determined the static parameters (Young's modulus and yield stress) as suggested in [12], kinematic variables were evaluated on experimental stabilized cycles at various strain amplitudes. This approach ensures that the estimated variables are suitable for being used over a wide range of strain amplitudes. Details on the estimation procedure (Young's modulus, yield stress and kinematic variables) is given in [5,20,21].

Figure 2 compares the simulated cycles (only kinematic model) with the first and stabilized experimental cycles, for $\varepsilon_a = 0.5\%$. For the considered material, the Armstrong and Frederick kinematic model (parameters C, γ) is selected, as it yields a better agreement with experiments, than does the Chaboche model (see [5]). For the first simulated cycle, the initial values of Young's modulus and the yield stress are assumed. As can be noticed, the model stabilizes in the first two cycles and agrees fairly well with the experimental cycle. In addition, Figure 2 also depicts the stabilized stress-strain cycle (black solid line) obtained with the stabilized parameters (E_s, σ_{0^*}). The change from the first cycle to the stabilized one confirms a softening behavior for the material.

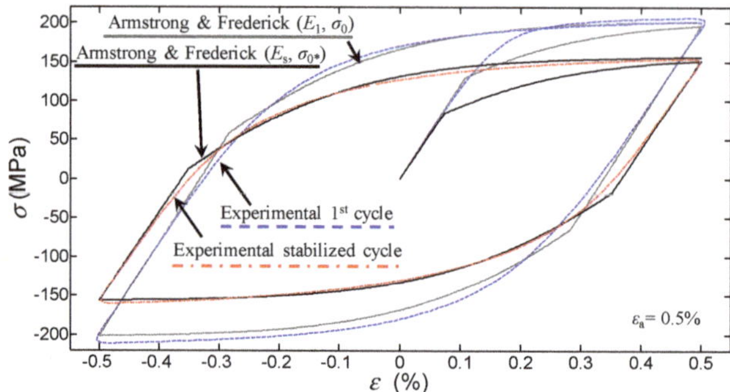

Figure 2. Comparison between simulated and experimental (first and stabilized) cycle.

Nonlinear isotropic parameters were then estimated according to Voce model. The maximum stress σ_{max} was measured for each stress-strain cycle. As the CuAg0.1 alloy never saturates completely, the stabilized stress $\sigma_{max,s}$ does not approach any asymptote and its value was then established by a conventional criterion, which considered the maximum stress at half the number of cycles to failure. The procedure was repeated for each strain amplitude. According to Equation (8), the saturation stress R_∞ was determined as:

$$R_\infty = \sigma_{max,\,s} - \sigma_{max,1} \tag{10}$$

In this case, $R_\infty < 0$, i.e., material exhibits softening (see also Figure 2). The speed of stabilization b was identified by fitting Equation (8) to experiments. Figure 3 compares the isotropic model curve and experiments for $\varepsilon_a = 0.5\%$. A satisfactory correlation is obtained, particularly at the smallest and the largest values of accumulated plastic strain (i.e., at the first and saturated cycle, respectively). Other than that, the model follows a trend that deviates significantly from that followed by experimental data. A similar inconsistency was also observed for other strain amplitudes. The aim to improve the fitting accuracy then motivates the attempt to modify the evolution rule of R in a new isotropic model, described in the next paragraph.

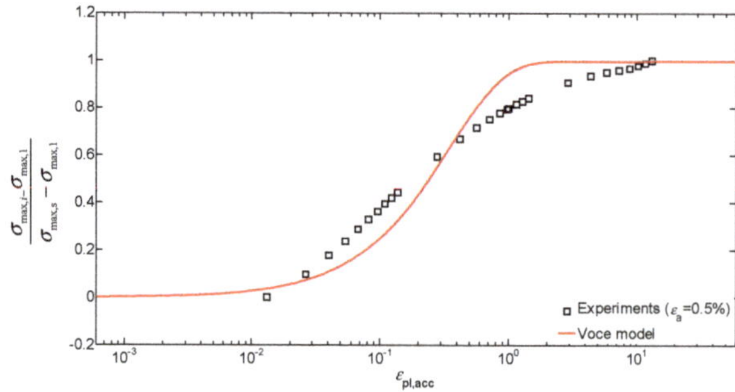

Figure 3. Experimental data for $\varepsilon_a = 0.5\%$ (markers) fitted by the isotropic model.

3. Proposed Isotropic Model

As previously observed, experimental results obtained for CuAg0.1 alloy do not fit perfectly well with the "S-shape" curve of the exponential law proposed by Voce. With the aim to get an improved fitting with experiments, a new model described by the following equation is thus proposed:

$$R = R_\infty \frac{\varepsilon_{pl,acc}^{s}}{a + \varepsilon_{pl,acc}^{s}} \tag{11}$$

in which a, s are material parameters that control the rate of cyclic hardening or softening. The proposed expression has a physical basis being it able to capture the two limiting cases: $R = 0$ for $\varepsilon_{\mathrm{pl,acc}} \to 0$ and $R = R_\infty$ for $\varepsilon_{\mathrm{pl,acc}} \to \infty$. Differentiation of Equation (11) gives the incremental relationship:

$$dR = \left[s \frac{R}{\varepsilon_{\mathrm{pl,acc}}} \left(1 - \frac{R}{R_\infty} \right) \right] d\varepsilon_{\mathrm{pl,acc}} \tag{12}$$

Similarly, to the Voce model, the evolution of R can be described by the relative change of maximum stress in each cycle:

$$\frac{\sigma_{\mathrm{max},i} - \sigma_{\mathrm{max},1}}{\sigma_{\mathrm{max},s} - \sigma_{\mathrm{max},1}} \approx \frac{R}{R_\infty} = \frac{\varepsilon^s_{\mathrm{pl,acc}}}{a + \varepsilon^s_{\mathrm{pl,acc}}} \tag{13}$$

Fitting of Equation (13) to experimental data yields the values of a and s parameters. The estimation procedure to evaluate the saturation stress R_∞ remains unchanged.

Figure 4 displays a sensitivity analysis of Equation (13), which is helpful to clarify the physical meaning of parameters a and s. At increasing values of a (keeping s constant), the speed of stabilization diminishes and the curve shifts to the right. In other words, a bigger amount of accumulated plastic strain is needed to reach the stabilized condition, see Figure 4a. By contrast, the exponent s (keeping a constant) controls the "slope" of the increasing portion of the curve (i.e., higher values of s give a steeper average slope), see Figure 4b.

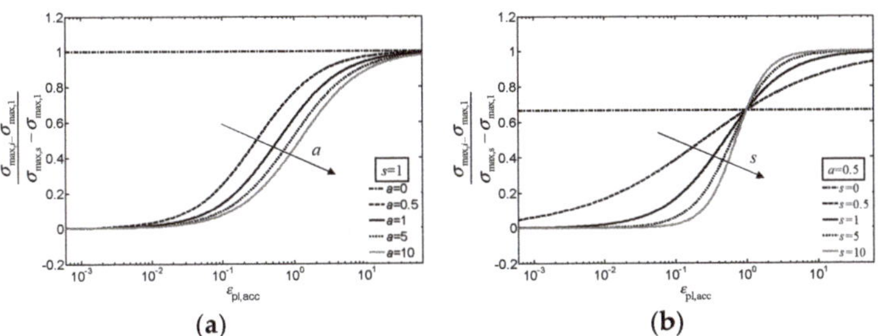

Figure 4. The sensitivity of the proposed Equation (10) to parameters (**a**) a and (**b**) s.

Please note that limiting values exist for both parameters. For example, no cyclic hardening/softening would actually occur if a becomes either infinite (for which $R = 0$ for any s) or zero (for which $R = R_\infty$ from the first cycle). A similar behavior also occurs when $s = 0$, for which $R = R_\infty/(a + 1)$ (in the example of Figure 4, $R \cong 0.667 R_\infty$). When s increases to infinite, $R = 0$ for $\varepsilon_{\mathrm{pl,acc}} < 1$, $R = R_\infty$ for $\varepsilon_{\mathrm{pl,acc}} > 1$ and $R = R_\infty/(a + 1)$ for $\varepsilon_{\mathrm{pl,acc}} = 1$.

4. Results and Discussion

The nonlinear kinematic model is able to capture quite accurately the first and the stabilized cycle, if the initial and stabilized static parameters are adopted, respectively, see Figure 2. However, the combined (kinematic and isotropic) model is needed, if the cyclic softening evolution of CuAg0.1 alloy has to be considered. It is then of interest to discuss in more detail on the results obtained with both the Voce and the new isotropic model proposed in this work.

4.1. Stress-Strain Behavior: Isotropic Model (Voce)

The solid lines in Figure 5 show how Equation (8) fits the cyclic data for $\varepsilon_a = 0.3\%$, 0.5% and 0.7%. The corresponding three estimated values of b are reported in Table 1 (along with the values estimated for other strain amplitudes). As it can be noticed, in all three cases examined the fitting curve does not match experimental data perfectly. In fact, experiments follow a quite smoother trend with respect to the "S-shaped" curve given by the exponential expression in Equation (8). On the other hand, it was already observed and shown in Figure 1 that the speed of stabilization b only makes the S-curve shift horizontally, without changing its shape.

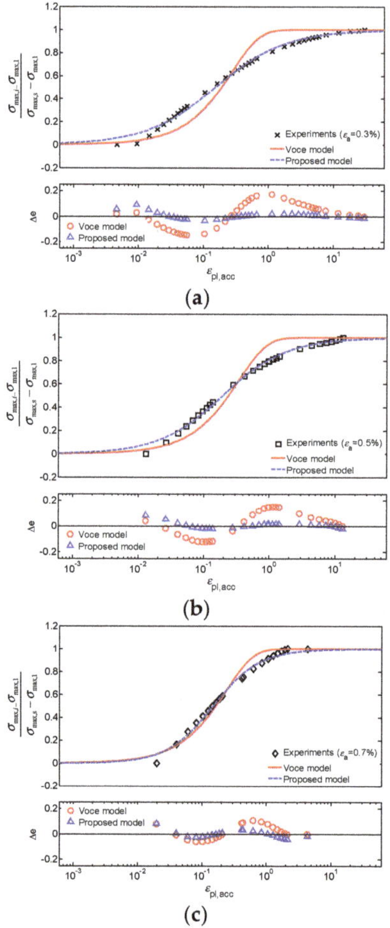

Figure 5. Isotropic model for different strain amplitudes ((**a**) $\varepsilon_a = 0.3\%$, (**b**) $\varepsilon_a = 0.5\%$ and (**c**) $\varepsilon_a = 0.7\%$); Voce model (red solid line) and proposed model (blue dashed line). Residual Δe is also reported.

Table 1. Isotropic parameters identified from experimental data for the Voce and proposed models. Goodness-of-fit examined in terms of sum of squares of residuals (SSE) for each model and strain amplitude.

Strain Amplitude		Material Parameters			Error Index (SSE)	
		Voce	Proposed		Voce	Proposed
ε_a	R_∞ (MPa)	b	a	s		
0.15%	−56	1.307	0.514	0.923	1.914	1.601
0.175%	−71	1.197	0.491	0.978	0.654	0.455
0.2%	−80	3.145	0.316	0.778	0.464	0.063
0.3%	−84	3.620	0.223	0.802	0.338	0.022
0.4%	−83	4.488	0.179	0.855	0.292	0.016
0.5%	−52	2.871	0.234	0.893	0.269	0.014
0.6%	−52	2.581	0.290	0.853	0.298	0.019
0.7%	−69	4.208	0.101	1.203	0.069	0.017
Single values	$R_{\infty,ave}$ [1] −68	b_{all} [1] 2.352	a_{all} [1] 0.199	s_{all} [1] 0.965	3.401	1.766

[1] parameters b_{all}, a_{all} and s_{all} are estimated from all experimental data merged together (whereas $R_{\infty,ave}$ is the average over all strain values).

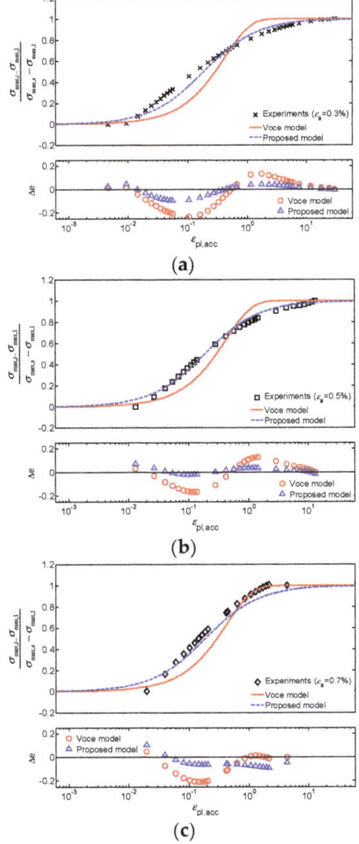

Figure 6. Isotropic model for different strain amplitudes ((**a**) $\varepsilon_a = 0.3\%$, (**b**) $\varepsilon_a = 0.5\%$ and (**c**) $\varepsilon_a = 0.7\%$) calculated with b_{all}, a_{all} and s_{all} parameters; Voce model (red solid line) and proposed model (blue dashed line). Residual Δe is also reported.

Different values of b and R_∞ characterize cycles with different strain amplitude values, see Table 1. This may represent a drawback, if commercial finite element codes have to be used, as they generally permit only one single value of the isotropic parameters to be input. On the other hand, a component may be subjected to a certain loading condition, for which the resulting plastic strain spans over a relatively wide range. A possible compromise for solving this issue could be to take an average value $R_{\infty,ave}$ and to identify a single value b_{all} by fitting Equation (8) to all the experimental data gathered together (see the last row in Table 1), as it was done in Figure 11 in [5]. A comparison between the fitted curve calculated with b_{all} and the experimental data is presented in Figure 6. As expected, this "averaging" procedure makes the fitting even worse with respect to the previous case.

4.2. Stress-Strain Behavior: Proposed Isotropic Model

Figure 5 also compares the proposed model (blue dashed line) with the same experimental data ($\varepsilon_a = 0.3\%$, 0.5% and 0.7%) considered previously for the Voce model. It is apparent how Equation (13) yields a more precise fitting with experiments.

Also for this model, a range of values for a (0.101÷0.514) and s (0.802÷1.203) characterize cycles with different strain amplitudes, see Table 1. Therefore, similarly to the case of Voce model, single values of a_{all}, s_{all} were estimated by fitting Equation (13) to all the experimental data gathered together. Table 1 then lists the values of a, s for each strain amplitude, as well as the "averages" a_{all}, s_{all} (see the last row). A comparison is presented in Figure 6. Even now, as it was for the previous comparison, for all three amplitudes examined the new isotropic model is far closer to experiments than Voce model is.

For the proposed model is not yet possible to prescribe a range of a and s values of general validity, as it would be desirable to have additional data from other materials.

4.3. Error Analysis

The previous comparison—clearly more qualitative—can be made more quantitative by the use of suitable error metrics through which the deviation of each model from experimental data can be quantified and compared.

Residual (Δe) and sum of squares of residuals (SSE) are used to measure the deviation between model and experiments. The residual provides a "local" measure of fitting at each point, whereas the SSE index gives a "global" measure of fitting. The residual measures the difference for each experimental point used in model calibration and it is defined as [22]:

$$\Delta e = y_{exp,\,i} - y_{Voce,\,i} \quad \text{and} \quad \Delta e = y_{exp,\,i} - y_{prop,\,i} \quad \text{for} \quad i = 1, 2 \ldots n \tag{14}$$

where:

$$y_{exp,\,i} = \frac{\sigma_{max,\,i} - \sigma_{max,1}}{\sigma_{max,\,s} - \sigma_{max,1}}; \quad y_{Voce,\,i} = 1 - \exp\!\left(-b\varepsilon_{pl,acc,\,i}\right); \quad y_{prop,\,i} = \frac{\varepsilon_{pl,acc,\,i}^{s}}{a + \varepsilon_{pl,acc,\,i}^{s}} \tag{15}$$

Symbol n denotes the number of experimental points used in calibration.

The bottom graph in Figures 5 and 6 show the values of residuals for both isotropic models, at three different strain amplitudes ($\varepsilon_a = 0.3\%$, 0.5%, 0.7%). Figure 5 refers to specific values (b, a, s) for the three amplitudes; Figure 6 refers to the "average" parameters (b_{all}, a_{all}, s_{all}), calculated as explained before. The residuals Δe are plotted on the vertical axis while the accumulated plastic strain, which is an independent variable, is plotted on the horizontal axis. In all considered cases, residual Δe obtained with the proposed model is significantly smaller with respect to Voce model. The maximum percentage residual for the considered strain amplitudes varies over a range of 44–51% and 5–18%, respectively, for the Voce and the proposed models.

To provide a single index value that quantifies the model accuracy for each strain amplitude, the sum of squares of residuals is also calculated [22]:

$$SSE = \sum_{i=1}^{n}\left(y_{exp,\,i} - y_{Voce,\,i}\right)^2 ; SSE = \sum_{i=1}^{n}\left(y_{exp,\,i} - y_{prop,\,i}\right)^2 \tag{16}$$

Values of SSE, calculated for each strain amplitude, are listed in Table 1. Both models give the biggest SSE for $\varepsilon_a = 0.15\%$ and $\varepsilon_a = 0.175\%$. These large values can be explained by considering that at those small strain amplitudes, the CuAg0.1 alloy showed a slight hardening in the first 5–10 cycles, after which softening occurred, as observed in [5]. Such a small hardening at the beginning of loading gives an increment of the maximum stress $\sigma_{max,i}$ with respect to the initial value $\sigma_{max,1}$. A positive difference $\sigma_{max,i}-\sigma_{max,1}$ results and thus a negative ratio in the left-hand side of Equations(8) and (13) appears. As Equations (8) and (13) are always positive and bounded by two horizontal asymptotes at both tails, such an initial hardening cannot be captured correctly and inevitably a small error is introduced in the fitting. At other strain amplitudes (0.2–07%), the proposed model gives from 75% up to 94% lower values of SSE with respect to the Voce model, see Table 1.

As expected, see Table 2, the fitting curves obtained by the parameters $(b_{all}, a_{all}, s_{all})$ are slightly less accurate than those obtained by the parameters (b, a, s) in Table 1 that were specifically calibrated for each strain amplitude. Nevertheless, it can be noticed that the proposed model always provides considerably better results in terms of SSE with respect to Voce model.

Table 2. Evaluation of fitted curves obtained with both models and calculated with b_{all}, a_{all} and s_{all}.

Strain Amplitude	Error Index (SSE)	
ε_a	Voce	Proposed
0.3%	0.493	0.080
0.5%	0.289	0.021
0.7%	0.340	0.094

4.4. Combined Armstrong-Frederick with Proposed Isotropic Model: Stress-Strain Evolution

A numerical algorithm is developed, see Appendix A, to implement the combined kinematic and proposed isotropic plasticity model, in the case of uniaxial stress and strain-controlled loading.

Figure 7 displays the stress-strain evolution at the 100th cycle for strain amplitude $\varepsilon_a = 0.5\%$. Although the difference between the three curves is relatively small, the proposed model fits more precisely the experimental stress-strain curve.

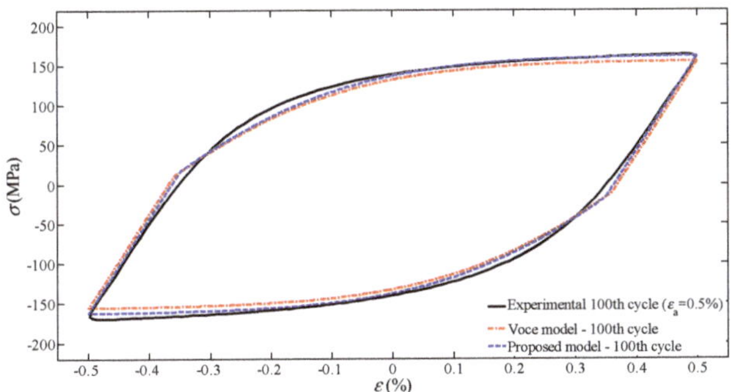

Figure 7. Stress-strain evolution of 100th cycle for strain amplitude $\varepsilon_a = 0.5\%$, comparison between experiments, combined model with Voce and with proposed isotropic models.

As a final remark, the cyclic hardening/softening evolution described with Equations (6) and (11) only depends on the accumulated plastic strain, but not on the plastic strain amplitude. This implies that all experimental data should collapse into one single curve independently on the imposed strain amplitude. On the other hand, the literature reports examples of materials—e.g., 316 stainless steel [12–14], nickel base superalloy [23] and 42CrMo4 steel [7]—for which the isotropic behavior depends on the strain amplitude. For the CuAg0.1 alloy considered in this work, such a dependence is not so evident, but cannot be excluded at all. Therefore, a further analysis in that sense would be desirable.

5. Conclusions

In this work, a new isotropic model is proposed for describing the cyclic hardening/softening behavior of metals. The model is an attempt to overcome the poor fitting observed in Voce isotropic model, when calibrated on experimental cyclic data of a CuAg0.1 alloy. Strain-controlled cyclic test data at different strain amplitudes, carried out in a recent study published by the authors, are used as a benchmark. Experimental data were used for identifying parameters of non-linear kinematic (Armstrong and Frederick, Chaboche) and isotropic (Voce) models. The calibration identified the one-pair kinematic model as the closest to experiments. By contrast, a poor agreement was observed when fitting the isotropic Voce model. The discrepancy is attributed to the fact that experiments follow a smoother trend with respect to the "S-shaped" curve defined by the exponential expression of Voce model equation. In fact, the Voce model is described by two parameters, i.e., the stabilized stress R_∞ and the speed of stabilization b, the latter only shifting the curve horizontally, while keeping the curve shape unaffected.

An improved isotropic model, aimed at providing a better fit with experimental data, is then proposed and calibrated to the same experimental data. In addition to the stabilized stress R_∞ also present in Voce model, the proposed model is governed by two parameters a, s, which both control the speed of stabilization and the shape of the hardening/softening evolution curve. An error analysis on calibration results confirms that the proposed model is much closer to experiments than Voce model in all cases examined.

The proposed isotropic model, when combined with the kinematic one, seems to better fit the experimental stress-strain curves than the Voce model combined with the same kinematic part. This permits the cyclic hardening/softening evolution of the material to be described quite accurately.

Finally, a numerical algorithm is developed to implement the combined kinematic and proposed isotropic plasticity model, in the case of uniaxial stress and strain-controlled loading. The framework of the presented algorithm permits convergence to be always achieved with low computational cost.

Author Contributions: Conceptualization: J.S.N. and D.B.; methodology: J.S.N., F.D.B. and D.B.; software: J.S.N.; validation: J.S.N., F.D.B. and D.B.; investigation: J.S.N., F.D.B. and, D.B., writing-original draft preparation: J.S.N., F.D.B. and D.B.; writing-review and editing: J.S.N., F.D.B. and D.B.; supervision: F.D.B. and D.B.

Funding: This research received no external funding.

Conflicts of Interest: The authors declare no conflict of interest. The research was conducted without personal financial benefit from any funding body, and no such body influenced the execution of the work.

Appendix A

A numerical algorithm is developed to implement the proposed model, currently not available in numerical codes. The algorithm applies to the case of uniaxial stress and strain-controlled loading (constant strain range) and it permits a comparison between the combined models (kinematic-Voce and kinematic-proposed model) and experimental data.

Figure A1 sketches the block diagram of the numerical algorithm. The calculation starts by defining the strain range $\Delta\varepsilon = 2\varepsilon_a$, the strain increment $d\varepsilon$ and the number of cycles N_f to be simulated. The number of iterations n_ε is obtained dividing $2\Delta\varepsilon N_f$ by $d\varepsilon$.

The stress increment is computed based on the strain increment assuming that the stress-strain relation is completely elastic. The elastic stress predictor σ_{el} is then evaluated simply adding to σ_i the stress increment $d\sigma$. In the same iteration, the actual yield stress σ_0^* remains unchanged and the plastic strain is set to be zero. If the elastic stress predictor indicates that yield stress has been exceeded (von Mises yield criterion is adopted), the increment becomes elasto-plastic. Consequently, the plastic strain increment $d\varepsilon_{pl}$ has to be calculated by adopting a flow rule. For von Mises criterion, the plastic multiplier $d\lambda$ corresponds to the increment of plastic strain; for uniaxial loading the accumulated plastic strain increment $d\varepsilon_{pl,acc}$ is equal to $d\varepsilon_{pl}$ [15] and therefore the accumulated plastic strain can be evaluated accordingly. Finally, the $(i + 1)$th value of stress is determined by taking into account the contribution of the back stress X and of the drag stress R (kinematic and isotropic) with Equations (4) and (11), respectively.

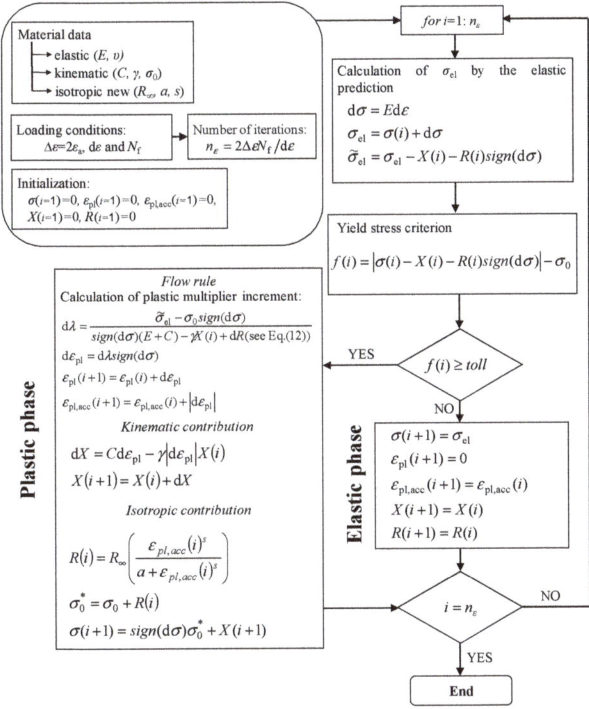

Figure A1. Calculation of stress-strain evolution considering uniaxial case with combined nonlinear kinematic and proposed isotopic models.

References

1. Moro, L.; Benasciutti, D.; De Bona, F. Simplified numerical approach for the thermo-mechanical analysis of steelmaking components under cyclic loading: An anode for electric arc furnace. *Ironmaking Steelmaking* **2019**, *46*, 56–65. [CrossRef]
2. Moro, L.; Srnec Novak, J.; Benasciutti, D.; De Bona, F. Thermal distortion in copper moulds for continuous casting of steel: Numerical study on creep and plasticity effect. *Ironmaking Steelmaking* **2019**, *46*, 97–103. [CrossRef]
3. You, J.H.; Miskiewicz, M. Material parameters of copper and CuCrZr alloy for cyclic plasticity at elevated temperatures. *J. Nucl. Mater* **2008**, *373*, 269–274. [CrossRef]
4. Hatanaka, K.; Yamada, T.; Hirose, Y. An effective plastic strain component for low cycle fatigue in metals. *JSME* **1980**, *23*, 791–797. [CrossRef]

5. Benasciutti, D.; Srnec Novak, J.; Moro, L.; De Bona, F.; Stanojevic, A. Experimental characterization of a CuAg alloy for thermo-mechanical applications. Part 1: Identifying parameters of non-linear plasticity models. *Fatigue Fract. Eng. Mater Struct.* **2018**, *41*, 1364–1377. [CrossRef]

6. Goodall, I.W.; Hales, R.; Walters, D.J. On constitutive relations and failure criteria of an austenitic steel under cyclic loading at elevated temperature. In *IUTAM Symp. Creep in Structures*; Ponter, A.R.S., Hayhurst, D.R., Eds.; Springer: Leicester, UK, 1980; pp. 103–127.

7. Basan, R.; Franulovic, M.; Prebil, I.; Kunc, R. Study on Ramberg-Osgood and Chaboche models for 42CrMo4 steel and some approximations. *J. Constr. Steel Res.* **2017**, *136*, 65–74. [CrossRef]

8. Chaboche, J.L. Constitutive equations for cyclic plasticity and cyclic viscoplasticity. *Int. J. Plast.* **1989**, *5*, 247–302. [CrossRef]

9. Kang, G.; Ohno, N.; Nebu, A. Constitutive modeling of strain range dependent cyclic hardening. *Int. J. Plast.* **2003**, *19*, 1801–1819. [CrossRef]

10. Chaboche, J.L.; Dang Van, K.; Cordier, G. Modelization of the strain memory effect on cyclic hardening of 316 stainless steel. In Proceedings of the Conference on Structural Mechanics in Reactor Technology, SMiRT 5, Berlin, Germany; 1979.

11. ASTM B 124-B 124M. *Standard Specification for Copper and Copper Alloy Forging Rod, Bar, and Shapes*; ASTM International: West Conshohocken, PA, USA, 2008.

12. Lemaitre, J.; Chaboche, J.L. *Mechanics of Solid Materials*; Cambridge University Press: Cambridge, UK, 1990.

13. Chaboche, J.L. A review of some plasticity and viscoplasticity constitutive theories. *Int. J. Plast.* **2008**, *24*, 1642–1693. [CrossRef]

14. Chaboche, J.L. Time-independent constitutive theories for cyclic plasticity. *Int. J. Plast.* **1986**, *2*, 149–188. [CrossRef]

15. Dunne, F.; Petrinic, N. *Introduction to Computational Plasticity*; Oxford University Press: New York, NY, USA, 2005.

16. Chen, W.F.; Zhang, H. *Plasticity for Structural Engineers*; Springer: New York, NY, USA, 1988.

17. Simo, J.C.; Hughes, T.J.R. *Computational Inelasticity*; Springer: New York, NY, USA, 1998.

18. Voce, E. A practical strain hardening function. *Metallurgia* **1955**, *51*, 219–226.

19. Chaboche, J.L.; Cailletaud, G. On the calculation of structures in cyclic plasticity or viscoplasticity. *Comput. Struct.* **1986**, *23*, 23–31. [CrossRef]

20. Novak, S.J.; Stanojevic, A.; Benasciutti, D.; De Bona, F.; Huter, P. Thermo-mechanical finite element simulation and fatigue life assessment of a copper mould for continuous casting of steel. *Procedia Eng.* **2015**, *133*, 688–697. [CrossRef]

21. Novak, S.J.; Benasciutti, D.; De Bona, F.; Stanojevic, A.; De Luca, A.; Raffaglio, Y. Estimation of material parameters in nonlinear hardening plasticity models and strain life curves for CuAg alloy. *Mater. Sci. Eng.* **2016**, *119*, 1–9. [CrossRef]

22. Montgomery, D.C.; Runger, G.C. *Applied Statistics and Probability for Engineers*, 6th ed.; John Wiley Sons: Hoboken, NJ, USA, 2014.

23. Zhao, L.G.; Tong, B.; Vermeulen, B.; Byrne, J. On the uniaxial mechanical behaviour of an advanced nickel base superalloy at high temperature. *Mech. Mater.* **2001**, *33*, 593–600. [CrossRef]

Article

Stress Relaxation Aging Behavior and Constitutive Modelling of AA7150-T7751 under Different Temperatures, Initial Stress Levels and Pre-Strains

Yixian Cai [1,2], Lihua Zhan [1,2,3,*], Yongqian Xu [2,3,*], Chunhui Liu [2,3], Jianguang Wang [4], Xing Zhao [2,3], Lingzhi Xu [2,3], Canyu Tong [2,3], Gengquan Jin [2,3], Qing Wang [2,3], Lan Hu [4] and Minghui Huang [2,3]

1   Light Alloy Research Institute of Central South University, Changsha 410083, China; caiyixian@csu.edu.cn
2   State Key Laboratory of High-Performance Complex Manufacturing, Central South University, Changsha 410083, China; chunhuiliu@csu.edu.cn (C.L.); xingzhao@csu.edu.cn (X.Z.); lexuswolf@csu.edu.cn (L.X.); tcyonns@csu.edu.cn (C.T.); 173812027@csu.edu.cn (G.J.); Qing.Wang@csu.edu.cn (Q.W.); meeh@csu.edu.cn (M.H.)
3   School of Mechanical and Electrical Engineering, Central South University, Changsha 410083, China
4   Shanghai Aerospace Equipment Manufacturing Co., Ltd., Shanghai 200245, China; wangjian.guang@outlook.com (J.W.); hulan.2019@outlook.com (L.H.)
*   Correspondence: yjs-cast@csu.edu.cn (L.Z.); yongqian.xu@csu.edu.cn (Y.X.); Tel.: +86-731-8883-6423 (L.Z.); +86-731-8883-0254 (Y.X.)

Received: 21 September 2019; Accepted: 7 November 2019; Published: 12 November 2019

Abstract: Age forming is an advanced manufacture technology for forming large aluminum panels. Temperature, initial stress level and pre-strains have a great effect on the formability and performance. The stress relaxation aging behavior of AA7150-T7751 under different temperatures, initial stress levels and pre-strains was studied through stress relaxation tests, tensile tests and TEM observations. The results show that the formability can be improved with the increase of temperature, initial stress levels and pre-strains. Deformation mechanisms during stress relaxation of the material were analyzed on the basis of creep stress exponent and apparent activation energy. The aging precipitates of the studied alloy were not sensitive to the age forming conditions, but drastically coarsened at over aging temperature, which decreased the mechanical properties. In addition, the relationship between stress relaxation behavior and aging strengthening is discussed. Based on the dislocation theory and the modified Arrhenius equation, a stress relaxation constitutive equation considering the initial mobile dislocation density and temperature dependent activation energy was established. This model can predict very well the stress relaxation behavior under various temperature, stress level and pre-strain conditions, with an average error of 2%.

Keywords: stress relaxation aging behavior; pre-strain; initial stress levels; temperature; constitutive modelling; AA7150-T7751

---

1. Introduction

Age forming (AF) technology was conceived in the 1970s by Textron Aerostructures to produce components with high strength and complex curvatures in the aeronautical industry [1]. The fundamental operational mechanism is based on stress relaxation and/or the creep phenomenon occurring during the course of artificial aging. Currently, the technology has been successfully applied to the upper wing of Airbus A340/A380 air planes and the upper wing skins of Gulfstreams [2,3]. Zhan et al. [4] and Zheng [5] outlined several important process parameters that affect the quality and precision of age forming, including aging time, aging temperature and initial stress level. Moreover, a simulation regarding the deformation behavior and mechanical property evolution of materials in age forming relies on the establishment of a unified constitutive equation on the basic of deformation and strengthening mechanism. Therefore, in order to realize the precise manufacturing of components, it is necessary to understand the comprehensive influence of process parameters and initial states on age forming so as to come up with an accurate prediction model.

Previous studies on the evolution of shape and properties in the age forming process have been carried out by simulating the creep phenomenon. Mostly, recent studies have outlined that when the work-piece is enforced to conform to the shape of the tool by atmospheric pressure acting on its upper surface, stress relaxation is the main phenomenon in the age forming process of the wing plate [6]. Stress relaxation of aluminum alloys at a certain temperature and initial stress is generally attributed to increased creep deformation. The power-low equations describe a function of steady-state creep rate on stress, temperature and grain diameter in pure metals [6]. However, for aluminum alloys, the grain structure and average size are not significantly altered, since the aging temperature is much lower than the recrystallization temperature [7,8]. Therefore, the subsequent studies on stress relaxation of aluminum alloys focused on stress, temperature. Considering the dislocation hardening, Kowalewski et al. [9] set creep damage constitutive equations to describe creep damage. Ho K C et al. [10] developed this and considered the solute hardening and aging hardening in the creep process. Their equations model the primary and secondary creep stage well. On this basis, Zhan et al. [11] further described the static and dynamic recovery effects in creep process by introducing relative dislocation density. Subsequently, this macro-micro modeling method based on the hardening and softening mechanism was generalized. Zhang et al. [12] and Ma et al. [13] extended the equation to a wider stress and temperature range and characterized the microscopic factors (precipitate size, volume fraction and aspect ratio) in more detail. Zheng et al. [14] established a deformation constitution model of the stress relaxation. Although the forms of the characteristic equation in their models of precipitates were different, the precipitated strengthening effect is introduced into the model by contributing to the yield strength of the material. However, Xu [7] et al. showed that a minor role was played by the precipitate size in the steady-state creep mechanism. Therefore, the physical relationship between the characteristics of precipitates and the strain rate in the process of stress relaxation still needs further study. Additionally, according to the Orowan equation, strain is a function of mobile dislocation density and its velocity; therefore, considering the evolution of mobile dislocation density, rather than total dislocation density, would be more suitable for physical mechanisms in the establishment of a stress relaxation model [15,16].

7150 Al alloy is a common application material of AF technology, the precipitate sequence of which can be summarized as: Solid solution $\rightarrow$ Guinier–Preston zones (GP zones) $\rightarrow$ Metastable η' $\rightarrow$ Stable η ($MgZn_2$) [17]. Since its development, its retrogression and re-aging treatment (RRA, commonly called T77 or T7751 temper) has been favored by the aerospace manufacturing industry due to its excellent resultant mechanical properties, which have been shown to not only guarantee the strength of a T6 temper, but also to offer the corrosion resistance of a T76 temper [18,19]. In this paper, AA7150-T7751, whose average grain size is generally more than 100 μm [20], was used as the experimental material to simulate the stress relaxation process on the relaxation testing machine. The control-variable method was used to explore the influence of the stress relaxation and age hardening at the initial stress level, aging temperature, pre-strain levels and the microstructure through TEM observation. The physical mechanism of stress relaxation behavior was investigated, and eventually, based on Orowan mechanism, a stress relaxation aging constitutive model AA7150-t7751 was initiated. It was evident that it accurately predicted different initial pre-strain levels, as well as complex thermal-mechanical conditions, and provided an essential theoretical basis for accurate prediction of spring-back in age forming and precision forming manufacturing.

2. Experiment

2.1. Material and Sample Preparation

In this study, the experiments were conducted on a commercial high-strength AA7150 whose chemical composition is listed in Table 1. Samples were machined into a dog-bone-shaped pattern with a parallel length of 50 mm along the rolling direction, whose specific geometry and dimensions are shown in Figure 1. The samples underwent a solution treatment at 470 °C for 1 h, water quenching for 1–2 min, natural aging for 2 days after pre-strain of 2% and 24 h of heat treatment at 120 °C. Afterward, retrogression and re-aging (RRA) heat treatment was conducted, where the samples were subjected to retrogression at 190 °C for 30 min and the re-aging treatment at 120 °C for 24 h. The samples were lastly air-cooled.

Table 1. Main compositional elements of AA7150-T7751, wt.%.

Zn	Mg	Cu	Mn	Fe	Si	Ni	Cr	Ti	Al
6.23	2.88	1.58	0.31	0.15	0.048	<0.01	0.16	0.025	Bal

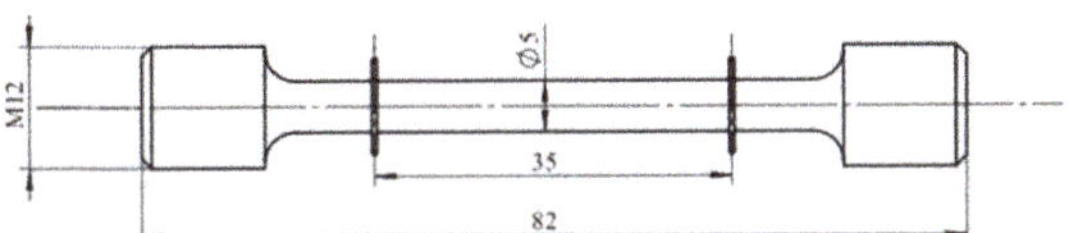

Figure 1. Specimen geometry (dimensions are in mm).

2.2. Experimental Procedure

To study the influence of different temperatures, initial stress levels, aging time, and pre-strains on the stress relaxation behavior of state AA7150-t7751, SRA experiments formulated as displayed in Table 2 were performed on RMT-D5, electronic stress relaxation testing machine. The sample was heated up to a certain temperature with a heating rate of 5 °C/min and then loaded to the predetermined stress level under a quasi-static condition with a loading rate of 15 N/s. After a period of aging, the sample was unloaded and cooled to room temperature inside the furnace. The aged samples were subjected to uniaxial tensile tests on a CMT-5105 test machine and the yield strengths were obtained as 0.2% offset yield stress. The tests were performed at room temperature with a strain rate of 7×10^{-4} s^{-1}. The microstructural evolution was characterized in detail by transmission electron microscopy (TEM). Slices for TEM observation were cut from the specimens and ground to about 70~120 μm in thickness and then punched out from the above slices to get foil discs of 3 mm in diameter. Finally, these foil discs were twin-jet electro-polished with a Struers TenuPol-5 machine using a mixture of 275 mL nitric acid and 825 mL methanol at −25 °C to further thinning down up to 40–60 nm. TEM images were acquired on a Titan G2 60–300 transmission electron microscope operated at 300 kV.

Table 2. Stress relaxation aging experiments of the AA7150-T7751.

Case ID	Pre-Stain (%)	Aging Stress (MPa)	Aging Temperature (°C)	Aging Time (h)
1	0	300	120/140/145/150/170	16
2	0	200/250/300/350/400	140	16
3	0/3/6	300	140	16
4	0	300	140	0/1/8/16

3. Results

3.1. Stress Relaxation Behavior of AA7150 under Different Temperatures

Figure 2 shows the stress relaxation behavior of AA7150 under different temperatures (120 °C, 140 °C, 145 °C, 150 °C and 170 °C). The stress relaxation curves under different temperatures are similar to the logarithmic-function curve, as shown in Figure 2a. A sharp stress drop period is observed at the start of the experiment, lasting about 1~1.5 h. After that, the stress declines at a slow rate, but no relaxation limit appears. Stress relaxation rate curves are used to characterize the stress relaxation resistance in the SRA process, as shown in Figure 2b. The stress relaxation rate curves of different temperatures also experience a steep drop at first and then transited to a stage in which relaxation rate gradually approaches zero. Interestingly, the stress relaxation curve at 170 °C shows an unusual phenomenon, which intersects the curves at 140 °C, 145 °C and 150 °C, as shown in Figure 2b. The aging time corresponding to the three intersections is about 0.52 h, 0.62 h and 0.98 h, respectively. Figure 2c represents the stress relaxation rate at different temperatures, $\dot{\varepsilon}_{120} < \dot{\varepsilon}_{170} < \dot{\varepsilon}_{140} < \dot{\varepsilon}_{145} < \dot{\varepsilon}_{150}$ within 1 h. After that, the higher the temperature is, the more the stress relaxation rate is. This unique phenomenon can be explained by the different mechanisms of SRA under 170 °C, which will be discussed in detail in Section 4.1. Relaxation efficiency is a measure of relaxation capacity and can be expressed by residual stress ratio:

$$\varphi = \frac{\sigma_t}{\sigma_i} \tag{1}$$

where σ_i and σ_t are the initial stress and the residual stress at a specific time. Higher temperature results in higher residual stress ratios at 16 h. These are 13.23%, 22.22%, 24.93%, 27.48%, and 38.70% respectively. This proves that temperature has a positive effect on stress relaxation.

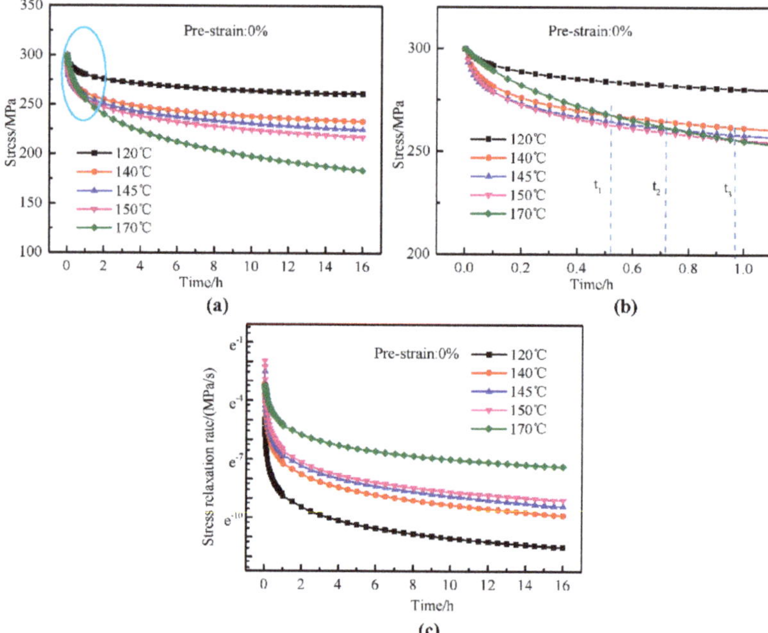

Figure 2. (a) Stress relaxation curves; (b) a partial enlargement of (a); (c) stress relaxation rate curves for 16 h under 120 °C, 140 °C, 145 °C, 150 °C and 170 °C, respectively.

3.2. Stress Relaxation Behavior of AA7150 under Different Initial Stress Levels

According to the experimental case 2 in Table 2, the effects of different initial stress levels (200, 250, 300, 350 and 400 MPa) on the stress relaxation behavior of AA7150 with non-pre-strain were studied, as exhibited in Figure 3a. Apparently, there is no difference in the trend of the stress relaxation curve under different initial stresses; both the fast and slow decline stages can be observed, which is similar to the curves under different temperatures in Section 3.1. This may be caused by same stress relaxation mechanism of different initial stress levels. By using Equation (1) to normalize the stress relaxation curves under different stresses, the evolution of the residual stress ratio over time can be obtained. Figure 3b indicates that the residual stress ratio at 16 h are 19.54%, 21.25%, 22.22%, 23.85% and 30.18%, respectively. The relaxation efficiency increases slightly with the increase of the initial stress levels from 200–350 MPa, but shows a significant rise at initial stress at 400 MPa. This is because 400 MPa is close to the elastic limit of AA7150-T7751 under 140 °C (the elastic limit of equivalent strain rate is about 430 MPa obtained by the hot tensile test under 140 °C). Although Figure 3c shows that the stress relaxation rate curves with different initial stresses are similar, the stress relaxation driving force increases with the increase of applied stress.

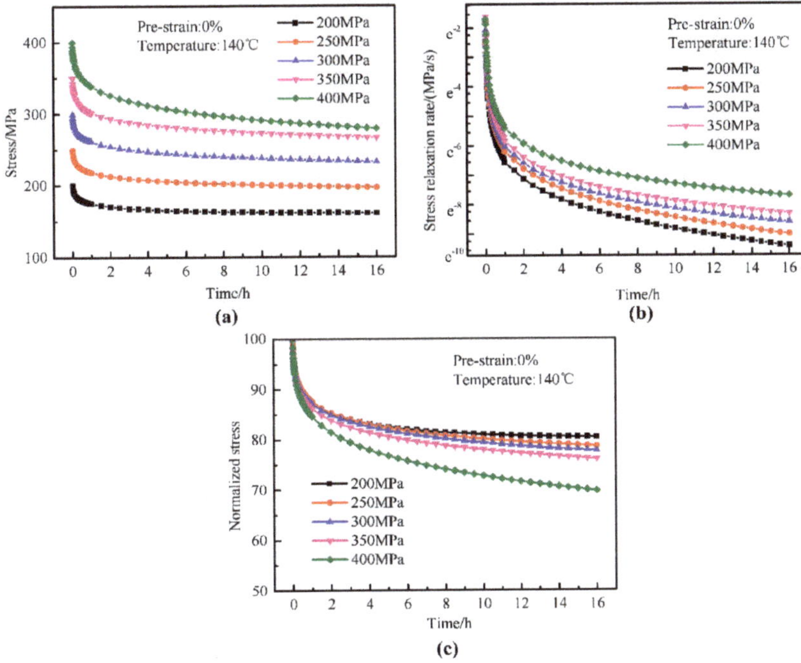

Figure 3. Stress relaxation curves (**a**), stress relaxation rate curves (**b**) and normalized stress curves (**c**) under 200 MPa, 250 MPa, 300 MPa, 350 MPa and 400 MPa, respectively.

3.3. Stress Relaxation Behavior of AA7150 under Different Initial Pre-Strains

Experiments at different pre-strains (0, 3% and 6%) were performed according to case 3 of Table 2. Figure 4a indicates that introducing pre-strain does not change the trend of stress relaxation curve but improves the relaxation efficiency, which is consistent with the results of previous researches [5,21]. The relaxation efficiency of three pre-strain levels are 22.22%, 29.96% and 39.94%, respectively. This increases with the enhancement of the pre-strain and the residual stress decreases by about 30 MPa for every 3% pre-strain. Similarly, there is still no relaxation limit, just like the stress relaxation behavior in the previous two sections. What we can observe from the different stress relaxation rate diagrams in Figure 4b is that the pre-strain augments the stress relaxation rate throughout the SRA phase. In the whole SRA process, it is evidently true that $\dot{\varepsilon}_0 < \dot{\varepsilon}_{3\%} < \dot{\varepsilon}_{6\%}$, and this means larger pre-strains lead to larger stress relaxation rates. Therefore, introducing pre-strains can improve the relaxation and forming efficiency.

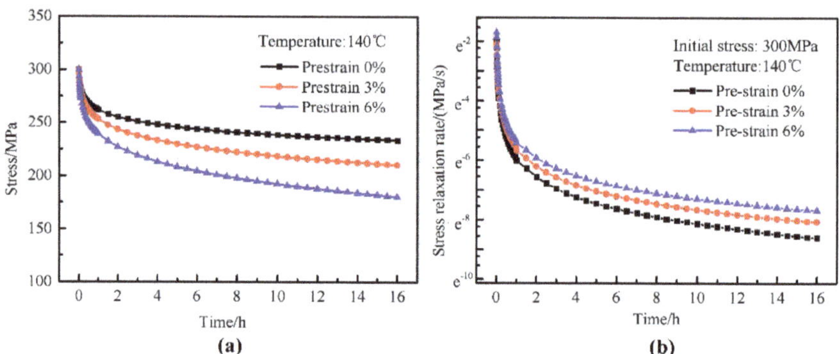

Figure 4. Stress relaxation curves (**a**) and stress relaxation rate curves (**b**) of 0%, 3% and 6% pre-strained samples, respectively.

3.4. Aging Behavior of AA7150 under Different Temperatures, Initial Stress Levels and Pre-Strains

Before studying the evolution of aging behavior of AA7150-T7751 during the SRA under different temperatures, initial stress levels and pre-strains, we need first to study the yield strength's change with time. Figure 5a shows the yield strength of stress-relaxation-aged samples at 0 MPa and 300 MPa under 140 °C for 0 h, 1 h, 8 h and 16 h. It is observed that the maximum yield strength of both stress-free aging and stress relaxation aging appears at 0 h, which are 589 MPa. The yield strength of stress-free aged samples fluctuates within 582 MPa to 589 MPa. When the initial stress is 300 MPa, the yield strength of stress-relaxation-aged samples has a small decrease trend with time. In addition, the yield strength after 16 h stress-aging is about 572 MPa, which is about 2.0% lower than that of initial state. With the increase of temperature, the yield strength gradually decreases, as shown in Figure 5b. The maximum drop is about 110 MPa, which occurs at 170 °C. When comparing the yield strength of the stress-relaxation-aged and stress-free–aged samples, the former is always lower than the later and the difference between them grows with temperature. Therefore, the reduction of property caused by stress relaxation is far less than caused by aging. After regression re-aging, SRA can be considered as an over-aging process. Higher initial stress level and higher temperature can deepen the degree of over-aging, which results in the decreasing of yield strength.

Figure 5d shows that the yield strength of 3% and 6% pre-strained sample are initially 30 MPa and 55 MPa higher than the non-pre-strained sample, respectively. This phenomenon is the strain hardening effect due to dislocations and entanglements [22]. The effects of initial pre-strain levels on the aging behavior have been studied by comparing its influence on the yield strength variation among non-pre-strained, 3% and 6% pre-strained samples stress-aged at 0 MPa and 300 MPa for 16 h, as shown in Figure 5d. Yield strength of both stress-free-aged and stress-relaxation-aged samples decreases with pre-strains. Moreover, the yield strength of stress-free-aged specimens is lower than that of stress-relaxation-aged specimens and its difference grows with pre-stains. The reason for this may be that the work hardening caused by pre-strains may not be completely recovered, which makes up for the performance weakening caused by the coarsening of precipitates.

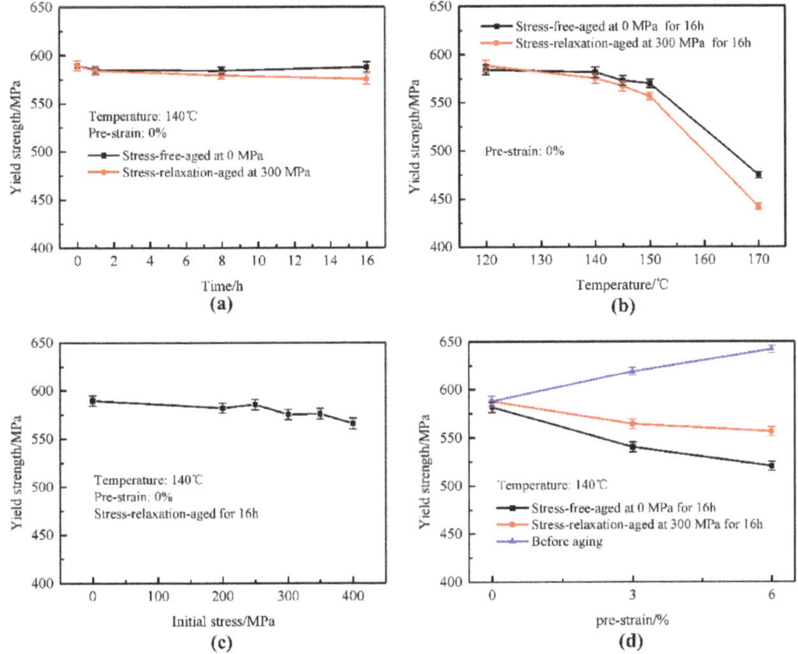

Figure 5. Yield strength for samples: (**a**) stress-free-aged at 0 MPa and stress-relaxation-aged at 300 MPa for different times; (**b**) stress-free-aged at 0 MPa and stress-relaxation-aged at 300 MPa under different temperatures; (**c**) stress-relaxation-aged at different initial stress levels; (**d**) before aging, stress-free-aged at 0 MPa and stress-aged at 300 MPa under different pre-strains.

To study the precipitation behavior under different temperatures, initial stress levels and pre-strains. The TEM bright field images of selected samples are shown in Figure 6. It is reported that the precipitations of the Al-Zn-Mg-Cu alloy in the T7751 state is dominated by the GP zones and the η′-precipitate. The former is a small cluster, and the latter is a small rod-shaped or platelet-like precipitate. TEM observation is mainly carried out along the <110>$_{Al}$ direction [23].

A large number of platelet-like and short rod-shaped precipitates can be observed in both initial materials, as shown in Figure 6a. According to the High Resolution Transmission Electron Microscopy (HRTEM) image of initial material, Figure 6f, they are GP zones and η′-precipitate. The size of the η′-precipitate is about 12 nm, and a small amount of GP zones have a size of about 2–5 nm. Based on precipitation hardening theories, the length of the η′-precipitate in the initial state is the same as the average critical length of the aluminum alloy, which can provide the maximum shear stress and bring the maximum hindrance to the dislocation motion, and the yield strength in this state may be the maximum.

Figure 6a,b shows that the types of precipitates after stress relaxation for 16 h at 300 MPa and 140 °C are unchanged, and the number of precipitates changes little. The average diameter of η′-precipitate is about 10 nm; thus, the mechanical properties before and after the stress relaxation aging are similar. When comparing Figure 6b,c, coarsening of the η′-precipitate is evident when the stress-relaxation aging temperature is 170 °C, where the average diameter of η′-precipitate is about 20 nm. In addition, a small amount of η′-precipitates appears, with an average size of about 50 nm. This explains why the yield strength drops sharply at 170 °C. In comparison with Figure 6b,d, when the initial stress is 400 MPa, the average diameter of the η′-precipitate is about 12 nm and a small amount of η′-precipitates is coarsened to 20 nm. The small GP zones disappear completely. Therefore, the yield strength in this state is slightly lower than initial state. Ma et al. and Zheng et al. [13,14] also demonstrated that the coarsening of the precipitates can be accelerated under higher stress. The comparison of Figure 6b,e shows that the average diameter of the η′-precipitate after stress relaxation aging of 6% pre-strain is about 20 nm. Pre-strains mainly affects the precipitation process by accumulating more dislocations of SRA, thus resulting in a noticeable coarsening response after SRA.

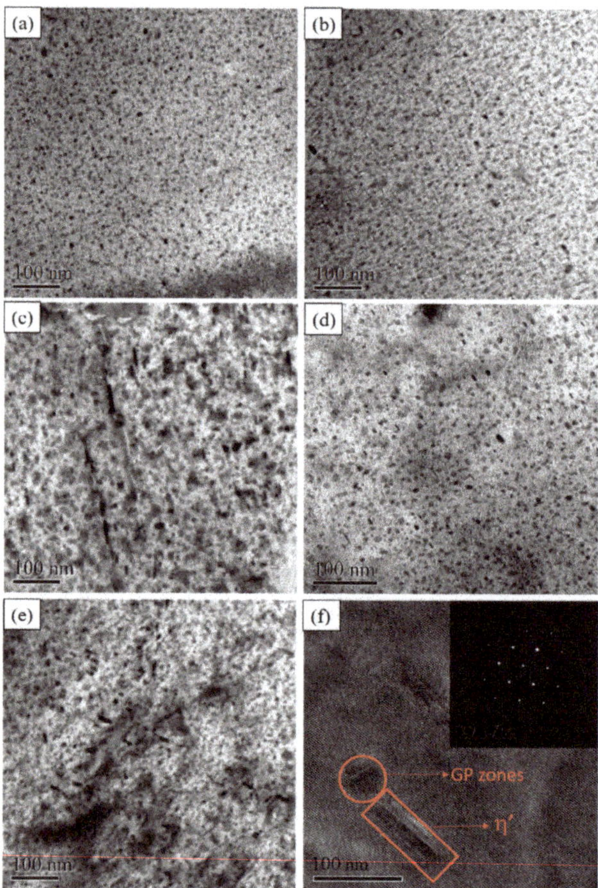

Figure 6. Bright field TEM images in <011> zone axis of Al matrix of samples after 16 h stress relaxation aging treatment: (**a**) T7751(initial); (**b**) 300 MPa + 140 °C; (**c**) 300 MPa + 170 °C;(**d**) 400 MPa + 140 °C; (**e**) 6% + 140 °C + 300 MPa; (**f**) HRTEM image of the GP-zones and η′-precipitate. Please note that, e.g., 6% + 140 °C + 300 MPa indicates the sample was pre-strained to 6% and the initial temperature and initial stress for SRA are 140 °C and 300 MPa.

4. Discussion

4.1. Deformation Mechanism during Stress Relaxation Aging Process under Three Parameters

A rational explanation of the stress relaxation phenomenon is that the total strain remains unaffected and a certain amount of creep strain is generated under the high temperature stress aging. This leads to a decrease in the force maintaining the elastic strain. The relation among the three strains can be described by Equation (2). The relationship between the stress relaxation rate ($\dot{\sigma}$) and the creep strain rate ($\dot{\varepsilon}_c$) can be directly determined by Equation (3).

$$\varepsilon_T = \varepsilon_e + \varepsilon_c \tag{2}$$

$$\dot{\sigma} = -E \times \dot{\varepsilon}_c \tag{3}$$

where ε_T, ε_e and ε_c are the total strain, elastic strain and creep strain, respectively. The deformation mechanism during the creep aging process at different initial stress levels can be determined by calculating the stress exponent at constant stress levels by Equation (4) [24]. There are two questions exist in stress relaxation aging process. One is that the steady creep strain rate of SRA is difficult to determine. Another is that the stress in SRA decreases gradually. However, we can calculate the "steady-state relaxation rate" through the linear fitting of the relaxation curve in the 13~16 h period. Then, Equation (5) is used to calculate the "stress exponent" between it and the initial stress level and the final residual stress, respectively as n_1 and n_2. The actual stress exponent in the stress relaxation process is between n_1 and n_2.

$$n = \frac{\ln(\dot{\varepsilon}_s)}{\ln \sigma} \tag{4}$$

$$n = \frac{\ln(-\dot{\sigma}_s/E)}{\ln \sigma_n} \tag{5}$$

where $\dot{\varepsilon}_s$ is the steady creep strain rate and $\dot{\sigma}_s$ is the "steady-state relaxation rate" obtained by fitting. σ_n represent substituting initial stress levels or finial residual stress. E is Young's modulus. The plots of $\ln(-\dot{\sigma}_s/E) - \ln \sigma_n$ are shown in Figure 7. Therefore, the value of n is found to range from 2.76 to 3.29. It is also known that n = 3 indicates a viscous dislocation glide creep mechanism; n = 5 represents the dislocation climb creep mechanism [25]. Consequently, we can infer that deformation mechanism of stress relaxation at 140 °C for different initial stress levels is principally controlled by dislocation glide.

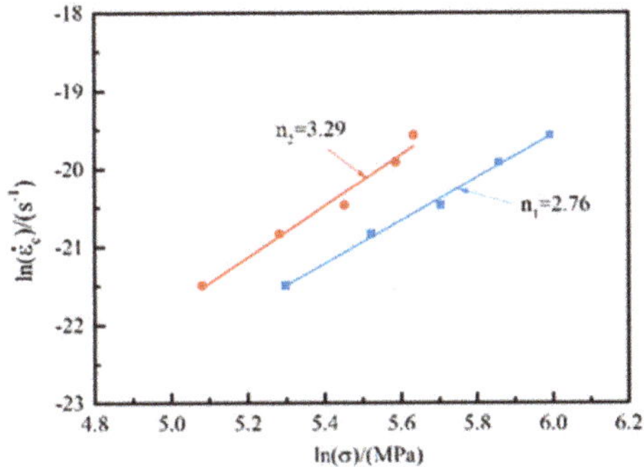

Figure 7. Variations of "steady-state creep strain rate" with initial stress levels and residual stresses, to determine the stress exponent range of stress relaxation aging with non-pre-strain samples at 140 °C.

From Section 3.1, it is evident that temperature has a great impact on stress relaxation behavior. We all known that the essence of deformation is the dislocation motion under thermal activation. Therefore, the deformation mechanism may change at different temperatures, which can be determined by calculating the value of the apparent activation energies. The Arrhenius equation (Equation (6)) has been shown to be applicable in describing the process of thermal activation [26,27]. The formula for calculating the apparent activation energy can be obtained by taking the logarithm on both the left and right sides of Equation (6).

$$\dot{\varepsilon}_s = A \times \exp(-\frac{Q_c}{RT}) \tag{6}$$

$$\ln\left(-\frac{\dot{\sigma}_s}{E(T)}\right) = \ln A - \frac{Q_c}{RT} \tag{7}$$

$$E(T) = 107452 - 119 \times T \tag{8}$$

where A is the material constant, R is the gas constant (8.314 J·mol^{-1}·K^{-1}) and T is the Kelvin temperature. $E(T)$ represents that Young's modulus is function of temperatures. The experimental data and fitting curves for Equation (8) are shown in Figure 8a Q_c is the apparent activation energy. Generally, when the temperature range is narrow, it is considered to be simply unaffected. However, some researchers had analyzed the relationship between the apparent activation energy (Q_c) and temperature of some pure metals such as aluminum and silver. They found that the apparent activation energy is the least and close to that for grain boundary diffusion energy (84 kJ/mol [28]) of pure aluminum when the temperature is about 0.4 T_m (T_m indicates that the melting point of aluminum, which is 993 K). At this time, the main deformation mechanism of creep is the grain boundary diffusion mechanism. When the temperature is > 0.7 T_m, Q_c is close to the self-diffusion activation energy of pure aluminum. The diffusion mechanism is dominant at this time. When the homologous temperature is between 0.4 and 0.7 (0.4 < T/T_m < 0.7), the activation energy increases with the temperature evidently. This change in the value of Q_c is usually explained by the contribution to strain by different deformation mechanisms. Dislocation motion is the main deformation mechanism during SRA. However, there are two main types of dislocation motion and they are dislocation glide and dislocation climb, and the energy barrier of dislocation climb is greater than that of dislocation slip, which causes the former is hard to occur at lower temperature [6,26]. Obviously, there are some differences in apparent activation energies between pure metals and alloys. Other researchers calculated the apparent activation energies of creep or stress relaxation in Al-Zn alloy and found out they lied within 91~134 kJ/mol, but little difference was found in the trend of creep or stress relaxation curves at different temperatures in their papers [29,30]. However, a significant difference appeared in stress relaxation process under different temperature. A special overlap phenomenon was observed in stress relaxation curves in this essay, as shown in Figure 2a. The apparent activation energies in two temperature ranges (120 °C, 140 °C, 145 °C and 145 °C, 150 °C, 170 °C) calculated by using Equation (7) are 109.0 kJ/mol and 135.2 kJ/mol, as shown in Figure 8b. It increases with the temperature. When the aging temperature rises from 120 °C to 170 °C, the dominant deformation mechanism in the stress relaxation process of AA7150-T7751 changes from dislocation slip to dislocation climbing. During dislocation climbing, the edge dislocation may generate positive and negative climbing. The former contribute to the multiplication of dislocation and the latter leads to dislocation annihilation. Therefore, the stress relaxation rate of stress-aged at 300 MPa with non-pre-strain sample under 170 °C is less than that under 140 °C to 150 °C in the initial stage of SRA, this may due to the fact that the rate of total dislocation multiplication under 170 °C decreases as a result of the presence of dislocation annihilation, which is lower than that under 140 °C to 150 °C.

Tamás Csanádi [22] reported that the introducing of pre-strain increases the initial dislocation density in the material. However, the introduction of pre-strains does not change the deformation mechanism in the stress relaxation process, which can be supported by the similar trend of stress relaxation and stress relaxation rate curves in Section 3.3. Another piece of evidence determining the mechanism of stress relaxation is that a large number of dislocations are observed in Figure 9, which shows the TEM images of 3% pre-strain samples after SRA treatment with an initial stress of 300 MPa. These dislocations become entangled with the precipitates and unwind at high temperatures to become mobile dislocations.

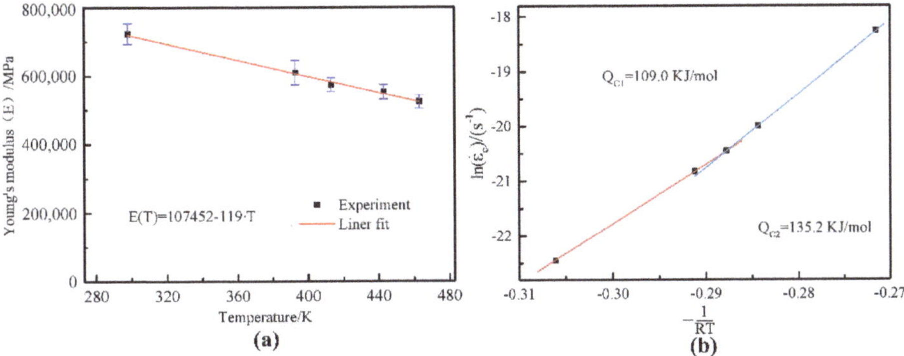

Figure 8. (a) Comparison of the predicted (line) and experimental (symbols) of the Young's modulus curves of AA7150 alloy under different temperatures; (b) apparent activation energy of SRA with temperatures.

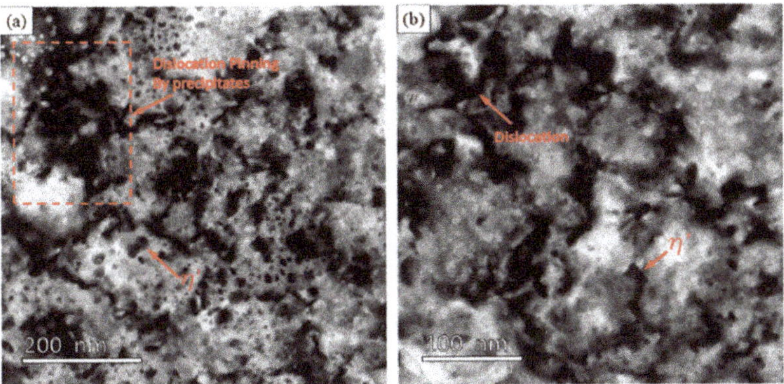

Figure 9. (a) and (b) are bright field TEM images of 3% pre-strained samples after 16 h stress relaxation aging treatment with an initial stress of 300 MPa under different magnifications, respectively.

Based on the above discussion, we can infer that the main deformation mechanism of stress relaxation is dislocation motion. At lower temperatures, the deformation mechanism of stress relaxation aging under different stress levels is primarily dislocation slip. At higher temperatures, the dominant mechanism of SRA changes from dislocation slip to dislocation climb. Moreover, the introduction of initial pre-strain cannot change the predominant mechanism of deformation mechanism in the SRA process. Therefore, we can infer the stress relaxation behavior and deformation mechanism of AA7150-T7751 by taking into account the three parameters as displayed in Figure 10.

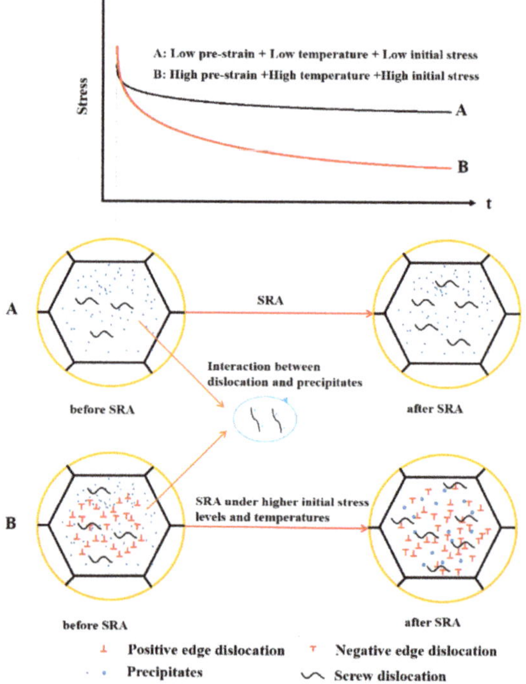

Figure 10. Diagram of stress relaxation behavior and microstructure evolution of AA7150-T7751, considering changes of temperature, initial stress level and pre-strain.

4.2. Constitutive Modeling and the Determination of Material Parameters

The mechanism controlling stress relaxation of AA7150-T7751 is a thermal activation dislocation motion, as was identified in Section 4.1. Equation (9), a formula combining the Orowan equation and the Arrhenius equation, is normally used to describe the dependencies of strain rate, $\dot{\varepsilon}$, mobile dislocation density, ρ_m, the average velocity of mobile dislocation, $\dot{\varepsilon}$ and temperature, T.

$$\dot{\varepsilon} = b\rho_m \mathrm{v}\left(-\frac{Q_c}{RT}\right) \tag{9}$$

where b is the absolute value of the Burgers vector. Q_c is the apparent activation energy of SRA and it changes with the aging temperature. Thus, a modified model needs to be introduced to define their relations, Equation (10), which was proposed by J.M. Montes et al. [31], is as follows:

$$Q_c = Q_{SD}\left(1 - \frac{a}{1 + \exp\left(b\left(\frac{T}{T_m} + c\right)\right)}\right) \tag{10}$$

where a, b and c are material constants, Q_{SD} is the self-diffusion activation energy, whose value for pure aluminum is142 kJ/mol [31,32].

ρ_m is the initial mobile dislocation density. The relationship between initial mobile dislocation density, initial dislocation density and stress (σ) is derived in Estrin and Mecking, as shown in Equation (11) [33,34]. The initial dislocation density and thus the dislocation hardening effect should increase with increasing pre-strain level. Equation (12) describes its increasing with pre-strains [35].

$$\rho_m = \rho_i \left(\frac{\sigma_s}{\sigma}\right)^{\mathrm{m}} \tag{11}$$

$$\rho_i = (1 + (\frac{C_1}{C_2} - \frac{C_2}{2})\varepsilon_p)\exp(-C_2\varepsilon_p) \times \rho_0 \tag{12}$$

where ρ_0, ρ_i and ρ_m are initial mobile dislocation density, dislocation density after introducing pre-stains and initial dislocation density, respectively. C_1 reflects material mobile dislocation proliferation, C_2 is on behalf of the mobile dislocation and mobile dislocation to capture and annihilate each other, ε_p is pre-strain. σ_s is the saturated stress and m is the material constant. With these two equations, the starting point of the stress relaxation process under the different pre-strains is determined.

v is the velocity of mobile dislocation and it is the time dependent and a strong function of stress in stress relaxation aging, and usually has three empirical equation forms as shown in Equation (13).

$$\bar{v} = \begin{cases} K\sigma^n, & \textit{low stress} \\ Cexp(\alpha\sigma), & \textit{high stress} \\ Asinh(B\sigma), & \textit{all stress level} \end{cases} \tag{13}$$

where K, C, A, n, α and B are material constants; σ is aging stress. The Norton equation, $K\sigma^n$, is suitable for lower stress levels; the Dorn equation, $Cexp(\alpha\sigma)$, is suitable for high stress aging; while the hyperbolic sine equation, $Asinh(B\sigma)$, is considered to be suitable for various initial stress levels.

In this paper, based on the $Asinh(B\sigma)$ model and the hardening and recovery theories, we proposed a stress relaxation model with considering the pre-strains, initial stress and temperature. Equations (14)–(19), as follows:

$$|\dot{\sigma}| = E(T) \times \dot{\varepsilon}_c \tag{14}$$

$$\dot{\varepsilon}_c = f_1 \times \overline{\rho_m} \times \sin h\{f_2\sigma(1-\bar{\rho})\}\exp(-\frac{Q_H}{RT}) \tag{15}$$

$$\overline{\rho_m} = (1 + (\frac{f_8}{f_9} - \frac{f_9}{2})\varepsilon_p)(\frac{f_7}{\sigma})^{f_5} \times \exp(-f_9\varepsilon_p) \tag{16}$$

$$\dot{\bar{\rho}} = \frac{f_3}{\sigma^{f_4} \times \exp(-\frac{Q_H}{RT})}(1 - \frac{\bar{\rho}}{f_6 \times (\frac{f_7}{\sigma})^{f_5} \times \exp(-f_9\varepsilon_p)})\dot{\varepsilon}_c \tag{17}$$

$$Q_c = Q_{SD}(1 - \frac{f_{10}}{1 + \exp(f_{11}(\frac{T}{T_m} + f_{12}))}) \tag{18}$$

$$Q_H = f_{13}(\frac{1}{1 + \exp(f_{14}(\frac{T}{T_m} + f_{15}))}) \tag{19}$$

where $f_1 \sim f_{15}$ are material constants. R is the gas constant (8.314 J·mol^{-1}·K^{-1}) and T is the Kelvin temperature. T and T_m are aging temperature and the melting temperature of pure aluminum respectively. ε_p is the initial pre-strain value. σ is the initial stress.

Equation (14) indicates that the decrease of stress relaxation rate is due to the slowdown of strain rate. $E(T)$ represents the change of Young's modulus with temperatures within experimental temperature.

$\overline{\rho_m}$ in Equation (16) is the normalized mobile dislocation density and means the mobile dislocation density ratio before stress relaxation between non-pre-strain and pre-strained samples, which is a combine of Equations (11) and (12).

The total density is composed in part of mobile dislocations and in part of network dislocations. The term $\sigma(1-\bar{\rho})$ in Equation (15) is used to describe the contribution of network mobile dislocations on strain rate under effective stress based on the assumption that only dislocations which are trapped contribute to the hardening of the material in stress relaxation process [36]. $\bar{\rho}$ is the relative newly generated network dislocation density fraction, which is defined as the density of the new network dislocation formed by the trapped mobile dislocation and its ratio of the instantaneous value to the possible peak value in the stress relaxation process. This peak value occurs when hardening and softening reach a dynamic equilibrium. Thus, $\bar{\rho}$ varies from 0 to 1. Equation (17) indicates that the transformation from mobile dislocations to network dislocations may be just related to dislocation multiplication caused by deformation in stress relaxation and dislocation annihilation caused by dynamic recovery. The rate of relative newly generated network dislocation density fraction is effected by stress, temperature, pre-strain and strain rate. The first term in Equation (17) represents the development of dislocation density due to deformation. The second term gives the effect of the dynamic recovery. With the increase of stress, temperature and pre-strain, $\bar{\rho}$ is smaller and thus is hard to reach steady state. Q_c is the apparent activation energy, whose relations between Q_c and aging temperature is given as Equation (18), which is the same form as Equation (10). Q_H is the recovery energy, which represents the energy required to convert mobile dislocations into network dislocations at specific temperatures and that decrease with increasing temperature, as shown in Equation (19), which is deduced on the basis of Equation (10).

The particle swarm algorithm was used for fitting the SRA curves. To prevent the program from producing no physical meaning constants and obtain more accurate material constants of Equation (14–19), the stress relaxation curves under the four experimental conditions (i.e., 0% + 300 MPa + 140 °C, 0% + 350 MPa + 140 °C, 0% + 300 MPa + 170 °C, 6% + 300 MPa + 140 °C) were fitted by manually modifying the range of material constants, as shown in Figure 11a. Standard deviation of the experimental data and the fitted data is used as a criterion for fitting accuracy. Table 3 shows the material parameters of stress relaxation aging constitutive equation of AA7150-T7751. Figure 11b gives the fitting results of the stress relaxation curves under different process parameters and the fitting curves of creep activation energy and recovery energy. The fitting error of the four stress relaxation curves is less than 1%. The evolution of Q_c and Q_H with the temperature is illustrated in Figure 11b. The apparent activation energy obtained by fitting at 170 °C and 120 °C is 134 kJ/mol and 117 kJ/mol, respectively, which are close to the apparent activation energy calculating results in Figure 8b. Figure 11c shows that the relative newly generated network dislocation density fraction obtained by using these models, which increase with the time and is closed to 1 when the hardening and recovery reach the dynamic balance. The higher temperature, initial stress levels and pre-stains are, the smaller $\bar{\rho}$ is.

Table 3. Material parameters of stress relaxation constitutive model of AA7150-T7751.

$f_1(-)$	$f_2(-)$	$f_3(-)$	$f_4(-)$	$F_5(-)$	$f_6(-)$
1.76 E+12	0.025	10.32	0.68	2.50	0.40
f_7(MPa)	$f_8(-)$	$f_9(-)$	$f_{10}(-)$	$f_{11}(-)$	$f_{12}(-)$
446.7	45	18.25	0.1618	205	0.4427
f_{13}(kJ/mol)	$f_{14}(-)$	$f_{15}(-)$	-	-	-
13.98	95.2	0.4521	-	-	-

Through modification of the variables such as initial pre-strain level, temperature and initial stress level, the stress relaxation curves under the other above mentioned process conditions were predicted as shown in Figure 12a. To further verify the accuracy of the model, predicted and experimental stress relaxation curves under two new experimental conditions (i.e., 2% + 275 MPa + 155 °C, 5% + 325 MPa + 160 °C) for 16 h was used to compare, as shown in Figure 12b. The standard deviation of each experimental curve and the predicted curve is less than 2%. Therefore, the stress relaxation constitutive model of AA7150-T7751 established in this paper has correct and accurate prediction ability for stress relaxation curves under different process conditions.

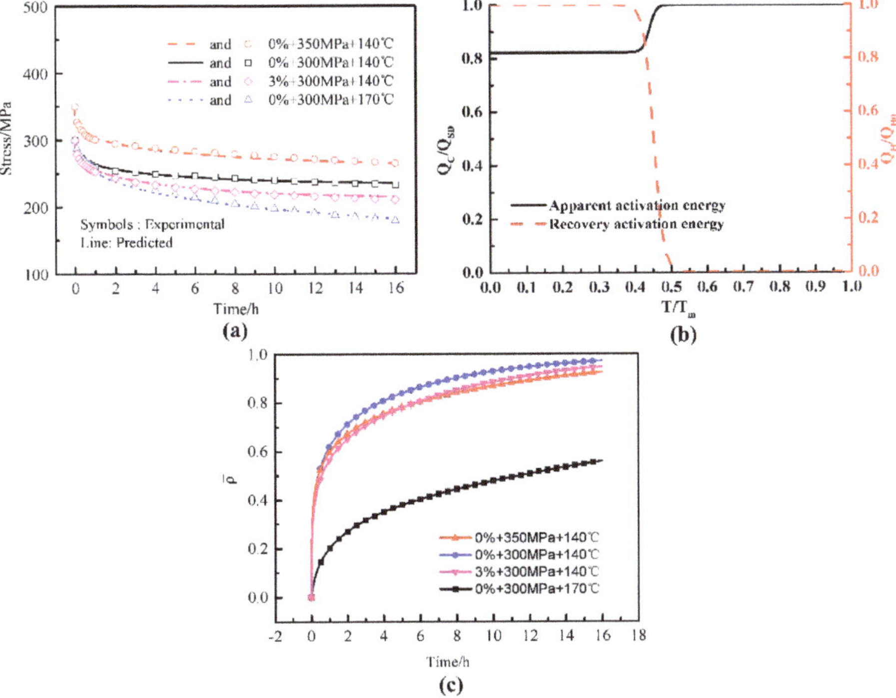

Figure 11. (**a**) Comparison of the predicted (lines) and experimental (symbols) of the SRA curves of AA7150-T7751 under different conditions; (**b**) The fitting curves of apparent activation energy and recovery activation energy. (**c**) The relative newly generated network dislocation density fraction with time. Please note that, e.g., 6% + 140 °C + 300 MPa + 16 h indicates the sample was pre-strained to 6% and the initial temperature and initial stress for 16 h SRA are 140 °C and 300 MPa.

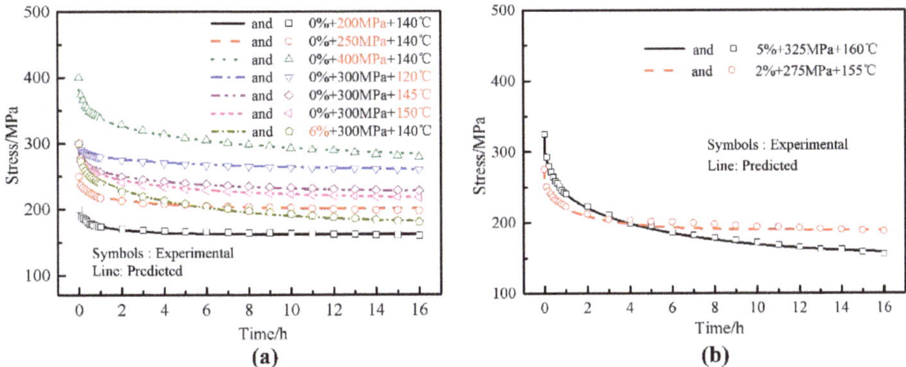

Figure 12. Comparison between the predicted (line) and experimental (symbols) stress relaxation curves of AA7050 during the stress relaxation aging process. (**a**) Other process conditions in this paper; (**b**) new conditions. Please note that e.g., 6% + 140 °C + 300 MPa + 16 h indicates the sample was pre-strained to 6% and the initial temperature and initial stress for 16 h SRA are 140 °C and 300 MPa.

5. Conclusions

By studying the stress relaxation behavior of AA7150-T7751 under different age forming conditions of temperature, initial stress levels and pre-strains, the following were concluded:

(1) Temperature, stress level and pre-strain have a great effect on stress relaxation behavior of AA7150-T7751. When the temperature rises by 30 °C, the initial stress level increases by 100 MPa and pre-strain value increases by 3%, relaxing efficiency increased by 16.48%, 7.96% and 17.72%, respectively, compared with SRA at 300 MPa under 140 °C for 16 h.

(2) Temperature, stress level and pre-strain considerably influence aging-strengthening during the stress relaxation aging behavior of AA7150-T7751. The improvement of these parameters can promote the coarsening of the η′-precipitate, but the decrease of yield strength is very small, especially compared with the stress-free aging samples. This means that the evolution of precipitates has no significant effect on the stress relaxation behavior in the complex process.

(3) The results of stress index and apparent activation energy show that the stress relaxation deformation mechanism of AA7150-T7751 is mostly dislocation slip at lower temperatures and dislocation climb at higher temperatures. This leads to the intersection phenomenon of stress relaxation curves at different temperatures.

(4) By introducing the correction formula of apparent activation energy and the formula of dislocation density changing with pre-strain, based on the Orowan model, a stress relaxation constitutive model considering the initial movable dislocation density and temperature-dependent activation energy is established. The average fitting error of a single curve is less than 2%. Not only does this promote the development of stress relaxation constitutive modeling, but it also provides a theoretical basis for the accurate prediction for spring-back of stress relaxation formed AA7150-T7751.

Author Contributions: Y.C. performed the stress relaxation curves and TEM experiments, the theoretical explanation of experimental curves, the establishment of the constitutive model and wrote the data; L.Z. performed the design of the work; Y.X. analyzed the data; C.L., J.W., X.Z., L.X., C.T., G.J., Q.W. and M.H., L.H. contributed reagents/materials/analysis tools; L.H. made effort to investigate.

Funding: This work is supported by National key R&D Program of China (2017YFB0306300); National Natural Science Foundation of China (No. 51601060); National Science and Technology Major Project of China (No. 2017ZX04005001); National Defense Program of China (JCKY2014203A001); the Fundamental Research Funds for the Central Universities of Central South University (2018zzts478).

Conflicts of Interest: The authors declare no conflict of interest.

References

1. Holman, M.C. Autoclave age forming large aluminum aircraft panels. *J. Mech. Work. Technol.* **1989**, *20*, 477–488. [CrossRef]

2. Ho, K.C.; Lin, J.; Dean, T.A. Modelling of spring-back in creep forming thick aluminum sheets. *Int. J. Plast.* **2004**, *20*, 733–751. [CrossRef]

3. Jeunechamps, P.P.; Ho, K.; Lin, J.; Ponthot, J.P.; Dean, T. A closed form technique to predict spring-back in creep age-forming. *Int. J. Mech. Sci.* **2006**, *48*, 621–629. [CrossRef]

4. Zhan, L.; Lin, J.; Dean, T.A.; Huang, M. Experimental studies and constitutive modelling of the hardening of aluminium alloy 7055 under creep age forming conditions. *Int. J. Mech. Sci.* **2011**, *53*, 595–605. [CrossRef]

5. Zheng, J.-H.; Pan, R.; Li, C.; Zhang, W.; Lin, J.; Davies, C.M. Experimental investigation of multi-step stress-relaxation-ageing of 7050 aluminium alloy for different pre-strained conditions. *Mater. Sci. Eng.* **2018**, *710*, 111–120. [CrossRef]

6. Ashby, M.F. A first report on deformation-mechanism maps. *Acta Metall.* **1972**, *20*, 887–897. [CrossRef]

7. Xu, Y.; Yang, L.; Zhan, L.; Yu, H.; Huang, M. Creep Mechanisms of an Al–Cu–Mg Alloy at the Macro- and Micro-Scale: Effect of the S′/S Precipitate. *Materials* **2019**, *12*, 2907. [CrossRef] [PubMed]

8. Lei, C.; Li, H.; Fu, J.; Bian, T.J.; Zheng, G.W. Non-isothermal creep aging behaviors of an Al-Zn-Mg-Cu alloy. *Mater. Charact.* **2018**, *144*, 431–439. [CrossRef]

9. Kowalewski, Z.L.; Hayhurst, D.R.; Dyson, B.F. Mechanism-based creep constitutive equations for an aluminium alloy. *J. Strain. Anal.* **1994**, *29*, 309–316. [CrossRef]
10. Ho, K.C.; Lin, J.; Dean, T.A. Constitutive modelling of primary creep for age forming an aluminium alloy. *J. Mater. Process Technol.* **2004**, *153–154*, 122–127. [CrossRef]
11. Zhan, L.; Lin, J.; Dean, T. A review of the development of creep age forming: Experimentation, modelling and applications. *Int. J. Mach. Tools Manuf.* **2011**, *51*, 1–17. [CrossRef]
12. Zhang, J.; Deng, Y.; Zhang, X. Constitutive modeling for creep age forming of heat-treatable strengthening aluminum alloys containing plate or rod shaped precipitates. *Mater. Sci. Eng. A* **2013**, *563*, 8–15. [CrossRef]
13. Ma, Z.; Zhan, L.; Liu, C.; Xu, L.; Xu, Y.; Ma, P.; Li, J. Stress-level-dependency and bimodal precipitation behaviors during creep ageing of Al-Cu alloy: Experiments and modeling. *Int. J. Plast.* **2018**. [CrossRef]
14. Zheng, J.H.; Lin, J.; Lee, J.; Pan, R.; Li, C.; Davies, C.M. A novel constitutive model for multi-step stress relaxation ageing of a pre-strained 7xxx series alloy. *Int. J. Plast.* **2018**, *106*, 31–47. [CrossRef]
15. Orowan, E. Problems of plastic gliding. *Proc. Phys. Soc.* **1940**, *52*, 8. [CrossRef]
16. Roters, F. A new concept for the calculation of the mobile dislocation density in constitutive models of strain hardening. *Phys. Status Sol.* **2003**, *240*, 68–74. [CrossRef]
17. Park, J.K. Influence of retrogression and reaging treatments on the strength and stress corrosion resistance of aluminum alloy 7075-T6. *Mater. Sci. Eng. A* **1988**, *103*, 223–231. [CrossRef]
18. Kanno, M.; Araki, I.; Cui, Q. Precipitation behaviour of 7000 alloys during retrogression and reaging treatment. *Mater. Sci. Technol.* **1994**, *10*, 599–603. [CrossRef]
19. Sha, G.; Cerezo, A. Early-stage precipitation in Al–Zn–Mg–Cu alloy (7050). *Acta Mater.* **2004**, *52*, 4503–4516. [CrossRef]
20. Fan, X.G.; Jiang, D.M.; Meng, Q.C.; Xian, H.Z.; Li, N.K. Influence of Ageing Treatment on Microstructure and Properties of 7150 Alloy. *Mater. Sci. Forum* **2007**, *546–549*, 849–854. [CrossRef]
21. Yang, Y.L. Effect of pre-deformation on creep age forming of AA2219 plate: Springback, microstructures and mechanical properties. *J. Mater. Process Technol.* **2016**, *229*, 697–702. [CrossRef]
22. Csanádi, T.; Chinh, N.Q.; Gubicza, J. Plastic behavior of fcc metals over a wide range of strain: Macroscopic and microscopic descriptions and their relationship. *Acta Mater.* **2011**, *59*, 2385–2391. [CrossRef]
23. Marlaud, T.; Deschamps, A.; Bley, F.; Lefebvre, W.; Baroux, B. Influence of alloy composition and heat treatment on precipitate composition in Al–Zn–Mg–Cu alloys. *Acta Mater.* **2010**, *58*, 248–260. [CrossRef]
24. Kassner, M.E. *Fundamentals of Creep in Metals and Alloys*; Butterworth Heinemann: Oxford, UK, 2015.
25. Poirier, J. *Creep of Crystals: High-Temperature Deformation Processes in Metals, Ceramics and Minerals*; Cambridge Earth Science Series; Cambridge University Press: New York, NY, USA, 1985.
26. Jeong, C.Y.; Nam, S.W.; Ginsztler, J. Activation processes of stress relaxation during hold time in 1Cr–Mo–V steel. *Mater. Sci. Eng. A* **1999**, *264*, 188–193. [CrossRef]
27. Sket, F.; Dzieciol, K.; Isaac, A.; Borbely, A.; Pyzalla, A.R. Tomographic method for evaluation of apparent activation energy of steady-state creep. *Mater. Sci. Eng. A* **2010**, *527*, 2112–2120. [CrossRef]
28. Davies, C.K. Introduction to creep. *Int. Mater. Rev.* **1994**, *3*, 123. [CrossRef]
29. Chen, J.F.; Jiang, J.T.; Zhen, L.; Shao, W.Z. Stress relaxation behavior of an Al–Zn–Mg–Cu alloy in simulated age-forming process. *J. Mater. Process Technol.* **2014**, *214*, 775–783. [CrossRef]
30. Mostafa, M.M.; Al-Ganainy, G.S.; El-Khalek, A.M.A.; Nada, R.H. Steady-state creep and creep recovery during transformation in Al–Zn alloys. *Physica B* **2003**, *336*, 402–409. [CrossRef]
31. Montes, J.M.; Cuevas, F.G.; Cintas, J.; Ternero, F.; Caballero, E.S. Phenomenological equation for the thermal dependence of the activation energy of creep. *Mater. Lett.* **2017**, *196*, 273–275. [CrossRef]
32. Frost, H.J.; Ashby, M.F. *Deformation-Mechanism Maps: The Plasticity and Creep of Metals and Ceramics*; Pergamon Press: Oxford, UK, 1982.
33. Estrin, Y.; Mecking, H. A unified phenomenological description of work hardening and creep based on one-parameter models. *Acta Metall.* **1984**, *32*, 57–70. [CrossRef]
34. Mecking, H.; Styczynski, A.; Estrin, Y. Steady State and Transient Plastic Flow of Aluminium and Aluminium Alloys. *Strength Metals Alloys* **1989**, *2*, 989–994.

35. Chinh, N.Q.; Illy, J.; Horita, Z.; Langdon, T.G. Using the stress-strain relationships to propose regions of low and high temperature plastic deformation in aluminum. *Mater. Sci. Eng. A* **2005**, *410–411*, 234–238. [CrossRef]
36. Basirat, M.; Shrestha, T.; Potirniche, G.P.; Charit, I.; Rink, K. A study of the creep behavior of modified 9Cr–1Mo steel using continuum-damage modeling. *Int. J. Plast.* **2012**, *37*, 95–107. [CrossRef]

Article

Flow Stress of bcc Metals over a Wide Range of Temperature and Strain Rates

Gabriel Testa, Nicola Bonora *, Andrew Ruggiero and Gianluca Iannitti

Department of Civil and Mechanical Engineering, University of Cassino and Southern Lazio, I-03043 Cassino, Italy; gabriel.testa@unicas.it (G.T.); andrew.ruggiero@unicas.it (A.R.); gianluca.iannitti@unicas.it (G.I.)
* Correspondence: nicola.bonora@unicas.it; Tel.: +39-0776-299-3693

Received: 25 November 2019; Accepted: 9 January 2020; Published: 13 January 2020

Abstract: A physical-based model for the flow stress of bcc metals is presented. Here, thermally activated and viscous drag regimes are considered. For the thermally activated component of the flow stress, the diffusion-controlled regime at elevated temperature is also taken into account assuming the non-linear dependence of the activation volume on temperature. The model was applied to A508 (16MND5) steel showing the possibility to accurately describe the variation of the flow stress over the entire temperature range (from 0 K to T_m) and over a wide strain-rate range.

Keywords: flow stress; activation volume; strain rate; temperature; bcc

---

1. Introduction

Today, numerical simulation tools have the potential to anticipate the performance of components and systems under varying operational scenarios, dramatically reducing the need for verification experiments and accelerating the engineering design process [1]. For what concerns the mechanical response is that the reliability of such predictive approach relies on the accuracy of material constitutive models used in the simulations. In many industrial sectors, such as defense, aerospace, oil and gas, automotive and manufacturing engineering applications, there is an increasing demand for models capable to describe material behavior under complex load paths involving large inelastic deformation and failure under different strain rate and temperature conditions. Over the past hundred years, modelling of deformation of metals and alloys has been largely investigated. A review of non-linear constitutive models can be found in [2,3]. For what concern the material yield stress is that the classical rate-independent plasticity theories represent idealizations which in general have limited applicability. Macroscopic constitutive formulations can be categorized into two main groups: phenomenological constitutive relationships and physical-based models. Models that fall in the first group are mathematical formulations for the material flow stress as a function of plastic strain, strain rate, and temperature, which are developed based on the empirical observations. Usually, these mathematical expressions fit the experimental data but not necessarily address any specific deformation mechanism. They require few material constants that can be determined easily by fitting or inverse calibration of simple tensile or compression test data. Since the mathematical expression of the flow stress is given in explicit form, its implementation in the finite element method (FEM) code is straightforward. Nowadays, many of these models are readily available in the material library of commercial FEM software. Examples of these type of models are Cowper and Symonds [4], Johnson and Cook [5], and Bodner and Partom [6]. The main limitation of these formulations is that they work well only over a limited range of variability of the constitutive variables (i.e., plastic strain, strain rate, and temperature) over which model parameters are determined. Plastic deformation in metals, which deform through dislocations motion and accumulation, is in general a rate- and temperature-dependent process [7]. The motion of dislocations through the crystals of a polycrystalline alloy is a complex phenomenon with various features which cannot be described by simple mathematical models. Physical-based models are derived considering the micromechanics of plastic deformation and are based on the thermally activated motion of dislocations [8]. These models usually can predict fairly well as material behavior over a much wider range of variability of strain, strain rate, and temperature, but they require a larger number of material constants that have physical meaning but are more challenging to be determined. Examples of such type of models can be found in Kocks et al. [9], Follansbee and Kocks [10], Nemat-Nasser and Li [7], and Voyiadjis and Abed [11].

Probably the feature that most distinguishes bcc pure metals and alloys from their fcc and hcp counterparts is the strong temperature dependence of the yield and flow stresses at low temperatures and its related effect on slip geometry [12]. Modelling the yield stress for these class of metals over a wide range of temperature and strain rates is a challenge. Cosrad [13] investigated the variation of the yield stress at low temperature in bcc polycrystals metals and alloys. In particular, he observed that the temperature sensitivity of the yield stress in iron increases with grain boundary and interstitial impurities. Kawata et al. [14] investigated strain rate effect on ductility in bcc and fcc metals. Zerilli and Armstrong [15] proposed a dislocation-mechanics-based constitutive relations for bcc and fcc metals based on the observation that in bcc metals the activation volume is essentially independent of plastic strain while in fcc metals thermal activation is strongly dependent on strain. Klepaczko [16] provided a comprehensive review of physical-based models for metals with different crystal lattice. More recently, Bonora and Milella [17] proposed a semi-empirical constitutive model incorporating damage for predicting both material response and fracture under varying strain rate and temperature conditions. Voyiadjis and Abed [18] developed a physically based yield function for bcc metals. Rusinek and Klepaczko [19] developed a visco-plastic, physically based model, later modified by Rusinek et al. [20] for application in a wide strain rate range. Later, Bonora et al. [21] extended the Rusinek-modified model to simulate deformation of OFHC (oxygen-free high-conductivity) copper under large strain, elevated temperature, and very high strain rate conditions. These are only few examples of an extensive literature. A detailed review of experimental testing and modelling for bcc metals can be found in [22,23] and more recently in [24–26].

However, most of the model formulations are limited to specific temperature and strain rate range. For instance, most of the models address the variation of the yield stress for temperature below 0.2 of the melting temperature T_m where experimental data show the largest temperature effect while very few consider the variation of the yield stress at elevated temperature where deformation process is controlled by diffusion. Similarly, the strain rate effect is often considered only in the range controlled by thermal activation while the regime of the very high strain rates, where viscous effects become dominant, is frequently not considered. Thus, the objective of this paper is to derive physically based constitutive relations for bcc metals over the whole temperature range, from 0 K to T_m, and for strain rate ranging from 10^{-4}/s up to 10^7/s considering thermally activated, diffusion controlled, and viscous regimes.

2. Materials and Methods

The temperature and strain rate sensitivity of bcc metals and alloys is considered. For this class of materials, the yield stress shows a strong dependence on the strain rate and temperature while that for the plastic strain hardening is weak or negligible. Such, sensitivity of bcc metals is generally attributed to the rate-controlling mechanisms of thermal component of the flow stress. The yield stress is mainly due to Peierls barriers, and dislocations slip is primarily controlled by high Peierls stress because of the non-planar structure of screw dislocations. Thermal activation helps dislocation glide by reducing the internal friction, facilitating the slip and decreasing the material flow stress. This results in a strong temperature dependence of the flow stress in bcc metals, which implies also a strong dependence on the strain rate. For bcc metals, experimental data indicates a substantial increase of the rate sensitivity at low temperature which is related to the thermally activated Peierls potential [27,28]. It follows that coupling between temperature and strain rate is critical for a correct definition of the transition between the athermal and thermally activated processes of plastic deformation at low temperature and at high strain rates above room temperature [29]. Based on these premises, a mathematical relationship for the yield stress is derived as follows.

The deformation of a metal beyond the elastic limit requires to activate and move dislocations present in the material through the crystal. Two types of obstacle oppose to dislocations motion: long-range and short-range barriers. The first are due to the structure and cannot be overcome by thermal energy. They supply the flow stress with a contribution that is not thermally activated, usually indicated as athermal stress component. The latter, which may include the Peierls stress, point defects, other dislocations that intersect the slip plane, or substitutional atoms, can be overcome by thermal energy and contribute to determining that part of the flow stress affected by the temperature [17]. At increasing strain rates or dislocation velocities viscous phonon drag becomes dominant [30,31] and the applied stress is high enough to overcome instantaneously the usual dislocation barriers without any aid from thermal fluctuations. Therefore, consistently with dislocation kink-pair theory, the yield stress can be expressed as the sum of three contributions,

$$\sigma_Y\left(\dot{\varepsilon}, T\right) = \sigma_{ath} + \sigma_{ta} + \sigma_{vd} \tag{1}$$

where the subscripts "*ath*," "*ta*," and "*vd*" indicate the athermal, thermally activated, and viscous drag components, respectively [7].

Athermal stress. The athermal stress component of the flow stress arises from the elastic interaction of the dislocations and depends on the temperature only though the weak temperature dependence of the shear modulus [32]. It is a function of the density and distribution of the dislocations, grain sizes and their distribution, as well as the density and distribution of precipitates, substitutional atoms, and other impurities and defects [14]. The athermal stress component of the yield stress can be determined experimentally by means of stress reduction and stress relaxation experiments [33] and it is usually given in the Hall-Petch form in order to account for the grain size effect,

$$\sigma_{ath} = \sigma_0 + \frac{K}{\sqrt{d}} \tag{2}$$

where d is the grain size.

Thermally activated stress. Plastic strain and dislocation density are related by the following simple relationship (Orowan's equation),

$$\dot{\varepsilon} = \rho_m b \bar{v} \tag{3}$$

where $\dot{\varepsilon}$ is the strain rate, b is the Burgers vector, ρ_m is the density of mobile dislocations, and $\bar{v}$ is the average of dislocation velocity. This equation is derived considering that when a dislocation moves, two atoms on sites adjacent across the plane of motion are displaced relative to each other by the Burgers vector b. However, the Orowan's equation holds also for screw and mixed dislocations. Johnston and Gilman [34] provided experimental evidences that the average velocity of mobile dislocations varies with temperature according to,

$$\bar{v} = \bar{v}_0 \exp\left(-\frac{G^*}{kT}\right) \tag{4}$$

Here, k is the Boltzmann's constant and T the temperature, that substituted in Equation (3) leads to:

$$\dot{\varepsilon} = \rho_m b \bar{v}_0 \exp\left(-\frac{G^*}{kT}\right) = \dot{\varepsilon}_0 \exp\left(-\frac{G^*}{kT}\right) \tag{5}$$

which is the rate equation for plastic flow controlled by thermal fluctuations as derived by Taylor [12] where $\dot{\varepsilon}_0$ is a parameter which depends on the dislocation density, vibrational frequency and strain. This equation is assumed as the starting point for the thermodynamical treatment of deformation [35]. G^* is the Gibbs free energy—assuming that the effects of dislocation character and slip on secondary systems are either subsumed in the pre-exponential term $\dot{\varepsilon}_0$ or ignored—which is strongly affected by the work done by the applied stress during activation. Thus, it can be written as:

$$G^* = G^*(\sigma) \tag{6}$$

Since the stress experienced by dislocation is a combination of the applied resolved stress and stresses from other sources, it is preferred to refer to an effective stress σ^* (i.e., σ_{ta}) defined as:

$$\sigma^* = \sigma_a - \sigma_i \tag{7}$$

where σ_a is the applied stress and σ_i is the internal, or athermal, stress arising from the elastic strain field of other dislocations [36]. Although, the internal stress can be positive or negative when considering the motion of dislocation over short barriers, it always subtract to the applied stress. Therefore, from the rate equation, Equation (6) can be rewritten as,

$$G^*(\sigma^*) = -kT \ln\left(\frac{\dot{\varepsilon}}{\dot{\varepsilon}_0}\right) \tag{8}$$

Therefore, differentiating with respect to effective stress the following expression is obtained:

$$\left(\frac{\partial G^*}{\partial \sigma^*}\right)_T = -kT\left(\frac{\partial \ln(\dot{\varepsilon}/\dot{\varepsilon}_0)}{\partial \sigma^*}\right)_T \tag{9}$$

This leads to the definition of the activation volume which is given as the negative stress derivative of the activation energy,

$$v^* = -\left(\frac{\partial G^*}{\partial \sigma^*}\right)_T = kT\left(\frac{\partial \ln(\dot{\varepsilon}/\dot{\varepsilon}_0)}{\partial \sigma^*}\right)_T \tag{10}$$

This form has the advantage with respect to other definitions of v^* of being obtained directly from experiments.

The activation energy cannot be obtained directly from experiments. Recalling that G^* can be written as:

$$G^* = H^* - TS^* \tag{11}$$

where H^* is the activation enthalpy that can be obtained explicitly in differential scanning calorimetry (DSC) tests, and S^* is the activation energy defined as,

$$S^* = -\left(\frac{\partial G^*}{\partial T}\right)_{\sigma^*} \tag{12}$$

Differentiating the rate equation with respect to temperature at constant stress we obtain,

$$\left(\frac{\partial G^*}{\partial T}\right)_{\sigma^*} = -k \ln\left(\frac{\dot{\varepsilon}}{\dot{\varepsilon}_0}\right) - kT\left(\frac{\partial \ln(\dot{\varepsilon}/\dot{\varepsilon}_0)}{\partial T}\right)_{\sigma^*} \tag{13}$$

At a given temperature, the effective stress and dislocation velocity are related. Therefore, the internal stress differential can be written as,

$$d\sigma^* = \left(\frac{\partial \sigma^*}{\partial T}\right)_{\dot{\varepsilon}/\dot{\varepsilon}_0} dT + \left(\frac{\partial \sigma^*}{\partial \ln(\dot{\varepsilon}/\dot{\varepsilon}_0)}\right)_T d\ln\left(\frac{\dot{\varepsilon}}{\dot{\varepsilon}_0}\right) \tag{14}$$

Similarly,

$$d\ln\left(\frac{\dot{\varepsilon}}{\dot{\varepsilon}_0}\right) = \left(\frac{\partial\ln(\dot{\varepsilon}/\dot{\varepsilon}_0)}{\partial T}\right)_{\sigma^*} dT + \left(\frac{\partial\ln(\dot{\varepsilon}/\dot{\varepsilon}_0)}{\partial\sigma^*}\right)_T d\sigma^* \tag{15}$$

Substituting Equation (15) in Equation (14), we get,

$$\left(\frac{\partial\ln(\dot{\varepsilon}/\dot{\varepsilon}_0)}{\partial T}\right)_{\sigma^*} = -\left(\frac{\partial\ln(\dot{\varepsilon}/\dot{\varepsilon}_0)}{\partial\sigma^*}\right)_T \left(\frac{\partial\sigma^*}{\partial T}\right)_{\dot{\varepsilon}/\dot{\varepsilon}_0} \tag{16}$$

and therefore,

$$H^* = kT^2\left(\frac{\partial\ln(\dot{\varepsilon}/\dot{\varepsilon}_0)}{\partial T}\right)_{\sigma^*} = -kT^2\left(\frac{\partial\ln(\dot{\varepsilon}/\dot{\varepsilon}_0)}{\partial\sigma^*}\right)_T \left(\frac{\partial\sigma^*}{\partial T}\right)_{\dot{\varepsilon}/\dot{\varepsilon}_0} \tag{17}$$

From this, the internal stress derivative with respect to the temperature at constant deformation rate can be obtained, and recalling the definition of the activation volume, we can finally write,

$$\left(\frac{\partial\sigma^*}{\partial T}\right)_{\dot{\varepsilon}/\dot{\varepsilon}_0} = -\frac{H^*}{kT^2}\bigg/\left(\frac{\partial\ln(\dot{\varepsilon}/\dot{\varepsilon}_0)}{\partial\sigma^*}\right)_T = -\frac{H^*}{T}\frac{1}{v^*} \tag{18}$$

According to Equation (18), the derivative of the internal stress with respect to temperature, which describes the temperature dependence of the flow stress at constant strain rate, is proportional to the ratio of the activation enthalpy and temperature and activation volume.

The activation volume is a nonlinear function of temperature. In the low temperature range, the activation volume increases with temperature while at elevated temperature, in the diffusion-controlled regime, it decreases with increasing temperature. In order to describe the variation of the activation volume over the whole temperature range, from 0 K to T_m, the following expression is proposed,

$$v^* = \frac{v_0^*}{T_m}[\Gamma_{TH}(T) + \Gamma_{DC}(T)]^{-1} \tag{19}$$

where the first term in the right square bracket is for the plasticity limited by lattice resistance (thermally activated regime, low temperature) while the latter is for the diffusion-controlled flow (elevated temperature), with

$$\Gamma_{TH} = \frac{A}{T_1}\exp\left(-\frac{T}{T_1}\right)\exp\left(-\left(\frac{T}{T_2}\right)^m\right)$$
$$\Gamma_{DC} = \frac{m}{T_2}\left(\frac{T}{T_2}\right)^{m-1}\left(1 + A\exp\left(-\frac{T}{T_1}\right)\right)\exp\left(-\left(\frac{T}{T_2}\right)^m\right) \tag{20}$$

where T_1 and T_2 are the mean temperature of the thermally activated and the diffusion controlled regime, respectively, A and m are dimensionless material constants, T_m is the melting temperature and v_0^* is the activation volume at T_m. Substituting Equation (20) in Equation (19) we finally obtain,

$$v^* = \frac{v_0^*}{T_m}\exp\left(\left(\frac{T}{T_2}\right)^m\right)\left[\frac{A}{T_1}\exp\left(-\frac{T}{T_1}\right) + \frac{m}{T_2}\left(1 + A\exp\left(-\frac{T}{T_1}\right)\right)\left(\frac{T}{T_2}\right)^{m-1}\right]^{-1} \tag{21}$$

Also the activation enthalpy is temperature dependent [37]. In the low temperature range, and for $T < T_1$, a linear relationship can be assumed,

$$H^* = cT \tag{22}$$

where c assumes different values for the thermally activated and for the diffusion-controlled regime. Then, substituting Equation (22) and Equation (21) in Equation (18) we obtain,

$$\left(\frac{\partial\sigma^*}{\partial T}\right)_{\dot{\varepsilon}/\varepsilon_0} = -\frac{cT_m}{v_0^*}\exp\left(-\left(\frac{T}{T_2}\right)^m\right)\left[\frac{A}{T_1}\exp\left(-\frac{T}{T_1}\right) + \frac{m}{T_2}\left(1 + A\exp\left(-\frac{T}{T_1}\right)\right)\left(\frac{T}{T_2}\right)^{m-1}\right] \tag{23}$$

Thus, Equation (14) can be rewritten as,

$$d\sigma^* = -\frac{H^*}{T}\frac{1}{v^*}dT + \frac{kT}{v^*}d\ln\left(\frac{\dot{\varepsilon}}{\varepsilon_0}\right) \tag{24}$$

that substituting the expression for the activation volume, integrating and grouping common factors leads to

$$\sigma^* = \frac{cT_m}{v_0^*}\Delta_1\Delta_2\left\{1 + \frac{k}{c}\ln\left(\frac{\dot{\varepsilon}}{\varepsilon_0}\right)\left[m\left(\frac{T}{T_2}\right)^m + \frac{T}{T_1}\left(1 - \frac{1}{\Delta_1}\right)\right]\right\} \tag{25}$$

where

$$\begin{aligned}\Delta_1 &= 1 + A\exp\left(-\frac{T}{T_1}\right) \\ \Delta_2 &= \exp\left(-\left(\frac{T}{T_2}\right)^m\right)\end{aligned} \tag{26}$$

In Equation (25) the stress contribution given by the second term in the square bracket is small compared to that of the first term, and in a first approximation, can be neglected. Therefore, we can write

$$\sigma^* = \sigma_{th}^0\Delta_1\Delta_2\left\{1 + \lambda\ln\left(\frac{\dot{\varepsilon}}{\varepsilon_0}\right)\right\} \tag{27}$$

where

$$\begin{aligned}\sigma_{th}^0 &= \frac{cT_m}{v_0^*} \\ \lambda &= \frac{k}{c}m\left(\frac{T}{T_2}\right)^m\end{aligned} \tag{28}$$

where σ_{th}^0 is a fraction of the thermal stress at 0 K (which is $\sigma_{th}^0(1 + A)$) and λ is the strain rate sensitivity parameter.

Viscous drag stress. At low strain rates in the thermally activated region, the viscous drag component is negligible. It becomes more and more relevant at increasing the strain rate. In the viscous drag-dominated regime the yield stress is a linear function of the strain rate. Here, the following simple expression is proposed,

$$\sigma_{vd} = \sigma_{vd}^0\left[\frac{\dot{\varepsilon}}{\varepsilon_0} - 1\right] \tag{29}$$

where σ_{vd} is a scale factor.

Finally, combining Equation (2), Equation (27), and Equation (29) the following expression for the constitutive equation is obtained,

$$\sigma^* = \sigma_0 + \frac{K}{\sqrt{d}} + \sigma_{th}^0\left(1 + A\exp\left(-\frac{T}{T_1}\right)\right)\exp\left(-\left(\frac{T}{T_2}\right)^m\right)\left\{1 + \lambda\ln\left(\frac{\dot{\varepsilon}}{\varepsilon_0}\right)\right\} + \sigma_{vd}^0\left[\frac{\dot{\varepsilon}}{\varepsilon_0} - 1\right] \tag{30}$$

Formally, the model requires ten parameters that can be identified from experimental data at different strain rate and temperature. However, they can be reduced to eight if the athermal stress is assumed as a single parameter. In the present formulation, the reference strain rate $\dot{\varepsilon}_0$ is the strain rate at which transition between the athermal and thermally activated regime occurs. In bcc metals, this is also temperature dependent as clearly shown by Campbell and Ferguson [38] for mild steel.

Here, the following expression for the temperature dependence of the transition strain rate in bcc is proposed,

$$\dot{\varepsilon}_0 = \dot{\varepsilon}_{vd}\left(\frac{T}{T_m}\right)^m \tag{31}$$

where $\dot{\varepsilon}_{vd}$ is the strain rate limit for the domain of validity of the linear viscous drag law [39].

3. Results

The proposed model has been validated predicting the yield stress at different strain rates and temperature for A508 steel (equivalent to 16MND5). This is a low alloyed steel with bcc atomic structure, developed for reactor pressure vessel applications. The microstructure is a restored bainite and the ferritic matrix is reinforced by carbides (of the order of 1 μm) resulting from the precipitation of cementite, upper and lower (spheroids) bainite together with several types of inclusions manganese sulphide (MnS). The reference composition is given in Table 1 [35].

Table 1. Chemical composition of A508 steel.

Element	C	S	P	Si	Mn	Ni	Cr	Mo	Cu	Co
wt.%	0.16	0.005	0.006	0.19	1.35	0.74	0.18	0.51	0.007	0.01

Experimental data at different temperature and strain rates were taken from different sources [40–42].

Identification of model parameters has been performed as follow. First, non-linear fitting of yield stress data as a function of temperature, for the nominal strain rate of 10^{-4}/s, was performed using Equation (30) neglecting the contribution of the viscous drag. Here, the athermal stress was treated as a single parameter. As a result of this procedure σ_{ath}, σ_{th}^0, A, T_1, T_2, and m were determined. Second, the remaining model parameters, λ, σ_{vd}^0 were determined by fitting yield stress data as a function of the strain rate for $T = 293$ K. T_m is given for the material under investigation and $\dot{\varepsilon}_{vd}^0$ is taken as 10^6/s.

In Figure 1, the comparison of the present model solution (fitted) for the yield stress as a function of temperature and experimental the data is given. In Figure 2, the predicted yield stress as a function of the strain rate at different temperature is shown.

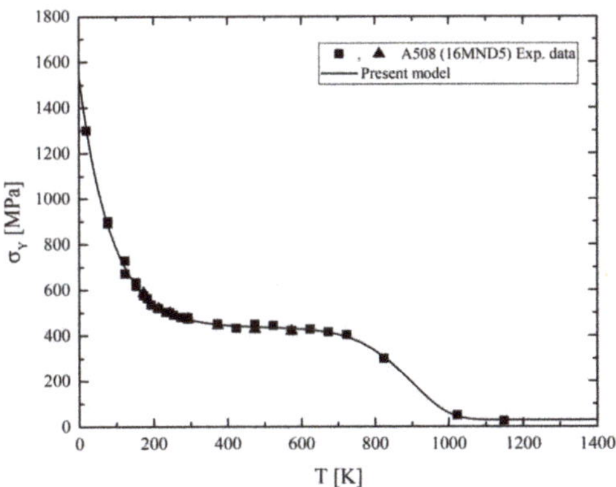

Figure 1. Yield stress as a function of temperature for A508 steel at nominal quasi-static strain rate (10^{-4}/s): result of fitting experimental data with Equation (30).

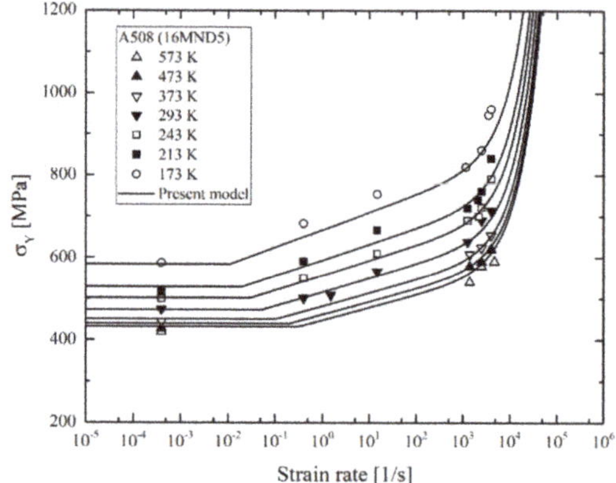

Figure 2. Yield stress of A508 steel as a function strain at different temperature. Here, only data at $T = 293$ K were used to fit the material model parameters while for other temperatures the behavior predicted by present model solution is shown.

Finally, material model parameters for A508 steel are summarized in Table 2.

Table 2. Model parameters for A508 steel.

Parameter	σ_{ath}[MPa]	σ_{th}^0[MPa]	A	T_1 [K]	T_2 [K]	m	λ	T_m [K]	$\dot{\varepsilon}_{vd}^0$[1/s]	σ_{vd}^0 [MPa]
Value	28.7	407.0	2.7	86.5	908.5	9.83	0.033	1623	10^6	5×10^{-3}

4. Conclusions

In this work a physically based model for the yield stress in bcc has been developed based on dislocations mechanics and according to a thermodynamical treatment of deformation. Although the approach for physically based model is well established in the literature, in the present paper the expression for the temperature activation volume, which is highly nonlinear over the whole temperature range, predicts the coupling between strain rate and temperature effect in the thermally activated regime of the yield stress. In addition, predicted strain rate effect, $1 + \lambda \ln(\dot{\varepsilon}/\dot{\varepsilon}_0)$, is found consistent with that of the Johnson and Cook of phenomenological derivation, providing a physical meaning to the scale factor $\dot{\varepsilon}_0$ that here is the strain rate at which the thermal activation effect becomes dominant over the athermal stress component, with $\dot{\varepsilon}_0$ being temperature dependent.

Author Contributions: Conceptualization G.T., N.B.; methodology, N.B.; validation, A.R.; data analysis, G.I. and G.T.; investigation, G.T.; writing—original draft preparation, N.B.; writing—review and editing, G.T. and N.B. All authors have read and agreed to the published version of the manuscript.

Funding: This research received no external funding.

Conflicts of Interest: The authors declare no conflict of interest.

References

1. Kuehmann, C.J.; Olson, G.B. Computational materials design and engineering. *Mater. Sci. Technol.* **2009**, *25*, 472–478. [CrossRef]
2. Allen, D.H.; Harris, C.E. *A Review of Nonlinear Constitutive Models for Metals*; Langley Research Center: Hampton, VA, USA, 1990.
3. Bodner, S.R. Review of a unified elastic—Viscoplastic theory. In *Unified Constitutive Equations for Creep and Plasticity*; Springer: Dordrecht, The Netherlands, 1987; pp. 273–301. [CrossRef]

4. Cowper, G.R.; Symonds, P.S. *Strain-Hardening and Strain-Rate Effects in the Impact Loading of Cantilever Beams*; Brown University: Providence, RI, USA, 1957.

5. Johnson, G.R.; Cook, W.H. Fracture characteristics of three metals subjected to various strains, strain rates, temperatures and pressures. *Eng. Fract. Mech.* **1985**, *21*, 31–48. [CrossRef]

6. Bodner, S.R.; Partom, Y. Constitutive equations for elastic-viscoplastic strain-hardening materials. *J. Appl. Mech.* **1975**, *42*, 385–389. [CrossRef]

7. Nemat-Nasser, S.; Li, Y. Flow stress of fcc polycrystals with application to OFHC Cu. *Acta Mater.* **1998**, *46*, 565–577. [CrossRef]

8. Surek, T.; Luton, M.; Jonas, J. Dislocation glide controlled by linear elastic obstacles: A thermodynamic analysis. *Philos. Mag. A J. Theor. Exp. Appl. Phys.* **1973**, *27*, 425–440. [CrossRef]

9. Kocks, U.; Argon, A.; Ashby, M. Thermodynamics and kinetics of slip. *Prog. Mater. Sci.* **1975**, *19*, 1–281.

10. Follansbee, P.; Kocks, U. A constitutive description of the deformation of copper based on the use of the mechanical threshold stress as an internal state variable. *Acta Metall.* **1988**, *36*, 81–93. [CrossRef]

11. Voyiadjis, G.Z.; Abed, F.H. Microstructural based models for bcc and fcc metals with temperature and strain rate dependency. *Mech. Mater.* **2005**, *37*, 355–378. [CrossRef]

12. Taylor, G. Thermally-activated deformation of BCC metals and alloys. *Prog. Mater. Sci.* **1992**, *36*, 29–61. [CrossRef]

13. Cosrad, H. Effect of temperature on yield and flow stress of BCC metals. *Philos. Mag.* **1960**, *5*, 745–751. [CrossRef]

14. Kawata, K.; Hashimoto, S.; Kurokawa, K. Analyses of high velocity tension of bars of finite length of BCC and FCC metals with their own constitutive equations. In *High Velocity Deformation of Solids*; Springer: Berlin/Heidelberg, Germany, 1979; pp. 1–15.

15. Zerilli, F.J.; Armstrong, R.W. Dislocation-mechanics-based constitutive relations for material dynamics calculations. *J. Appl. Phys.* **1987**, *61*, 1816–1825. [CrossRef]

16. Klepaczko, J.R. Constitutive modeling in dynamic plasticity based on physical state variables-A review. *Le Journal de Physique Colloques* **1988**, *49*, C3-553–C3-560. [CrossRef]

17. Bonora, N.; Milella, P.P. Constitutive modeling for ductile metals behavior incorporating strain rate, temperature and damage mechanics. *Int. J. Impact Eng.* **2001**, *26*, 53–64. [CrossRef]

18. Voyiadjis, G.Z.; Abed, F.H. A coupled temperature and strain rate dependent yield function for dynamic deformations of bcc metals. *Int. J. Plast.* **2006**, *22*, 1398–1431. [CrossRef]

19. Rusinek, A.; Klepaczko, J. Shear testing of a sheet steel at wide range of strain rates and a constitutive relation with strain-rate and temperature dependence of the flow stress. *Int. J. Plast.* **2001**, *17*, 87–115. [CrossRef]

20. Rusinek, A.; Rodríguez-Martínez, J.A.; Arias, A. A thermo-viscoplastic constitutive model for FCC metals with application to OFHC copper. *Int. J. Mech. Sci.* **2010**, *52*, 120–135. [CrossRef]

21. Bonora, N.; Testa, G.; Ruggiero, A.; Iannitti, G.; Mortazavi, N.; Hörnqvist, M. Numerical simulation of dynamic tensile extrusion test of ofhc copper. *J. Dyn. Behav. Mater.* **2015**, *1*, 136–152. [CrossRef]

22. Liang, R.; Khan, A.S. A critical review of experimental results and constitutive models for BCC and FCC metals over a wide range of strain rates and temperatures. *Int. J. Plast.* **1999**, *15*, 963–980. [CrossRef]

23. Khan, A.S.; Liang, R. Behaviors of three BCC metal over a wide range of strain rates and temperatures: Experiments and modeling. *Int. J. Plast.* **1999**, *15*, 1089–1109. [CrossRef]

24. Huh, H.; Ahn, K.; Lim, J.H.; Kim, H.W.; Park, L.J. Evaluation of dynamic hardening models for BCC, FCC, and HCP metals at a wide range of strain rates. *J. Mater. Process. Technol.* **2014**, *214*, 1326–1340. [CrossRef]

25. Salvado, F.C.; Teixeira-Dias, F.; Walley, S.M.; Lea, L.J.; Cardoso, J.B. A review on the strain rate dependency of the dynamic viscoplastic response of FCC metals. *Prog. Mater. Sci.* **2017**, *88*, 186–231. [CrossRef]

26. Gray III, G.T. High-strain-rate deformation: Mechanical behavior and deformation substructures induced. *Annu. Rev. Mater. Res.* **2012**, *42*, 285–303. [CrossRef]

27. Conrad, H. Thermally activated deformation of metals. *JOM* **1964**, *16*, 582–588. [CrossRef]

28. Klepaczko, J. A general approach to rate sensitivity and constitutive modeling of FCC and BCC metals. *Impact Eff. Fast Transient Load.* **1988**, 3–35.

29. Rusinek, A.; Zaera, R.; Klepaczko, J.R. Constitutive relations in 3-D for a wide range of strain rates and temperatures–application to mild steels. *Int. J. Solids Struct.* **2007**, *44*, 5611–5634. [CrossRef]

30. Kumar, A.; Hauser, F.; Dorn, J. Viscous drag on dislocations in aluminum at high strain rates. *Acta Metall.* **1968**, *16*, 1189–1197. [CrossRef]

31. Vreeland, T., Jr.; Lau, S. *Study of Dislocation Mobility and Density in Metallic Crystals*; Technical report; WM Keck Lab. of Engineering Materials; California Institute of Technology: Pasadena, CA, USA, 1974.
32. Nabarro, F.R.; Duesbery, M.S. *Dislocations in Solids*; Elsevier: Amsterdam, The Netherlands, 2002; Volume 11.
33. Kruml, T.; Coddet, O.; Martin, J. About the determination of the thermal and athermal stress components from stress-relaxation experiments. *Acta Mater.* **2008**, *56*, 333–340. [CrossRef]
34. Johnston, W.G.; Gilman, J.J. Dislocation velocities, dislocation densities, and plastic flow in lithium fluoride crystals. *J. Appl. Phys.* **1959**, *30*, 129–144. [CrossRef]
35. Vineyard, G.H. Frequency factors and isotope effects in solid state rate processes. *J. Phys. Chem. Solids* **1957**, *3*, 121–127. [CrossRef]
36. Tanibayashi, M. A Method to Determine the Internal Stress and the Stress Exponent m of the Dislocation Velocity. *Phys. Status Solidi (A)* **1990**, *120*, K19–K23. [CrossRef]
37. Arsenault, R.; Li, J.C. Thermally activated dislocation motion in a periodic internal stress field. *Philos. Mag. A J. Theor. Exp. Appl. Phys.* **1967**, *16*, 1307–1311. [CrossRef]
38. Campbell, J.; Ferguson, W. The temperature and strain-rate dependence of the shear strength of mild steel. *Philos. Mag.* **1970**, *21*, 63–82. [CrossRef]
39. Gurrutxaga-Lerma, B.; Balint, D.; Dini, D.; Sutton, A. The mechanisms governing the activation of dislocation sources in aluminum at different strain rates. *J. Mech. Phys. Solids* **2015**, *84*, 273–292. [CrossRef]
40. Tanguy, B.; Besson, J.; Piques, R.; Pineau, A. Ductile to brittle transition of an A508 steel characterized by Charpy impact test: Part II: Modeling of the Charpy transition curve. *Eng. Fract. Mech.* **2005**, *72*, 413–434. [CrossRef]
41. Chapuliot, S.; Lacire, M.; Marie, S.; Nédélec, M. Thermomechanical analysis of thermal shock fracture in the brittle/ductile transition zone. Part I: Description of tests. *Eng. Fract. Mech.* **2005**, *72*, 661–673. [CrossRef]
42. Reytier, M.; Chapuliot, S.; Marie, S.; Ferry, L.; Nedelec, M. Study of cleavage initiation under thermal shock by tests on cracked rings and thermomechanical calculations. *Nucl. Eng. Des.* **2006**, *236*, 1039–1050. [CrossRef]

Article

Probabilistic Modeling of Slip System-Based Shear Stresses and Fatigue Behavior of Coarse-Grained Ni-Base Superalloy Considering Local Grain Anisotropy and Grain Orientation

Benedikt Engel [1],*, Lucas Mäde [2], Philipp Lion [3], Nadine Moch [4], Hanno Gottschalk [4] and Tilmann Beck [3]

[1]   Gas Turbine and Transmission Center Research Center (G2TRC), University of Nottingham, Nottingham NG7 2RD, UK

[2]   Department for Technology & Innovation, Gas and Power Division, Siemens AG, Huttenstraße 12, 10553 Berlin, Germany

[3]   Institute of Materials Science and Engineering, TU Kaiserslautern, 67663 Kaiserslautern, Germany

[4]   School of Mathematics and Science, Bergische Universität Wuppertal, 42119 Wuppertal, Germany

*   Correspondence: Benedikt.Engel@nottingham.ac.uk

Received: 26 June 2019; Accepted: 20 July 2019; Published: 24 July 2019

Abstract: New probabilistic lifetime approaches for coarse grained Ni-base superalloys supplement current deterministic gas turbine component design philosophies; in order to reduce safety factors and push design limits. The models are based on statistical distributions of parameters, which determine the fatigue behavior under high temperature conditions. In the following paper, Low Cycle Fatigue (LCF) test data of several material batches of polycrystalline Ni-base superalloy René80 with different grain sizes and orientation distribution (random and textured) is presented and evaluated. The textured batch, i.e., with preferential grain orientation, showed higher LCF life. Three approaches to probabilistic crack initiation life modeling are presented. One is based on Weibull distributed crack initiation life while the other two approaches are based on probabilistic Schmid factors. In order to create a realistic Schmid factor distribution, polycrystalline finite element models of the specimens were generated using Voronoi tessellations and the local mechanical behavior investigated in dependence of different grain sizes and statistically distributed grain orientations. All models were first calibrated with test data of the material with random grain orientation and then used to predict the LCF life of the material with preferential grain orientation. By considering the local multiaxiality and resulting inhomogeneous shear stress distributions, as well as grain interaction through polycrystalline Finite Element Analysis (FEA) simulation, the best consistencies between predicted and observed crack initiation lives could be achieved.

Keywords: LCF; René80; Probabilistic modeling; slip system-based shear stresses; probabilistic Schmid factors; polycrystalline FEA; anisotropy; Ni-base superalloy

---

1. Introduction

Due to the worldwide increase in fluctuating renewable energy generation, flexibly operating gas turbine power plants are necessary to secure stable power supply and grid frequencies. However, frequent start-ups and shut downs as well as load changes lead to high requirements to the materials used in the hot gas section components, foremost the turbine blades. Due to their outstanding properties at high temperatures, Ni-base superalloys are appropriate as turbine blade material. While single crystal and directionally solidified components are nowadays used to accommodate the highest demands in creep and oxidation resistance, components made of polycrystalline solidified nickel-base superalloys from conventional cast are still more often used for economic reasons. The remarkable Low Cycle Fatigue (LCF) crack initiation life scatter observed in lab tests at polycrystalline Ni-base superalloys samples is well known and usually considered by material safety factors for engine part designs. Deterministic models to predict the LCF lives of the components have been applied successfully for decades, but the increasing demand for risk-based business decision making and the reduction of conservatism in design (safety factors) have led to the development of probabilistic fatigue prediction models. In the early days of probabilistic modeling for LCF, defects were considered as sources of randomness in fatigue behavior. Accordingly, [1,2] utilized measured defect size distributions for probabilistic modeling. This approach is also used for fatigue prediction of turbine disks [3,4]. Other probabilistic fatigue models are based on the weakest link concept which is often described with a Weibull distribution in fatigue limit [5–10] but also in fatigue life [11–15]. The probabilistic fatigue limit and fatigue life modeling often resulted in integral approaches to model the probabilistic size effect, which are already put down in design guidelines [16–18], but are up to now still under research [19–23]. Particularly in [22], a local probabilistic approach based on a Weibull distribution in LCF life arising from inherent material scatter was validated with Ni-base superalloy specimens. While the previously mentioned approaches use mostly parametric distributions to describe the stochastic nature of fatigue mechanisms and the statistical size effect, more physics and micro-mechanics-based approaches are favorable to improve the model accuracy. Therefore, it is necessary to determine the distributional aspects of the high temperature fatigue mechanisms in the investigated materials and the influence of those factors to the cyclic life and life scatter.

Coarse-grained polycrystalline Ni-base superalloys such as IN 738 LC or René80 are used for turbine blades in the rear stages of gas turbines. Due to the vacuum casting process and geometry-related cooling conditions, they tend to form grains of up to 3 mm and more in diameter. As a result, only a few grains in highly stressed areas of limited size are formed within the component, such as the transition from root to the airfoil. Due to the low grain numbers, the orientation of the crystal lattice of each grain has a major influence on the mechanical behavior of the component. On the one hand, the orientation has a direct influence on the elastic material behavior. Due to the pronounced elastic anisotropy of nickel and Ni-base superalloys (anisotropy factor of up to 3), the effective value of Young's modulus for uniaxial loading of a single grain varies over a wide range. For room temperature it can vary between 130 GPa and 330 GPa and between 95 GPa and 260 GPa at 850 °C [24–26]. This aspect in combination with low grain numbers, leads to high scatter in the determined mechanical properties, which makes high safety factors in design inevitable. On the other hand, the crystal orientation in addition to the elastic behavior has an influence on the onset of plastic behavior. As determined by [27], the resolved shear stress at a slip system can be calculated by projecting the applied normal stress at the slip system orientation. The modified Schmid factor is defined as the quotient of maximum resolved shear stress and equivalent stress and compares the propensity of shear glide between crystals of different orientation. The effects of Schmid factor on the crack initiation and fatigue behavior have been investigated in several studies. Seibel et al. [28,29] could show for the coarse grain Ni-base superalloy René80 at 850 °C and low total strain amplitudes, that fatigue cracks initiate on slip bands at grains with high Schmid factors. A plot of the LCF life against the resulting shear stresses in the crack-initiating slip system led to a significant reduction in life scatter compared to a strain Wöhler plot. Gottschalk et al. [28,30] showed, that there is a correlation between the distribution of the Schmid factor and the lifetime of the Ni-base superalloy René80 tested in high temperature LCF tests. Engel [25,31] could also show coarse grained René80 that, besides the Schmid factor, also the local Young's modulus of the crack initiating grain has a huge influence on the resulting shear stress within the slip systems. With the proposed E·m model, it could be shown, experimentally and analytically, that cracks predominantly initiate in grains with Schmid factors between 0.35–0.45 and high corresponding Young's moduli and therefore high values in the product E·m. Cracks in grains can also initiate if these are surrounded by grains with high E·m values (neighborhood effect). The presented models all are used to estimate the resulting shear stress in the slip system of the grains, to provide predictions about the LCF life distribution of Ni-base superalloys. However, the following simplifying assumptions are applied in these models:

- Uniaxial stress states in each grain with global uniaxial load
- No influence of the deformation behavior of the surrounding grains
- Only Young's modulus and Schmid factor in direction of a uniaxial stress were considered
- Homogeneous resolved shear stress field at slip system within the grain.

In order to improve the model for probabilistic Schmid factor distribution modeling from [30], several polycrystalline finite element models are generated by 3D Voronoi tessellation using the software Neper [32–34] and solved in ABAQUS® 2017; considering the elastic anisotropy of the crystals. These simulations allow approximating the local mechanical properties in a coarse-grained polycrystalline, uniaxially loaded LCF specimen where the local grain orientations and their interaction lead to various multiaxial stress states. The polycrystalline Finite Element Analysis (FEA) simulations were carried out for the case of randomly and for preferentially oriented grains. These preconditions led to distinct distributions of local resolved shear stresses and therefore Schmid factors. Deterministic formulae are used to derive statistical distributions for the LCF crack initiation life of both cases. In order to verify the differences seen in the simulation results, LCF test data of two René80 batches (isothermal 850 °C) was generated in strain-controlled experiments. One batch has had coarse grains with random orientation, while the other batch has had smaller grains with preferential orientation. Hence, polycrystalline FEA simulations according to this grain orientation were carried out. The thereof derived life distribution was shifted to a higher median compared to the case of random grain orientation. Using the large LCF test data set of Seibel [29] (coarse, randomly oriented grains) for calibration of the Schmid factor based crack initiation life model, it was possible to predict the observed crack initiation lives of the René80 batch with preferential grain orientation.

The presented work is subdivided into three major sections. The examined material and applied testing and simulation methods are described in Section 2. Experimental and simulation outcomes are presented in Section 3 and discussed in Section 4. Sections 2 and 3 follow an equivalent substructure. It starts by describing the investigated material René80 and its microstructural examination (Sections 2.1 and 3.1). The LCF testing procedure and its outcomes are detailed in the Sections 2.2 and 3.2. Sections 2.3 and 3.3 switch to the simulation part of the presented work specifying how polycrystalline anisotropic FEA models of the LCF specimen were set up and how they can capture their global stiffness. These FEA models are partly the basis for probabilistic Schmid factor modeling, which is introduced in Section 2.4 and presented in Section 3.4. Probabilistic Schmid factors are the basis for the microstructure-based crack initiation life models that are described in Section 2.5 and calibrated with the test data in Section 3.5. These microstructure based probabilistic approaches for LCF are the highlight of the presented work and are furthermore compared to the Weibull approach from [19] in Section 4. It is concluded that the LCF life prediction based on the "modified Schmid factor" distribution derived from polycrystalline FEA simulations has the best prediction accuracy.

2. Materials and Methods

Two batches of René80 LCF specimens from the same melt, but different solidification processes are investigated in the present work. The mechanical properties of the specimens are simulated with FEA models mimicking the polycrystalline microstructure and the anisotropic stiffness of the grains. From those FEA solutions, the distribution of normalized maximum shear stresses at the slip systems (modified Schmid factors) are derived. It is then explained how the distribution of Schmid factors leads to a distribution in crack initiation life.

2.1. Material

Cylindrical bars with 150 mm length made of the polycrystalline Ni-base superalloy René80 [35–39], were conventionally casted and heat treated by Doncasters/Germany. To achieve the final microstructure, a two-step heat treatment consisting of solution annealing and aging was applied to the material. From the same melt, bars with diameters of 20 mm (coarse grained batch) as well as 12 mm were produced in order to generate a different grain size and grain orientation distribution.

These bars were machined to the final specimen geometry shown in Figure 1 and the gauge section was polished to a surface roughness of N5.

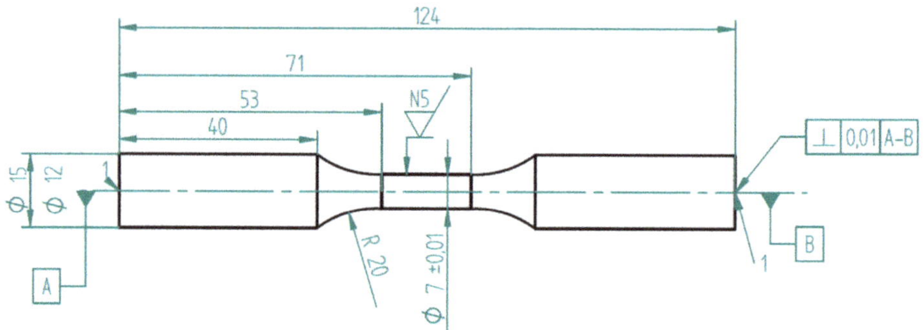

Figure 1. Specimen geometry. Only the cylindrical gauge section is considered in Finite Element Analysis (FEA) simulations.

Table 1 shows the chemical compositions of the René80 melt, determined by the manufacturer, as well as the composition of Inconel 738 with low amount of carbon (IN 738 LC).

Table 1. Chemical composition of René80 and IN 738 LC (taken from Hermann, W. (2014) [26]) in wt.%.

Element	Ni	Cr	Co	Ti	Mo	W	Al	C	B	Zr	Ta	Nb	Fe
René80	Bal.	14.04	9.48	5.08	4.03	4.02	2.93	0.17	0.015	0.011	-	-	-
IN 738 LC	Bal.	16	8.3	8.7	3.4	1.8	2.7	3.4	0.11	-	1.9	0.9	0.1

2.2. Experimental Isothermal LCF Testing

Isothermal LCF tests were carried out at 850 °C on an MTS 810 servo hydraulic test rig with a maximum load capacity of 100 kN. The specimens were heated by a Huttinger TruHeat generator MF5000 with an induction coil and temperature controlled by a ribbon thermocouple type K attached to the middle of the gauge length. The temperature gradient was measured before the testing campaign and was below ±8 °C across the entire gauge length. Total strain control condition with $R = -1$ was realized by a 12 mm MTS high temperature extensometer Type MTS 632.53. Depending on the value of total strain, the test frequency varies between 1 Hz for low total strain and 0.1 Hz for high total strains. Cycles to failure were determined by a load drop of 2.5% from the stabilized measured stress amplitude in order to reach a crack surface of 0.962 mm^2 and to compare the results to [29].

2.3. FEA Models for Polycrystalline Microstructure Modelling

In order to simulate polycrystalline material behavior, the open source software NEPER (Version 3.3, by Romain Quey, MINES Saint-Étienne, Saint-Étienne, France) was applied to generate random grain morphologies using the 3D Voronoi tessellation method [40,41]. Since only the gauge lengths of the specimens were of interest in the simulation, a cylinder with the dimensions $r = 7$ mm, $h = 18$ mm were created. The specification of 49 Voronoi seeds leads to 49 grains with an average grain diameter $d_{grain} \approx 3$ mm for the given volume. This is in good accordance to metallographic analyzes of the tested coarse grain René80 batch (random orientation) presented in Section 3.1. Equivalently, the model for the other batch which solidified to grains with an average grain size of $d_{grain} \approx 1.3$ mm and preferential direction in orientation contains 500 grains. Within the meshing tool of NEPER a relative characteristic length value of $rcl = 0.3$ defines the size of the elements relative to the average cell size. As a result, the mesh consists of a uniform distribution of approximately 50,000 quadratic tetrahedral elements for the coarse grain model and approximately 90,000 quadratic tetrahedral elements for the fine grain model. Both polycrystalline specimen models are shown in Figure 2.

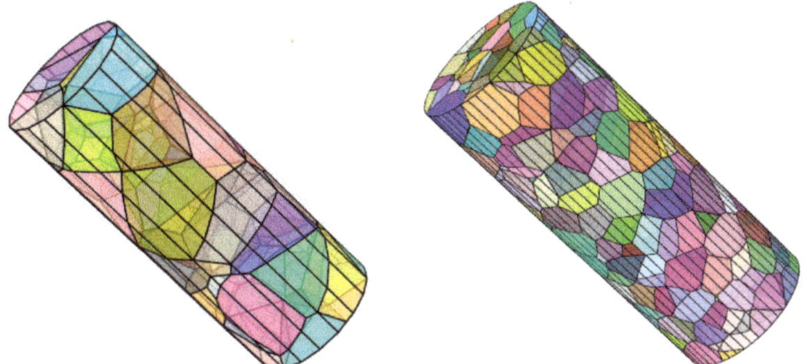

Figure 2. Polycrystalline model for the cylindrical gauge section with 49 and 500 grains.

Additionally, to the morphology, Neper also creates a consistent mesh at the boundary interfaces of the grains such it can be used as an input for ABAQUS® (Dessault systemes, Vélizy-Villacoublay, France). In order to account for anisotropic stiffness of René80, a global, anisotropic, linear-elastic material law was defined in ABAQUS®.

$$\underline{\sigma} = \underline{\underline{C}} \cdot \underline{\varepsilon} \text{ where } \underline{\underline{C}} = \begin{bmatrix} C_{11} & C_{12} & C_{12} & & & \\ C_{12} & C_{11} & C_{12} & & & \\ C_{12} & C_{12} & C_{11} & & & \\ & & & C_{44} & & \\ & & & & C_{44} & \\ & & & & & C_{44} \end{bmatrix} \tag{1}$$

As the authors are not aware of any values for the elastic constants of René80 at high temperatures, elastic constants for IN 738 LC were taken from Hermann, W. (2014) [26]. Since both, composition and content of the γ' phase are very similar in IN 738 LC and René80, it is assumed that the elastic behavior of both alloys are qualitatively comparable. A linear interpolation from 800 °C and 898 °C to 850 °C results in $C_{11} = 225.83$ MPa, $C_{12} = 161.45$ MPa and $C_{44} = 98.79$ MPa. In order to create local material models, each grain was rotated in a preprocessing step and the respective data written to the input file. The grains in the first René80 batch (coarse-grained) are assumed to have no preferential direction in orientation, which is why the rotational matrices U used in this preprocessing step are distributed according to the isotropic measure, mathematically given by the Haar measure at the SO3 group of rotations [42]. For the second batch (fine grains) however, the grains are assumed to have a non-isotropic orientation distribution (see Section 3.1) and the respective FEA models are set up accordingly. During the calculation, ABAQUS® transfers the globally defined material law by means of tensor rotation with U into the local coordinate system of the individual grains. With this procedure, the grains interact according to their orientation dependent stiffness. However, grain boundary interactions are not explicitly modeled by physics-based considerations. In order to have high comparability to the LCF experiments, all FEA simulations were modeled with given displacements using a material model for $T = 850$ °C. Since the latter was linear-elastic and a stable cyclic behavior for low total strain amplitudes was observed [43], only one load case was evaluated. All nodes of the cylinder top face were displaced by 0.045 mm which is equal to a total strain of 0.25%. The nodes at the bottom face were fixed but allowed transverse contraction.

2.4. Derivation of the Schmid Factor Distribution

Plastic deformation of a metallic grain begins if one resolved shear stress within the slip systems of the crystal exceeds a critical value τ_{CRSS}. As the E·m model in [25,31] proposed for coarse grained Ni-base superalloys, the resulting shear stress in the slip systems depends on the grains local Young's moduli and its Schmid factor $\widetilde{m}$. According to [28,30] $\widetilde{m}$ is defined as

$$\widetilde{m} = \max_{i,j}\left(\frac{\tau_{i,j}}{\sigma_{vM}}\right) \text{ where } \tau_{i,j} \text{ is the resolved shear stresses at slip systems } j \text{ of slip plane } i \tag{2}$$

The 12 possible values for $\tau_{i,j}$, representing the slip systems <111>{110} in a fcc crystal are obviously dependent on the crystal orientation towards the load. Let U be the rotation matrix describing the orientation from the coordinate system of the stress tensor $\underline{\underline{\sigma}}$ and the lattice coordinate system. The resolved shear stresses $\tau_{i,j}(U)$ in a crystal rotated by U are calculated by

$$\tau_{i,j}(U) = \vec{n}_i \cdot \underline{\underline{\sigma}}_{aniso}(U) \cdot \vec{s}_{i,j} \text{ where } U \in SO(3) \tag{3}$$

$$\underline{\underline{\sigma}}_{aniso}(U) = \underline{\underline{C}}(U) \cdot \underline{\underline{\varepsilon}}_{iso} \tag{4}$$

$\vec{n}_i$ are the normal vectors of the slip planes and $\vec{s}_{i,j}$ are the slip vectors on each slip plane. Equation (3) was simplified exploiting the orthogonality of U [30]. In order to also account for the stiffness anisotropy of the Ni lattice cell the anisotropic stress tensor $\underline{\underline{\sigma}}_{aniso}(U)$ is calculated by the tensor product of isotropic strain tensor $\underline{\underline{\varepsilon}}_{iso}$ and rotated $\underline{\underline{C}}(U)$ stiffness tensor. The randomness of crystal grain orientations in a polycrystal is the root cause why the Schmid factor as defined in (2) is a stochastic variable with the cumulative distribution function $F_{SF}(\widetilde{m}) = P(SF \leq \widetilde{m})$.

Repeated plastic deformation of the crystal causes lattice dislocations to move along the slip systems towards the surface eventually creating persistent slip bands (PSB's) which are manifested as intrusions and extrusions and initiate surface fatigue cracks. Grains with an orientation leading to high $E \cdot \widetilde{m}$ values under given load are creating PSB's faster than those with a lower value of $E \cdot \widetilde{m}$. That is why it could be shown that LCF cracks preferably initiate in grains with such an orientation [29,31]. A high stiffness in direction of the load leads to high anisotropic stresses in the crystal and a high Schmid factor leads to high resolved shear stresses at the slip system.

In order to include the stochastic character of the Schmid factor in a fatigue life prediction, it is necessary to quantify the Schmid factor distribution $F_{SF}(\widetilde{m})$ in a first step. One approach uses only the geometric considerations for a single crystal at isotropic strain, which is reflected in Equations (3) and (4). A Monte-Carlo sampling of lattice orientations described by rotation matrices U creates a distribution of resolved shear stresses $\tau_{i,j}(U)$ according to (3) which translates into the Schmid factor distribution $F_{SF}(\widetilde{m})$ via (2.2) [30]. Note that $U = U(\vartheta, \varphi_1, \varphi_2)$ is conveniently defined as a function of the Euler angles $\vartheta, \varphi_1, \varphi_2$ in this publication.

This approach is computationally inexpensive but suffers from a major drawback. The derived single crystal Schmid factor distribution has limited meaning for reflecting the true distribution of maximum shear stresses in a polycrystal since no grain interaction and therefore no complex grain distortion is considered.

Hence, a second approach to simulate the Schmid factor distribution was carried out in order to overcome the drawback of simplification to only single crystal behavior. The polycrystalline FEA simulations described in Section 2.2 were repeated three times with different grain orientations. These were realized using Monte-Carlo sampled rotation matrices U. Following these simulations, Schmid factor values were calculated from all nodal stress tensors according to Equations (2) and (3) during the post-processing routine [30]. The thereby received distribution will be denoted as modified Schmid factor $\widetilde{m}_{mod}$ distribution in the following. Section 3.3 describes the differences from the single grain Schmid factor distribution. They arise from the distinct distribution of stress tensors with different multiaxiality, even within single grains, which develops due to the mutual grain interactions which is discussed in Section 3.2.

Note, that both approaches do not consider elastic-plastic deformation models and therefore only linear-elastic shear stresses at the slip systems are calculated. Still, their distribution at the model surface give an indication about the expected onset of plasticity. Table 2 summarizes the description of the two types of Schmid factor distributions.

Table 2. Overview of Schmid factor distribution generation.

Distribution	Modeling Approach
$F_{SF}(\widetilde{m})$	Monte-Carlo sampling of statistically distributed orientations of a single crystal, i.e., single grain. Maximum normalized resolved shear stresses calculated at global uniaxial stress state. Explicit consideration of elastic stiffness anisotropy.
$F_{SF}(\widetilde{m}_{mod})$	Monte-Carlo sampling of statistically distributed orientations of single crystals in polycrystalline FEA simulations. Maximum normalized resolved shear stresses calculated at all FEA nodes from local (multiaxial) stress states. Explicit consideration of elastic stiffness anisotropy.

2.5. Calibration of the Probabilistic LCF Fatigue Model and the Cyclic Material Strength Model

Three probabilistic models are chosen for modeling the strain Wöhler curves of the LCF test results. The first model combines deterministic LCF life modeling using the Coffin-Manson-Basquin equation $CMB(N_i)$

$$\varepsilon_a = \frac{\sigma_f - \sigma_m}{E}(2N_i)^b + \varepsilon_f(2N_i)^c \tag{5}$$

with the Weibull distribution in crack initiation life.

$$F_N(n|\eta, m) = 1 - e^{-\left(\frac{n}{\eta}\right)^m} \tag{6}$$

The deterministic Coffin-Manson-Basquin CMB life N_i is considered as the distributions median value and hence determines the Weibull scale

$$\eta = N_i \cdot (\ln 2)^{-1/m} \tag{7}$$

The Weibull distributed crack initiation life is also the basis for the approach in [19]. There, it is furthermore combined with a surface integration approach to cover the statistical size effect. The latter was not carried out in the presented work. In order to calibrate the CMB parameters for this Weibull based model, a Maximum-Likelihood estimation (MLE) is used. The other two approaches combine the Schmid factor distribution with the CMB equation to derive a life distribution. This is based on the hypothesis that crack initiation life scatter observed in the experiments emanates from the underlying statistical distribution of shear stresses in the crystallites which itself originates in the statistical distribution of grain orientation (see Section 2.4). In order to project the distributional behavior of crack initiating shear stresses to an actual cycle distribution, the approach by [28,30] is followed. The averaged stress in the specimen is calculated from the applied strain with the Ramberg-Osgood model, $RO(\sigma_a)$.

$$\varepsilon_a = \frac{\sigma_a}{E} + \left(\frac{\sigma_a}{K'}\right)^{1/n'} \tag{8}$$

$RO(\sigma_a)$ is inverted such that $\sigma_a = RO^{-1}(\varepsilon_a)$

A large sample set (>200,000 samples) is drawn from the Schmid factor distribution $F_{SF}(\widetilde{m})$ and its median $\widetilde{m}^{50\%}$ renormalized to one. The specimen stress σ_a is then multiplied with this sample set and all values are subsequently transferred back to strain values with the Ramberg-Osgood equation.

$$\varepsilon_a(\widetilde{m}) = RO(\sigma_a(\widetilde{m})) \tag{9}$$

This set of strain samples is then calculated into a set of life samples with the Coffin-Manson-Basquin relationship.

$$N_i(\widetilde{m}) = CMB^{-1}(\varepsilon_a(\widetilde{m})) \tag{10}$$

Equation (10) is the numerical representation for the crack initiation life distribution $F_N(n(\widetilde{m}))$ based on probabilistic Schmid factors. Note, that the procedure described above is applied identically for the distributions $F_{SF}(\widetilde{m})$ and $F_{SF}(\widetilde{m}_{mod})$. For model calibration with MLE, the likelihood summands are calculated from the probability density function.

$$f_N(n(\widetilde{m})) = \frac{d}{dn}F_N(n(\widetilde{m})) \tag{11}$$

Since $F_{SF}(\widetilde{m})$ is only available as a numerical sample set, $f_N(n(\widetilde{m}))$ is differentiated numerically from the empirical cumulative distribution function.

Furthermore, fine and coarse-grained metals have different strengths due to the different intensity of dislocation pile up at the grain boundaries (Hall-Petch relation) [44]. In order to account for this effect, the cyclic stress-strain relation was calibrated by estimating the Ramberg-Osgood model parameters in Equation (8) by means of Maximum-Likelihood. Though the cyclic stress response was assumed to be statistically distributed in a log-normal fashion for simplicity in the MLE, only the median values of the cyclic stress prediction is used in (9).

3. Results

The results of metallographic examination of the specimen material (Section 3.1) and its implications on the FEA modelling (Section 3.2) are described in this section. It follows that different specimen stiffness (Section 3.2) and Schmid factor (Section 3.3) distributions are calculated for the two material batches (coarse grains with random grain orientation and fine grains with preferential grain orientation). By considering the different grain orientation distributions in both batches it is possible to explain the observed crack initiation life differences qualitatively, as well as quantitatively (Sections 3.4 and 3.5). Note that LCF test data from Seibel [29] was used for probabilistic model calibration and its predictions are compared to the fine grain test data generated by Engel et al. [25,31].

3.1. Microscopic Material Examinations and Orientation Distributions

Metallographic investigations using scanning electron microscope (SEM) reveal the typical γ,γ' microstructure with an averaged γ' content of about 35 vol.%, as Figure 3 illustrated for both batches. The average size of the primary cuboidal γ' is 0.4 μm whereas the secondary spherical γ' shows diameters between 10–50 nm.

Due to the different cooling conditions, especially thin cross sections of components made of polycrystalline, Ni-base superalloys tend to form a crystallographic texture on edge layers, due to the preferred grain growth of <100> near orientations in direction of the temperature gradient [45]. The specimen slugs for the data from Engel [25,31] were casted as cylindrical bars of 12 mm in diameter for fine grain realization and 20 mm for coarse grain realization. It is important to note, that the fine grain batch, with an average grain size of about 1 mm, is still coarse with regard to material science, but in order to differentiate it from the material with an average grain size of about 3 mm it is called fine grained batch.

Based on the metallographic etching of the vertical cross section of the 20 mm bar in Figure 4, it is assumed, that a radial and an axial temperature gradient were present during the solidification. The edge layer shows a dendritic solidification with small elongated grains. The 20 mm specimens show large, randomly oriented grains towards the bar center. It is assumed that rapid cooling rates occurred in the edge areas, resulting in small grains. The preferential grain orientation of those might have originated from an axial temperature gradient, as the melt started to cool at the bottom of the upright standing mold. However, it could have originated from crystal seeds at the mold wall as well. With a decrease of the cooling rate towards the bar center, the grains have more time to grow resulting in large grains with no preferential direction. Within the gauge length, they have an averaged diameter of approximately $d_{coarse} = 3$ mm, determined as an equivalent circular diameter of the grain area. This size allows roughly 49 grains in the gauge section. By machining the specimens gauge section geometry (see Figure 1) most of the dendritically solidified material is removed. The mechanical behavior of the specimen is then mainly determined by coarse randomly oriented grains in the gauge length.

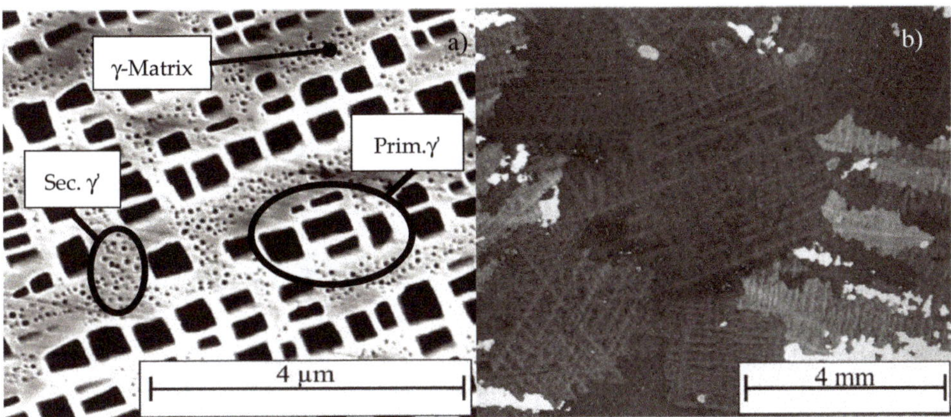

Figure 3. (**a**) Scanning electron microscope (SEM) image of the γ,γ′ microstructure; (**b**) light microscope image of cross section shows dendritic grain growth.

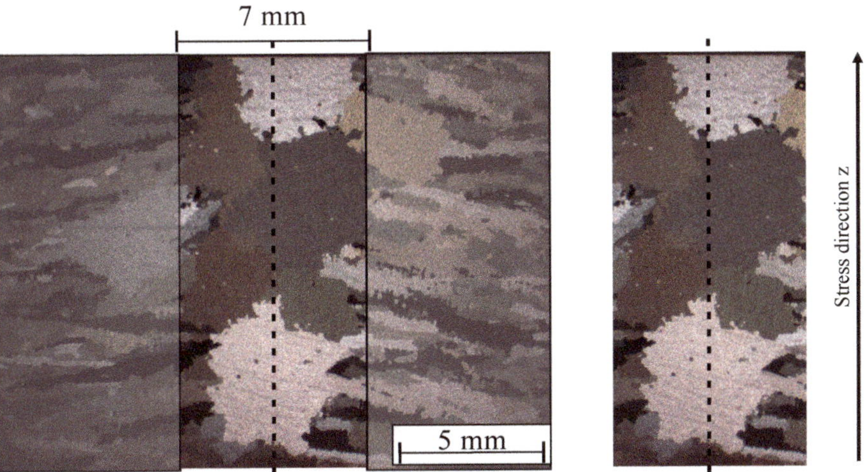

Figure 4. Light microscopy image of the vertical cross section of a 20 mm bar of René80. Coarse grains with random orientation solidified in the center of the gauge section but also fine, dendritically solidified grains are also visible in the edge area. The right-hand side shows the section that remained after specimen machining.

Furthermore, bars of René80 with a length of 150 mm and diameter of 12 mm were cast. The metallographic etching of the vertical cross section in Figure 5 also shows small, dendritically solidified, elongated grains but throughout the entire cross section of the bar. While the axial and radial temperature gradients were most likely comparable to those in the 20 mm bar, the faster cooling in 12 mm bar prevented the formation of larger grains without preferential direction in the center. Due to the large grain size, the determination of the orientation distribution of the material using electron backscatter diffraction (EBSD) was not suitable as only an insufficient number of grains could be examined, which provided inadequate statistics. Therefore, the alignment within the gauge length were measured by light microscope and image processing. On average, the grains' <100> direction is aligned to the specimen horizontal in an angle ϑ of approximately 25°.

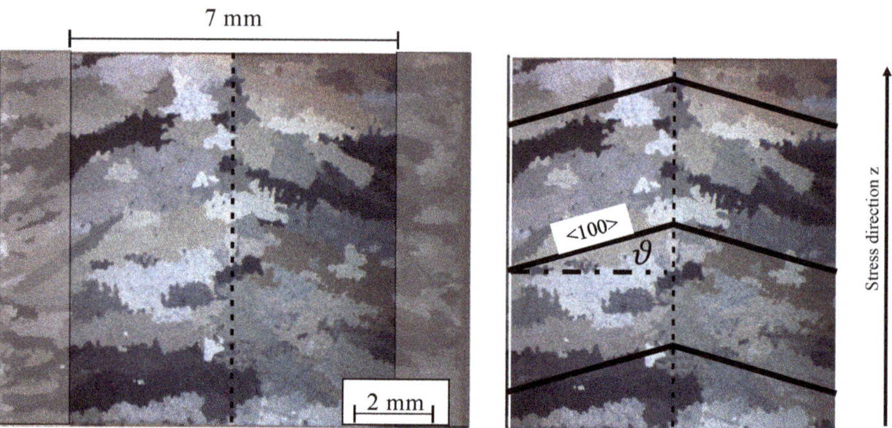

Figure 5. Light microscopy image of the vertical cross section of a 12 mm bar of René80. A preferential orientation of the grains was likely caused by the temperature gradients in the mold during solidification.

The dendritic character of the grains is present in the entire gauge length. Thus, it is assumed that the mechanical behavior is significantly determined by this texture. In contrast to the coarse-grained material, the grains appear rather lengthy in the cross section with an average grain size of $d_{fine} = 1.3$ mm. This relates to approximately 500 grains in the gauge section.

The LCF test data from cylindrical bar slug specimens is shown in Section 3.5. However, the probabilistic model calibration presented there uses only LCF test data from Seibel [29] where the specimen slug was casted as a plate with a thickness of 20 mm. The materials microstructure is comparable to the just described coarse-grained bar material regarding grain size distribution and morphology. Therefore, a similar mechanical and fatigue behavior can be assumed. The consistency of properties is proven in Sections 3.2 and 3.5.

3.2. Results of the Isothermal LCF Tests at 850 °C

Figure 6 shows the cyclic stress-strain data in the stabilized regime and the calibration curves of the respective Ramberg-Osgood model.

It is found that the average cyclic Young's moduli of the test specimens (global values from hysteresis using extensometer) are different. The value E_{fine} from the fine-grained batch with preferential grain orientation is 9.8% lower than the value of E_{coarse} with approximately random grain orientation ($E_{coarse}/E_{fine} \approx 1.11$). The Ramberg-Osgood model calibration using the cyclic stress strain data was conducted using only the cyclic strength coefficient K' as a degree of freedom. The Young's moduli from the experiment were set according to the experimental data of the respective batch. Since the amount of data points in the plastic deformation regime was insufficient for a reliable description of the hardening slope determined by the exponent n', an n' value for René80 was taken from the Siemens proprietary material data base. The derived K' values differ in the ratio $K'_{coarse}/K'_{fine} \approx 1.07$ meaning the coarse-grained material has also a higher plastic strength than the fine-grained. The results for the experimentally derived Young's moduli and the calibrated cyclic strength coefficient are discussed in Section 4.1.

The cyclic stress-strain plot in Figure 6 indicates that the coarse grain batches from both data sets (Engel and Seibel) show comparable mechanical behavior as assumed in Section 3.1. The crack initiation life data derived from the 2.5% load drop as crack initiation criterion is presented as strain-life Wöhler plot in Figure 14 for all data sets from Engel [25,31] and Seibel [29]. It is found that the fine grain batch specimen withstands a significantly higher number of cycles at equal strain level. The detailed examination of this findings and graphs of the respective data are presented in Section 3.5.

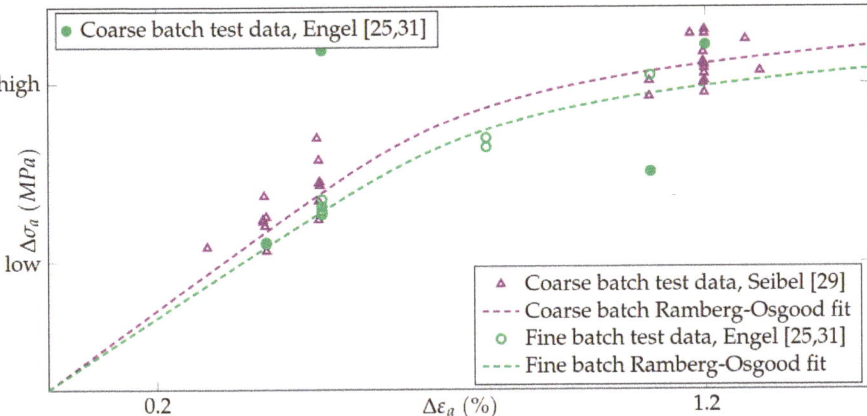

Figure 6. Stabilized cyclic stress response data and respective Ramberg-Osgood calibration curve for coarse grain batch and fine grain batch.

3.3. Results of the Finite Element Simulation

Linear elastic simulations of polycrystalline FEA models were carried out to model the mechanical interactions during an LCF test in a coarse-grained Ni-base superalloy. Therefore, an anisotropic elasticity model of IN 738 LC at 850 °C was applied. Figure 7 shows the computed stress and strain distribution in z-direction for the coarse-grained batch model (49 grains with random orientation distribution) for a globally applied strain of 0.25%. The stress distribution indicates maximal values with more than 600 MPa near some grain boundaries as seen in (a). Most parts of the surface show a significantly lower stress of about 300 MPa. Considering a homogenized Young's modulus of IN 738 LC of $E = 152$ GPa a stress of $\sigma_{iso} = 379$ MPa is expected for isotropic modeling.

In contrast to an isotropic material behavior, high stresses do not simultaneously lead to high strains in this anisotropic simulation, as obvious for area (b) in Figure 7. Area (c) shows both, high stresses and high strains. The local strains can vary compared to the globally applied strain by nearly a factor of 2. Hence, local plastic deformation may occur in some grains, despite the globally elastic response. The yield of these grains is however not modelled explicitly. Furthermore, also inhomogeneous stress and strain states within individual grains are observed. Figure 8 shows this for a free cut grain of the previously described specimen.

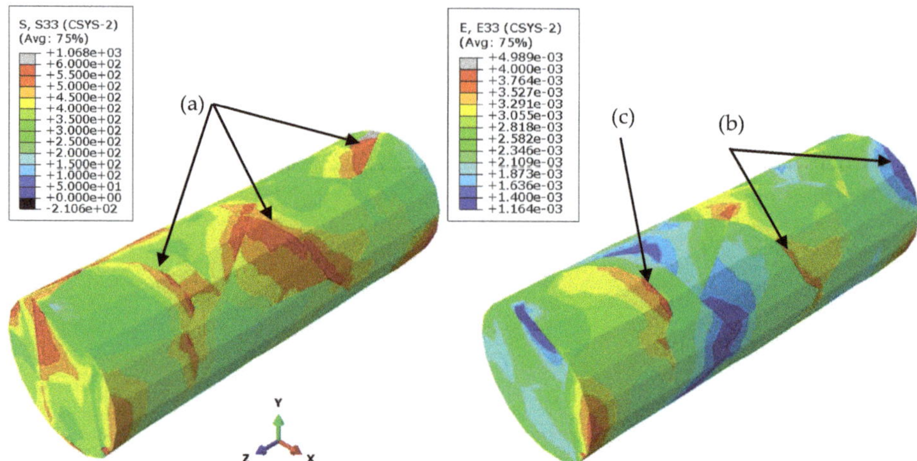

Figure 7. Stress and strain distribution in loading direction for a specimen with 49 grains and random orientation at 0.25% total strain. An anisotropic elasticity model of IN 738 LC at 850 °C was applied.

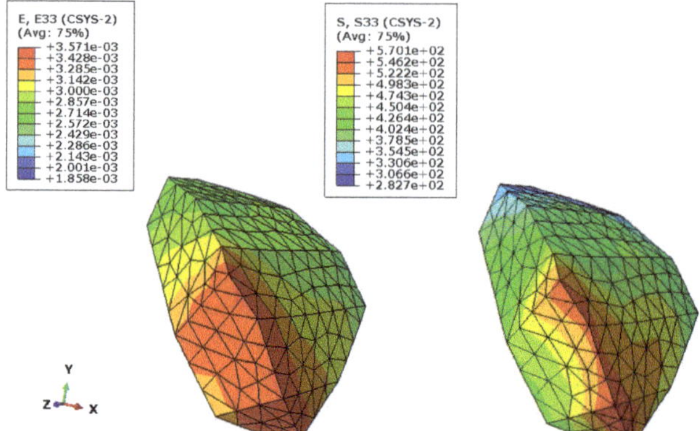

Figure 8. Stress and strain distribution in loading direction of a free cut grain from the specimen model with 49 grains at 0.25% total strain and 850 °C.

The examination of the individual nodes in Figure 8 clearly shows that stress and strain are also inhomogeneously distributed within the grains. An evaluation of the stress tensors at each node shows that a globally applied uniaxial stress leads to local multiaxial stress and strain states within the grains. The unidimensional parameter κ is introduced for describing the multiaxiality of the stress tensor using the principal components of the stress deviator $\underline{\underline{\sigma}}'$.

$$\text{For } |\sigma'_I| \geq |\sigma'_{II}| \geq |\sigma'_{III}|, \ \kappa = \frac{|\sigma'_{III} - \sigma'_{II}|}{|\sigma'_I|} \tag{12}$$

For $\kappa = 0$, the stress state at the node is equal to a uniaxial load, while $\kappa = 1$ is equivalent to a stress state where $|\sigma'_I| = |\sigma'_{III}|$ which is equal to a full torsional loading on the node. Figure 9 shows the histograms of all κ values calculated from the nodes of both FEA models, the coarse grain model (random grain orientation) and the fine grain model (preferential grain orientation).

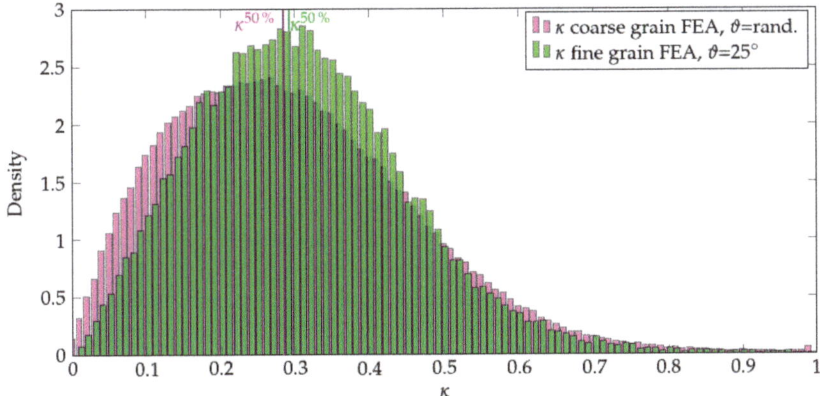

Figure 9. Frequency distribution of the parameter κ to evaluate the stress state on every simulated node of both specimen models.

The histograms indicate a continuous distribution of κ values across the entire range. No node in the globally uniaxial loaded specimen has a real uniaxial stress state with $\kappa = 0$ but also none with $\kappa = 1$. Although the κ distributions differ in dispersion, the median quantiles are similar. $\kappa^{50\%} = 0.285$ for the coarse grain model and $\kappa^{50\%} = 0.293$ for the fine grain model.

As described in Section 2.3, a preferential orientation of the grains was simulated in the fine-grained FEA model (500 grains) in order to accommodate the corresponding observation at the tested material described in Section 3.1. Hence, the rotation matrices were set up such that all grains align their <100> direction at an angle of $\vartheta = 25\,^\circ$ towards the specimen horizontal. This correlates approximately to the grain alignment observed in the metallographic etchings (see Figure 5). The Euler angles φ_1, φ_2 however were assumed to be uniformly distributed as no further distribution information could be derived from the metallographic examination. Figure 10 shows the stress and strain distribution in z-direction computed for the fine-grained batch model for a globally applied strain of 0.25%.

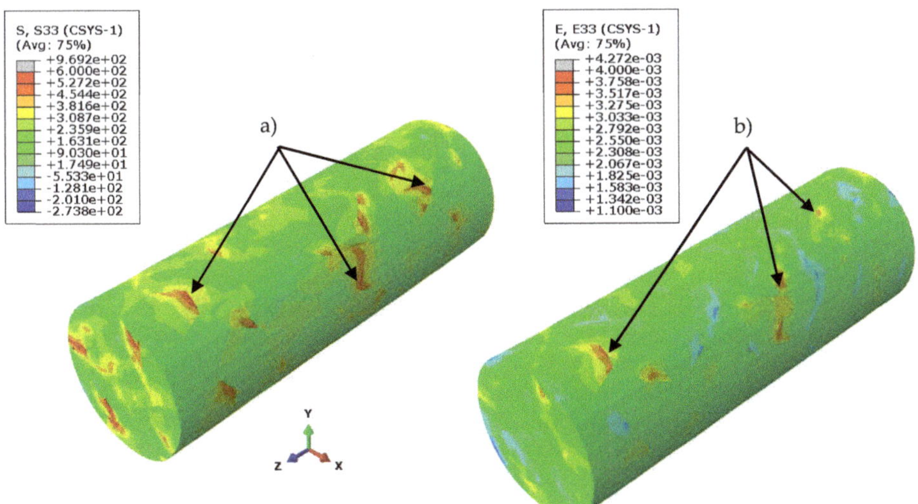

Figure 10. Stress and strain distribution in loading direction for the specimen model with 500 grains and directed orientation distribution at 0.25% total strain. An anisotropic elasticity model of IN 738 LC at 850 °C was applied.

It can be observed, that the surface distribution of stresses is generally more homogeneous than in the coarse-grained material (cf. Figure 7). Small maxima are located at the grain boundaries, as seen in area (a). Large areas of the surface show a stress in loading direction close to the average value of 200 MPa. Furthermore, the strains in loading direction are also very homogeneously distributed and close to the globally applied strain of 0.25%. Occasionally higher values occur in small areas at the grain boundaries as seen in (b). There, they can reach values up to 0.4%. The examination regarding the local stress state multiaxiality is also presented as a histogram for κ in Figure 8.

Comparing the stress fields of both specimen simulations (coarse-grained, isotropically distributed grain orientation with $\vartheta = rand.$ and fine-grained, preferential grain orientation, $\vartheta = 25°$) it becomes apparent, that peak stress levels in the fine-grained model are comparable to the coarse-grained model. The explanation for both observations is discussed in Section 4.1. A global Young's modulus E_{global} is introduced for comparing the effective stiffness of the specimen models in loading direction.

$$E_{global} = \frac{\overline{\sigma_{zz,node}}}{\varepsilon_t} \tag{13}$$

Here, $\overline{\sigma_{zz,node}}$ is average of all nodal stresses in loading direction z (see coordinate system in Figure 7.) of all simulated specimens and ε_t is the applied total strain. The following Table 3 shows the results for E_{global} in dependence of the orientation distribution.

Table 3. Comparison of the determined global Young's moduli from polycrystalline Finite Element Analysis (FEA).

Value	Random Orientation $\vartheta=rand.$	Preferential Orientation $\vartheta=25°$	Shift
E_{global}	160 GPa	142 GPa	−11%
Standard deviation	±1.5 GPa	±0.3 GPa	-

The global Young's modulus of the textured material model is 18 GPa lower than for the model with randomly oriented grains.

3.4. Results of the Schmid-Factor Distribution Calculations

The resulting shear stresses at the slip systems $\tau_{i,j}$ were calculated for both FEA models, using Equation (3). Figure 11 shows the distribution of the maximum resulting shear stresses $\max_{i,j}\left(\tau_{i,j}\right)$ at each node of the FEA models. The one with large, randomly orientated grains to the left and the one with smaller, preferentially directed grains to the right. As in Section 3.3, the total applied strain was 0.25% and an anisotropic elasticity model of IN 738 LC at 850 °C was applied.

The spots marked with (a) clearly indicate local minima of the resulting shear stress distribution in the slip systems of the grains. Here, a purely elastic behavior can be assumed because $\tau_{res} < \tau_{crit.}$. These spots occur much more frequent at the surface of the textured specimen model. In addition, the spot in area (b) shows an example of clearly increased shear stresses in the slip systems. There, dislocation movement, i.e., local plastic deformation can be expected. These areas with significantly increased shear stresses were not found at the surface of the fine-grained textured FEA model. Most regions at the coarse-grained model show an average value of $\tau_{rss} \approx 250$ MPa, while the resolved shear stresses of the fine-grained FEA model fluctuate between $\tau_{rss} \approx (180–200)$ MPa. Values for the critical resolved shear stresses of René80 are not known to the authors and would have to be determined in further experiments. However, Nitz and Nembach [46] present values of the critical resolved shear stress (crss) for different crystal orientations for the Ni-base superalloy Nimonic 105 (single-crystalline) under compression loads. The highest crss at 850 °C was measured with in [110] direction (325 MPa), the crss for the [111]-orientation is slightly lower (319 MPa) and in [100] it was measured to approximately 290 MPa. Österle, et al. showed for the Ni-base superalloy single crystal

SC16 that besides octahedral slip, which is generally assumed in this paper, especially for grain orientations near [111] cubic slip can occur at very high temperatures. The reason is a higher Schmid factor of 0.46 occurring at the [011](100)-slip system compared to 0.293 at the [111](110) slip systems [47]. Due to the statistical rarity, as shown in [31], [111]-orientations with low corresponding octahedral Schmid factors of 0.29 only occur with a probability of <0.5% and are therefore neglected in this work.

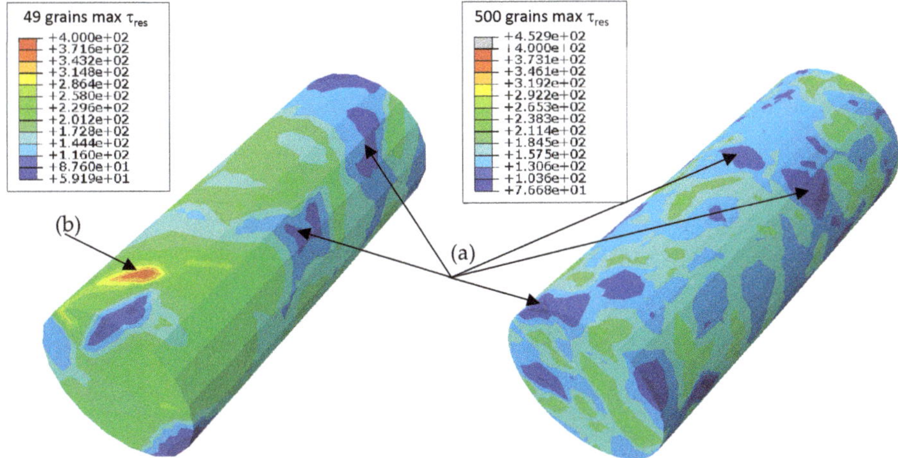

Figure 11. Distribution of maximum resulting shear stress in the <111>{110} slip systems at total strain of 0.25%. Coarse grain morphology and random orientation to the left and fine grain morphology with preferential orientation ($\vartheta = 25°$) to the right.

Figure 12 shows the z-component of the stress (middle) and maximum resolved shear stress (left) distribution for three grains of the coarse-grained FEA model (49 grains, random orientation). Grain 1 and 2 show the previously discussed high stresses near the grain boundaries (area a). However, the maximum resolved shear stresses at these nodes are low, due to the local grain orientation with corresponding low values for the local Schmid factors. Grain 3 shows high stress and high corresponding resulting shear stress. Therefore, a high local Schmid factor can be observed for this grain.

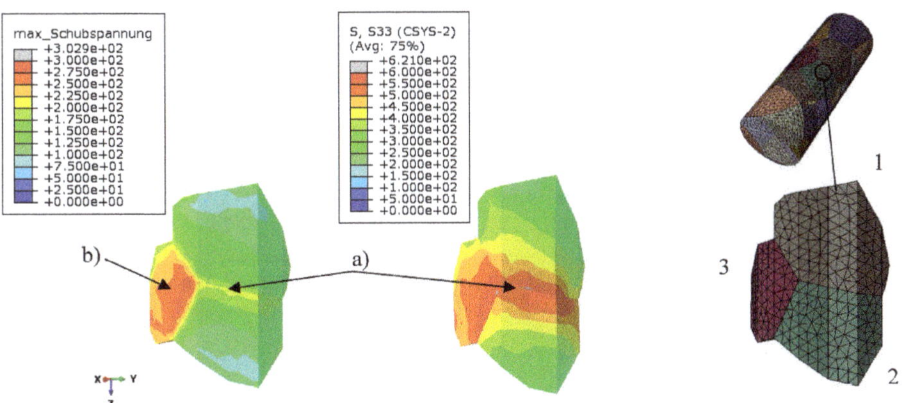

Figure 12. Distribution of maximum resolved shear stress at free cut grains in the [111] (110) slip systems for globally applied strain of 0.25%.

Figure 12 also shows that the inhomogeneous stress field at the surface nodes leads to an inhomogeneous shear stress distribution, i.e., pronounced shear stress gradients across the grain. As presented in Section 3.3, the polycrystalline FEA simulations have confirmed the observed differences in the global Young's modulus between the two specimen models (randomly oriented grains and preferentially oriented grains). It was also of interest whether a systematic effect of grain orientation could be found for the crack initiation life. For that reason, the Schmid factor distributions for both polycrystal realizations were computed with the two approaches explained in Section 2.4.

In the first approach two Schmid factor distributions $F_{SF}(\widetilde{m})$ are created by rotating a single crystal lattice using Monte-Carlo sampled rotation matrices $U(\vartheta, \varphi_1, \varphi_2)$ from the isotropic distribution of orientations and calculating the Schmid factor according to Equations (2)–(4). This corresponds to the coarse-grained material batch. The effect of the preferential grain orientation in the fine grain batch (see Section 3.1) was modelled by creating a distribution of rotation matrices $U(\varphi_1, \varphi_2, \vartheta = 25°)$ with fixed Euler angle ϑ to 25° while the other Euler angles φ_1 and φ_2 were uniformly distributed. This corresponds to the observation of the <100> alignment to the specimen horizontal and the assumption of no preferential orientation of the other crystal directions.

In the second approach, the Schmid factors were calculated from the nodal stress states in the three different realizations (different grain rotations) of the two polycrystalline FEA models. The resulting Schmid factor distribution densities of the two modeling approaches are compared to each other in Figure 13.

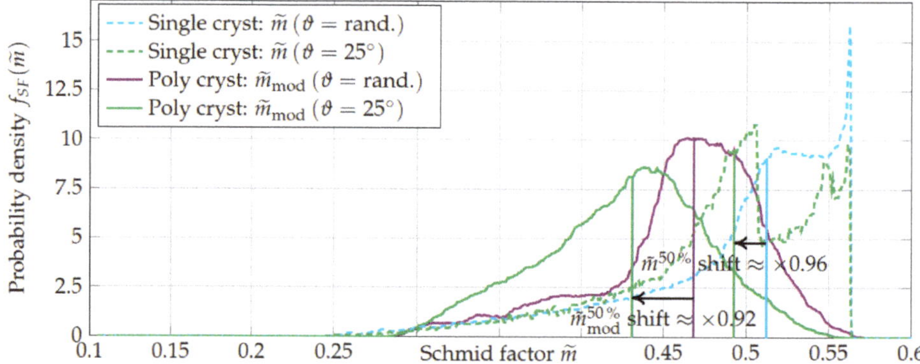

Figure 13. Comparison of the Schmid factor distribution density functions for a single crystal $f_{SF}(\widetilde{m})$ and a polycrystal $f_{SF}(\widetilde{m}_{mod})$ (from FEA). The random Euler angle ϑ (coarse grain batch) refers to isotropically and $\vartheta = 25°$ (fine grain batch) refers to preferentially distributed grain orientations. A uniaxial stress was applied to all models. Vertical lines indicate the median values.

Figure 13 visualizes that the Schmid factor distribution densities derived from polycrystalline FEA simulations are less structured, resulting in lower median values than the densities from the first, purely geometric approach (see Table 4). The distributions from nodal Schmid factors are also broadened, especially towards the upper tail. Particularly the distribution for a randomly oriented single crystal at purely uniaxial load has a concentration at high Schmid factors. Values larger than 0.5 are reached because of the elastic anisotropy. This is discussed in more detail in Section 4.2. Figure 13 further shows that the median values of the Schmid factor distributions are shifted to lower values for the cases where a preferential orientation was modeled. The median shift for $\widetilde{m}^{50\%}_{mod}$ (FEA model approach) is more significant.

Table 4. Comparison of the median values of the Schmid factor distributions $f_{SF}(\widetilde{m})$ (geometric approach) and $f_{SF}(\widetilde{m}_{mod})$ (from FEA).

Schmid Factor	Random Orientation $\vartheta = random$	Preferential Orientation $\vartheta = 25°$	Shift
$\widetilde{m}^{50\%}$	0.512	0.493	−4%
$\widetilde{m}_{mod}^{50\%}$	0.468	0.431	−8%

These differences will affect the calibration and prediction result presented in the following Section 3.5.

3.5. Procedure of the LCF Life Calibration and Prediction

Three different probabilistic fatigue models were created and calibrated only with the LCF data of the coarse-grained batch (random grain orientation). They are then used for prediction the LCF life distribution of the fine-grained batch (preferential grain orientation). All values commonly use the CMB model (2.5) as deterministic baseline for the median curve modeling but use different distributions. They are listed Table 5 below.

Table 5. Comparison of crack initiation life distribution densities for the applied probabilistic crack initiation model approaches.

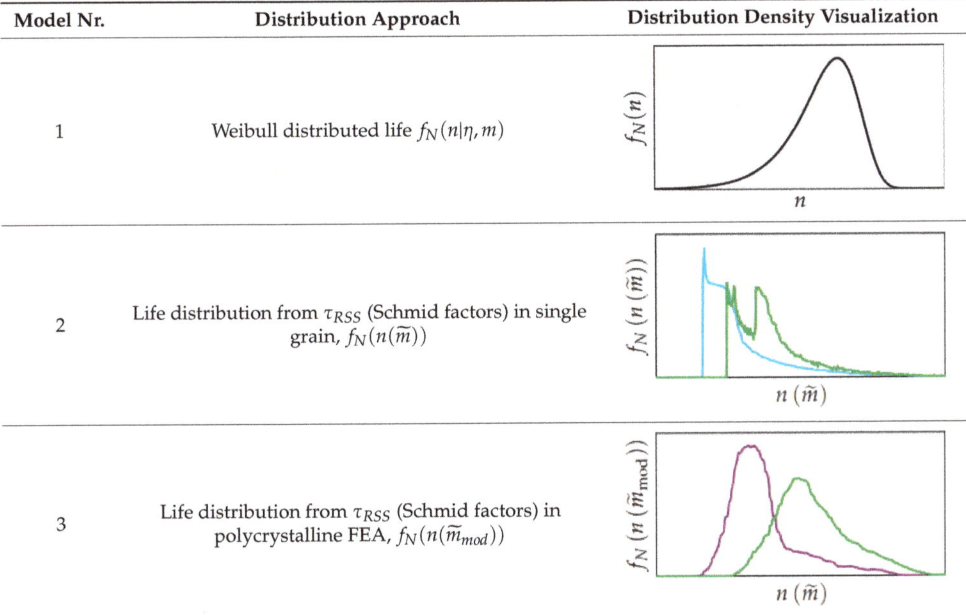

Model Nr.	Distribution Approach	Distribution Density Visualization
1	Weibull distributed life $f_N(n|\eta, m)$	
2	Life distribution from τ_{RSS} (Schmid factors) in single grain, $f_N(n(\widetilde{m}))$	
3	Life distribution from τ_{RSS} (Schmid factors) in polycrystalline FEA, $f_N(n(\widetilde{m}_{mod}))$	

The aim of creating these three models is to evaluate their applicability for predicting the observed differences in crack initiation life between the two tested batches. The latter two Schmid factor-based models are taking the grain orientation into account. The applicability of each model is tested by visual comparison of calibration and prediction curves, and the associated negative log-Likelihood. The life differences of the strain-Wöhler curves are quantified at a strain range $\Delta\varepsilon_{comp} = 0.6\%$, which is common for all three analyses. Equivalently to the Ramberg-Osgood model calibration, all CMB model calibrations use fixed exponents b and c as not enough test points in the plastic deformation regime exist to confidently calibrate a slope for the elastic-plastic branch c which in return largely influences the slope b of the elastic branch when calibrated. Hence, Siemens proprietary values for b and c for René80 are used here as well and only the cyclic fatigue strength coefficient σ'_f and the cyclic fatigue ductility coefficient ε'_f are calibrated with the coarse-grained batch data and the same values are used for predicting the fine-grained batch data. Furthermore, all calibrations and predictions were conducted using the respective average value for the Young's modulus of each batch. During the calibration of the model CMB with Weibull distribution, the Weibull shape value m is also calibrated at the test data. The presented strain-Wöhler curves are drawn by interpolating the median values of the respective LCF life distribution along the strain axis.

Prediction of the Wöhler Curve from a Weibull Distribution

The strain-life Wöhler plot in Figure 14 contains all LCF data from the present work and the LCF data set from Seibel [29]. As mentioned in Section 3.2, the data of the coarse-grained batch generated by Engel [25,31] and the presented data of Seibel [29] show comparable fatigue behavior due to the similar microstructure and grain size. Moreover, the grains in both material batches (Seibel and Engel-coarse) seemed to be oriented randomly, according to the isotropic distribution of orientations. Since the data set from Seibel contains many more values, only those are used for calibrating the probabilistic LCF models. The other coarse grain batch test points from Engel [25,31] are therefore not shown in the further strain-life Wöhler plots. Figure 14 shows the test points as well as the calibrated and predicted strain Wöhler curves given by the median of the Weibull LCF life distribution.

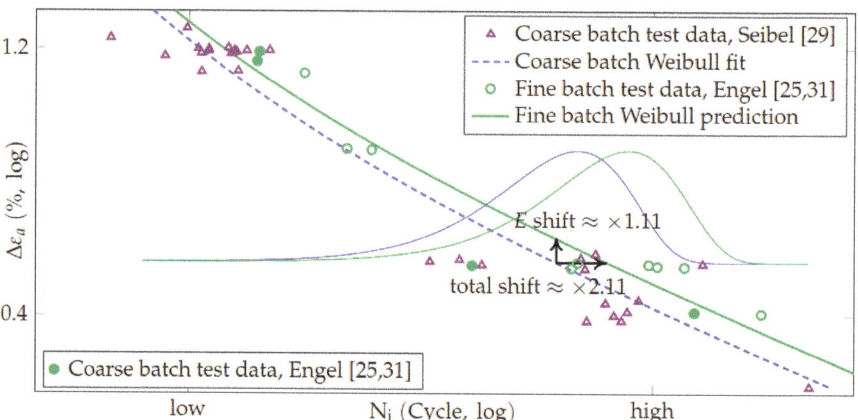

Figure 14. Fit and prediction curve for the model: Weibull-distributed Coffin-Manson-Basquin life.

The CMB model was calibrated with the coarse-grained batch test data. The LCF life distributions are qualitatively equal (relative dispersion) for both data sets since the Weibull shape value calibrated with the coarse grain batch data set is also used for the prediction curve of the fine grain batch. The difference between fit and prediction curve results only from the different values of global Young's modulus. The lower value of E_{fine} compared to E_{coarse} ($E_{coarse}/E_{fine} \approx 1.11$) causes a positive shift of the fine batch prediction curve in $\Delta\varepsilon_a$ direction and effectively results in shift towards higher LCF life by a factor of 2.11 at $\Delta\varepsilon_{comp}$. Hence, the Weibull model explains the increased crack initiation life observed in the strain controlled LCF experiments only with the decreased stiffness of the fine grain batch material (preferential grain orientation).

$$N^{50\%}_{fine} \approx 2.11 \cdot N^{50\%}_{coarse}$$

Prediction of the Wöhler Curve from the Single Grain Schmid Factor Distribution

Figure 15 shows the strain Wöhler curves given by the median values of the LCF life distributions $F_N(n(\widetilde{m}))$ calculated according to Equations (9) and (10) from the Schmid factor distributions $F_{SF}(\widetilde{m})$. One was computed for random single grain orientation (for coarse batch) and the other for preferential single grain orientation (for fine batch) as described in Section 3.4.

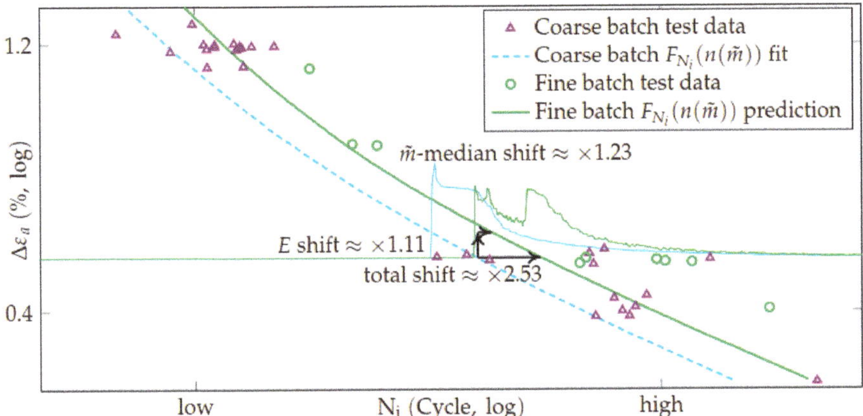

Figure 15. Fit and prediction curve for the CMB with single grain Schmid factor distribution model.

The CMB model was also calibrated only with the coarse batch test data. The lower E value of the fine grain batch causes the prediction to shift in positive $\Delta\varepsilon_a$ direction. Additionally, the effect of the grain orientation becomes apparent. For the modeled preferential orientation, the median value of the Schmid factor distribution $\widetilde{m}^{50\%}$ is 4% lower than for an entirely random grain orientation (see Figure 13). This is visible as a second, positive shift of the life distribution $F_N(n(\widetilde{m}))$. Altogether, the total effective life shift factor between coarse and fine batch median is 2.53 at $\Delta\varepsilon_{comp}$. It is already apparent that both curves are located too far left of the test point clouds. The reason for that is the shape of the underlying distribution which is discussed in more detail in Section 4.3.

$$N_{fine}^{50\%} \approx 2.53 \cdot N_{coarse}^{50\%}$$

Prediction of the Wöhler Curve from the Modified Schmid factor Distribution

Figure 16 shows the strain Wöhler curves given by the median values of the LCF life distributions $F_N(n(\widetilde{m}_{mod}))$ calculated according to Equations (9) and (10) from the modified Schmid factor distributions $F_{SF}(\widetilde{m}_{mod})$ derived from the coarse FEA model (random grain orientation) and fine FEA model (preferential grain orientation) described in the Sections 3.3 and 3.4.

As before, the CMB model was calibrated with the coarse batch test data. The difference between fit and prediction curve originates on one hand from the different values of global Young's modulus between both batches and on the other hand significantly from the different median values $\widetilde{m}_{mod}^{50\%}$ of the modified Schmid factor distributions $F_{SF}(\widetilde{m}_{mod})$. The Schmid factor distribution from the fine-grain FEA model where a preferential orientation of <100> aligning with $\vartheta = 25°$ to the specimen horizontal, the median value of the Schmid factor distribution $\widetilde{m}_{mod}^{50\%}$ is 8% lower than for an entirely random grain orientation (see Figure 13). The total median life is shifted by a factor of 3.6 at $\Delta\varepsilon_{comp}$.

$$N_{fine}^{50\%} \approx 3.6 \cdot N_{coarse}^{50\%}$$

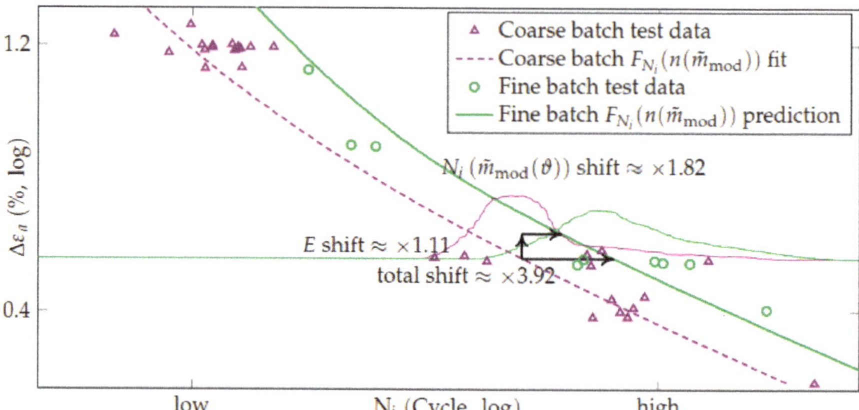

Figure 16. Fit and prediction curve for the CMB with modified Schmid factor distribution model.

4. Discussion

4.1. Influence of the Grain Orientation Distribution on the Mechanical Properties

The tested René80 batches were different with respect to grain size and grain orientation distribution. It was observed that specimens of the fine-grained material batch which had a preferential grain orientation (see Section 3.1) showed lower global Young's moduli (in loading direction) in average. An equivalent observation was made for the polycrystalline FEA models whose grain orientations were sampled according to the assumed distributions in the two batches (isotropically distributed orientations for the coarse-grained batch and preferentially oriented for the fine-grained batch). Both orientation distribution density functions (ODF) are visualized as inverse pole plot in the left column of Figure 17. The right column shows the histograms of local stiffnesses in z-direction calculated by Equation (13). For both orientation cases, σ_{zz} was once derived from the respective polycrystalline specimen FEA results and once from a single crystal Monte-Carlo-simulation. For these, a single crystal with the elastic anisotropy of IN 738 LC at 850 °C was rotated according to the respective ODF (100,000 times) and an isotropic strain load was applied to it.

The global Young's moduli averaged from nodal FEA stresses (z-components only) have a difference of 18 GPa between the batch with (coarse) and without preferential grain orientation (fine). The ratio $E_{coarse}/E_{fine} = 1.13$ of the Young's moduli averaged from FEA results is slightly larger for the experimental results where $E_{coarse}/E_{fine} = 1.11$. Although also the absolute values of the tested Young's moduli are slightly lower than the averaged computed values the comparison of the ratios indicates a good qualitative consistency. That is a satisfying outcome as it confirms the applicability of the elasticity model from IN 738 LC for René80 with respect to the anisotropies at 850 °C. Nevertheless, the uncertainties in the determination of the Young's moduli must not be neglected. The computed average value is based on FEA solutions considering just three different grain orientation realizations at one grain morphology. The simulated grain orientation distributions are also an approximation derived from the results of a limited microstructure evaluation. Particularly approving the assumption of uniformly distributed Euler angles φ_1, φ_2 would need further investigations. Moreover, the scatter in experimentally determined Young's moduli is significant, particularly for large grain realizations. Only the grains between the extensometer tips (distance of 12 mm) are influencing on the measurement results of Young's moduli. As investigations in [31] reveal, the measured Young's moduli strongly depend on the extensometer position. Different measurements positions along the gauge length lead to results in a window of ±18 GPa for a coarse-grained specimen. Further polycrystalline FEA simulations in combination with a virtual 12 mm extensometer revealed differences of up to 20 GPa between the Young's modulus derived from this virtual extensometer strain and the globally averaged

Young's modulus considering the entire volume. The reason for that is the large stiffness anisotropy of the René80 lattice. Many grains are required to homogenize the materials global stiffness (Young's modulus) over a certain volume of interest [48]. Due to the lower standard deviation of local stiffnesses in the case of preferentially oriented grains (see green histogram in Figure 17b), homogenization and quasi-isotropic mechanical behavior is more pronounced at lower grain numbers already.

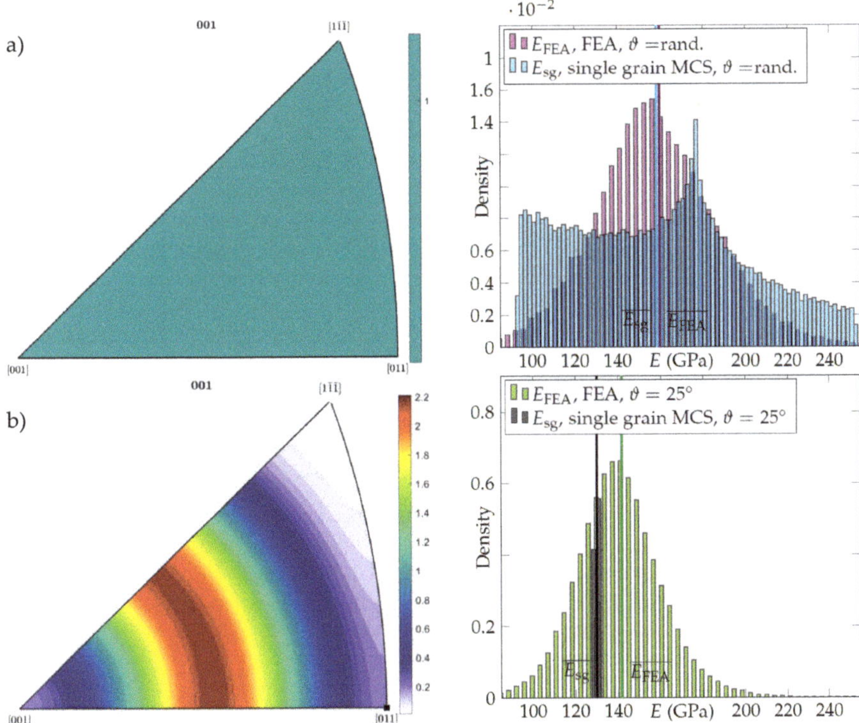

Figure 17. Orientation distribution functions and resulting local stiffness distributions (z-direction). (**a**) Isotropically distributed grain orientation: Resulting local stiffness distribution from a single crystal Monte-Carlo simulation (E_{sg}, cyan) and from polycrystalline FEA results (E_{FEA}, violet). (**b**) Preferential grain orientations: Resulting local stiffness distribution from a single crystal Monte-Carlo simulation (E_{sg}, gray) and from polycrystalline FEA results (E_{FEA}, green).

A Monte-Carlo simulation on different orientations of a single-grain was conducted to study the pure effect of orientation at the stiffness distributions. Figure 17a shows the resulting z-stiffness distribution (in z-direction) of a single crystal with random orientation (as for the coarse-grained material). The values vary between 100–260 GPa because orientations leading to low and high stiffness in loading direction are equally likely. The equivalent distribution of local stiffnesses from FEA in comparison is less dispersed but centered around a similar average. The single grain z-stiffness distribution in subfigure Figure 17b) however shows a very different result. The ODF was constructed as described in Section 3.5. This results in a slim peak band in the ODF with values > 2 (left column). Hence, also the distribution of z-stiffness from all single grain rotations is hardly dispersed and varies only between 126–129 GPa. Especially orientations near [111], which correlate to high stiffness and orientations close to [001], with the lowest stiffness are unlikely for this orientation distribution. The equivalent distribution of local stiffnesses from FEA in is now much more dispersed, has a higher average and resembles a normal distribution. The latter is also reflected in the stress fields shown in Figure 7; Figure 10 (Section 3.3). Despite discriminated contact stresses (gray) at the top and bottom surface comparable surface stresses are observable in Figure 7 at the coarse-grained model and at the fine-grained model in Figure 10. All boundary conditions (material law, temperature, applied strain load) were identical. The mutual mechanical grain interaction which is considered in the polycrystalline FEA leads to different multiaxial stress tensors across the specimen and even within single grains (Figure 8). Due to the high elastic anisotropy of the material (anisotropy up to a factor of 3 [31]), agglomerations of grains with different stiffnesses in load direction undergo strain restrictions near the grain boundaries, which result in the observed inhomogeneities and high peak stresses. These peak stresses emerging from grain interactions are the reason why the resulting local stiffnesses from the FEA model with preferential grain orientation are so different from the single grain simulation.

Besides the elastic anisotropy and the related consequences, which were observed in the experiments and FEA simulations, also a difference in the elastic-plastic strength of the examined material batches was observed. Calibration of the Ramberg-Osgood models returned cyclic strength coefficients with $K'_{coarse}/K'_{fine} \approx 1.07$. At first sight, this contradicts the typically occurring hardening effects of grain refinement, known as Hall-Petch relationship [44]. However, the difference in average grain size ($d_{coarse} = 3$ mm, $d_{fine} = 1.3$ mm) is not significant for such effects and the grain orientation distribution is more likely to play a role, similarly as for the elasticity properties. The onset of plastic deformation is largely determined by the critical resolved shear stress τ_{crss}. Nitz [46] and Österle [47] presented orientation dependent values for τ_{crss} of Ni-base superalloys at high temperature but the investigated materials were not sufficiently comparable to René80. Therefore, no simulation study using an orientation-dependent elastic-plastic material model could be conducted equivalently to the procedure presented for the elastic mechanical behavior.

4.2. Influence of Grain Orientation Distribution on the Fatigue Behavior

All Schmid factor distributions shown Figure 13 have non-zero likelihood values $\widetilde{m} > 0.5$, even for the simulation of a single grain under uniaxial stress. The reason for that is definition (2) identifying the maximum resolved shear stress at the slip systems as Schmid factor combined with anisotropic elasticity. Uniaxial external stresses can then translate into multiaxial stress states in the crystal leading to quotients $\tau_{i,j}/\sigma_{vM} > 0.5$. Note, that σ_{vM} is always calculated from the external stress tensor. Consider the following example case:

$$\varphi_1 = 304.9°, \ \varphi_2 = 341.6°, \ \vartheta = 112.0° \text{ and } \vec{\sigma}_{ext} = (1,0,0,0,0,0)^T \text{ in Voigt notation}$$

In the case of isotropic elasticity, the rotated stress tensor deviator $\underline{\underline{\sigma}}'_{iso}(\boldsymbol{U})$ results to

$$\vec{\sigma}'_{iso}(\boldsymbol{U}) = (0.076, 0.172, -0.248, 0.208, -0.187, -0.454)^T \ (\text{Voigt notation})$$

$\vec{\sigma}'_{iso}(\boldsymbol{U})$ is the deviator of a uniaxial stress state ($\kappa = 0$) and leads to $\widetilde{m} = 0.43$. Considering the anisotropic elasticity of IN738 LC at 850 °C however yields to the multiaxial stress state deviator

$$\vec{\sigma}'_{aniso}(\boldsymbol{U}) = (0.044, 0.099, -0.143, 0.368, -0.331, -0.805)^T$$

The multiaxial stress state $\vec{\sigma}_{aniso}(\boldsymbol{U})$ in the material yields a Schmid factor $\widetilde{m} = 0.56$. In both cases, the same slip system (1–11) [011] is activated. Furthermore the Schmid factor distributions shown in Figure 13 differ significantly for the two modeling approaches polycrystalline FEA and single grain Monte-Carlo simulation (MCS). The higher dispersion of the distributions $F_{SF}(\widetilde{m}_{mod})$ calculated from nodal FEA stress tensors is attributed to the effect of grain interactions. The thereof created multiaxial stress states create the observed, broadened Schmid factor distributions. The major difference to those created from single grain MCS, the shallow, not steeply descending right tail, is consistent with the studies of Moch [30]. There, it was shown that the steep right tail of the Schmid factor distribution for a single grain under uniaxial load decreases and flattens with increasing stress multiaxiality. The steep right tail of $F_{SF}(\widetilde{m})$ leads to a steep left tail of $F_N(n(\widetilde{m}))$ since $F_N(n(\widetilde{m})) = 1 - F_{SF}(\widetilde{m})$ due to the inverse proportionality of shear stresses and life cycles (compare Figure 13 and Table 5).

Furthermore, the inhomogeneous resolved shear stress distributions exemplarily shown in Figures 11 and 12 lead to different probabilities of crack initiation within the grains in a polycrystal compound. Hence, the polycrystalline FEA based modelling approach even indicates the location within a grain while the E·m model only determines which grain shows the highest shear stresses under a given stress condition.

Even more important, the determined grain alignment in the fine-grained batch (preferential orientation distribution) also aligns the slip systems such that lower median values result for the distributions $F_{SF}(\widetilde{m}_{mod})$ and $F_{SF}(\widetilde{m})$. Therefore, higher local stresses or more load cycles are in average required to trigger shear glide with the same intensity as for random grain orientation. Additionally, it was explained how the observed preferential grain orientation also lead to lower global Young's moduli. That again decreases the resolved shear stress at the slip systems and retards shear glide and PSB cracking. Both effects are combined in the case of strain-controlled LCF testing and the subsequent strain-life Wöhler curve representation of the data points. In a stress-life representation only the decreased probability of high Schmid factors obviously takes effect. An example is shown in Section 4.3. The grain size does not have an influence on the Schmid factor $F_{SF}(\widetilde{m}_{mod})$ distribution and the thereof derived life distribution.

As already mentioned in Section 4.1, the simulated grain orientation distributions are approximations motivated by metallographic analyses of few specimen cuts. Both, isotropic distribution of orientations as well as the full texturing with fixed Euler angle $\vartheta = 25°$ are rather special cases of a real material.

4.3. Comparison of Fit and Prediction Quality of Probabilistic Models

Table 6 shows the average negative Log-Likelihood (*nLL*) per test point for all probabilistic models. The *nLL* was chosen as an appropriate coefficient of determination since the fitted relationship is non-linear and the residual distribution is not normal.

Table 6. Values of negative Log-Likelihood for the different distribution approaches.

Model Combination: Scale Model + Life Distribution	Neg. Log-Likelihood per Data Point at Calibration	Neg. Log-Likelihood per Data Point at Prediction
CMB + Weibull distribution	10.7	8.7
CMB + Schmid factor-corrected life distribution	0.87	12.38
CMB + modified Schmid factor-corrected life distribution	0.57	0.61

Since low negative Log-Likelihood values in Table 6 indicate good accordance of the test data with the estimated distribution it becomes apparent that the Weibull distribution is not well suited to describe the statistical behavior of the tested LCF lives compared to the microstructure-based approaches. Out of those, the modified Schmid factor-corrected life distribution performs significantly better in predicting the LCF test points of the fine grain material batch. The comparison of the *nLL* values supports the visual impression of Figure 16 where the prediction curve goes through the test point cloud splitting it approximately into halves.

It was not expected that this could be achieved using the single grain Schmid factor-corrected life distribution $F_N(n(\widetilde{m}))$ since the underlying simulation comprises the strong simplification of no grain interaction and the same homogeneous stress state at all grains. Although the calibration curve for this case has low *nLL*, Figure 15 shows that its position is dominated by the position of the far-left curve point at $\Delta\varepsilon_a = 0.5\%$. The reason for that is the sharp left flank of the distribution density which is zero, allowing no point before the first visible ascent. But not only the missing consideration of grain interaction in $F_N(n(\widetilde{m}))$ is a simplification. $F_{SF}(\widetilde{m})$ is the cumulative distribution function for the probability $P(SF \leq \widetilde{m})$ of for a single grain. However, there are many grains in the gauge section of the specimen in between the extensometer tips. Hence, the maximum value distribution for $\tilde{m}$ in the observed section would have to be calculated. It combines the probabilities of all grains for reaching a Schmid factor $SF \leq \tilde{m}$. Gottschalk et al. [28] have shown that such an approach assuming independently distributed grain orientations at uniaxial stress leads to distributions with unrealistically small scatter bands. If such an approach would be followed, the nLL values would be ∞ for the given data set. That is why further grain interaction and dependencies have to be taken into account, e.g., with a crack percolation model [28,30]. The modified Schmid factor distribution from polycrystalline FEA solutions is a step towards that as it considers the intergranular dependencies of the local stress states but is also computationally expensive.

Comparing the Weibull LCF life distribution and the modified Schmid factor based LCF life distribution it becomes apparent that both are very different. Specifically, the lower probability tails which are important for reliable design of components, e.g., hot gas parts of gas turbines in the case of René80, differ largely. The modified Schmid factor based N_i distribution implicates that there is a safe life cycle range with zero probability of crack initiation from PSB formation. Of course, care must be taken, and further evaluations and tests must be conducted before transferring such outcomes to component design. Note in this context that considerable efforts are necessary to achieve a sufficiently high number of LCF tests to verify the distribution tail shape. Additionally, LCF tests are always subject to experimental uncertainties, such as limited crack initiation detection accuracy [49].

All in all, the microstructure-based modeling approach shows that there is potential for increasing a parts time in service at the same risk for crack initiation at persistent slip bands. Note, that the approach presented here is not suitable for quantitative design assessments since it does not utilize the local approach for LCF life evaluation and therefore does not consider the statistical size effect such as the local probabilistic model for LCF which is presented and validated in [19,20,22,23].

Furthermore, it was observed that a preferential direction in the polycrystal grain orientation has a beneficial impact on the LCF life. The previously shown E-N-plots visualize two root causes. On the one hand, the material with preferential grain orientation has a lower stiffness and therefore, the total stress at the grains and hence slip systems is lower compared to the material with random grain orientation. On the other hand, also the distribution of the Schmid factors has a lower expectation value than in the case of randomly oriented grains. That itself, again results in lower resolved shear stresses at the slip systems in average for the material with preferential grain orientation. Both effects are present in strain controlled LCF experiments but only the latter plays a role for engineering components which are typically under a certain force load. Figure 18 shows the stress Wöhler curve predictions for both material batches using the modified Schmid factor concept.

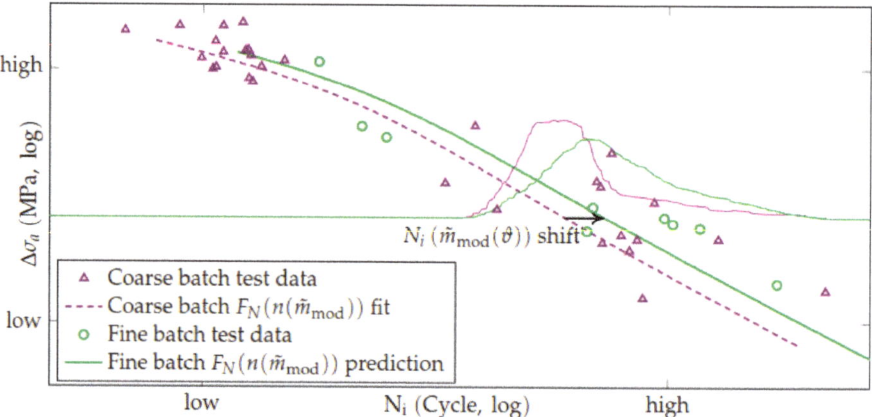

Figure 18. Stress-life plot of test data and fit and prediction curves for the modified Schmid factor distribution model.

5. Conclusions

Probabilistic LCF life models for the high temperature fatigue behavior of the Ni-base superalloy René80 (CC) were derived. The models are based on the statistical distributions of resulting shear stress, respectively Schmid factors, depending on the grain orientation distribution. One the one hand geometrical concepts, Monte-Carlo sampling of a single crystal orientations under uniaxial load and resolved shear stress evaluations were carried out. One the other hand polycrystalline FEA models were computed for different grain morphologies, i.e., randomly oriented coarse and preferentially oriented fine grains. It turned out that grain interaction has a crucial influence on the local stress and strain states and thus the shear stresses in the slip systems. The fine-grained material batch with a preferential grain orientation showed higher crack initiation life in the experiment. The respective simulations revealed lower Schmid factors and lower Young's modulus in average. Both effects combined lead to higher crack initiation life in the strain-controlled tests compared to the coarse-grained material with random grain orientation. The Schmid factor based LCF life model was calibrated with test data from a batch with isotropic grain orientation distribution (random) using a Maximum-Likelihood approach. The thus calibrated model was then used to predict the LCF life for the other material batch with the preferential grain orientation. The LCF life predictions based on the Schmid factor distributions derived from FEA show the best coincidence with the experimentally determined lifetimes and is superior to the well-established Weibull-approach and the Schmid factor distributions calculated from geometric considerations. However, it has the drawback of not covering the statistical size effect and being computationally expensive. That makes it difficult for direct fatigue-risk assessment tool integration, e.g., as an FEA post-processor.

While the conducted studies shed light on the elasticity anisotropy of René80 and comparable materials like IN 738 LC, no anisotropic plasticity models were available in the literature. Hence, the lower yield strength of the preferentially oriented material batch could not be explained.

The presented work covers two cases of grain orientation distribution. It is favorable to conduct validation studies with further, different, more precisely determined orientation distributions. But it is already possible to draw conclusions from few polycrystalline FEA simulations and in certain cases extrapolate the influence of preferential grain orientations on the fatigue behavior. It is of course also desirable to study more component relevant geometry and loading conditions and their effect on the modified Schmid factor distribution such as notches or torsional loads.

Author Contributions: B.E. and L.M. wrote the present paper, and did the experimental work as well as the simulation and lifetime modeling under the supervision and support of H.G. and T.B. P.L. and N.M. have supported the results of the paper with their work.

Funding: The investigations were conducted as part of the joint research program COOREFLEX TURBO in the frame of AG Turbo. The work was supported by the Federal Ministry for Economic Affairs and Energy as per resolution of the German Federal Parliament under grant number 03ET7041K. N.M. acknowledges financial support under the same scheme with the grant number 03ET7041J. H.G. acknowledges partial financial support from the Federal Ministry of Education under the Grant-Nr. 05M18PXA.

Acknowledgments: The authors gratefully acknowledge AG Turbo and Siemens AG for their support and permission to publish this paper. The responsibility for the content lies solely with its authors.

Conflicts of Interest: The authors declare no conflict of interest.

References

1. Bussac, A.; Lautridou, J. A probabilistic model for prediction of lcf surface crack initiation in pm alloys. *Fatigue Fract. Eng. Mater. Struct.* **1993**, *16*, 861–874. [CrossRef]
2. Grison, J.; Remy, L. Fatigue failure probability in a powder metallurgy Ni-base superalloy. *Eng. Fract. Mech.* **1997**, *57*, 41–55. [CrossRef]
3. Brückner-Foit, A.; Jäckels, H.; Quadfasel, U. Prediction of the lifetime distribution of high-strength components subjected to fatigue loading. *Fatigue Fract. Eng. Mater. Struct.* **1993**, *16*, 891–908. [CrossRef]

4.	Pineau, A.; Antolovich, S.D. Probabilistic approaches to fatigue with special emphasis on initiation from inclusions. *Int. J. Fatigue* **2016**, *93*, 422–434. [CrossRef]

5.	Böhm, J.; Heckel, K. Die Vorhersage der Dauerschwingfestigkeit unter Berücksichtigung des statistischen Größeneinflusses. *Materialwissenschaft und Werkstofftechnik* **1982**, *13*, 120–128. [CrossRef]

6.	Zheng, X.-L.; Lü, B.; Jiang, H. Determination of probability distribution of fatigue strength and expressions of P-S-N curves. *Eng. Fract. Mech.* **1995**, *50*, 483–491. [CrossRef]

7.	Hertel, O.; Vormwald, M. Statistical and geometrical size effects in notched members based on weakest-link and short-crack modelling. *Eng. Fract. Mech.* **2012**, *95*, 72–83. [CrossRef]

8.	Okeyoyin, O.A.; Owolabi, G.M. Application of Weakest Link Probabilistic Framework for Fatigue Notch Factor to Turbine Engine Materials. *World J. Mech.* **2013**, *03*, 237–244. [CrossRef]

9.	Zok, F.W. On weakest link theory and Weibull statistics. *J. Am. Ceram. Soc.* **2017**, *100*, 1265–1268. [CrossRef]

10.	Weibull, W. *A Statistical Theory of the Strength of Materials*; Generalstabens Litografiska Anstalts Foìrlag: Stockholm, Sweden, 1939.

11.	Schijve, J. A normal distribution or a weibull distribution for fatigue lives. *Fatigue Fract. Eng. Mater. Struct.* **1993**, *16*, 851–859. [CrossRef]

12.	Schijve, J. Fatigue predictions and scatter. *Fatigue Fract. Eng. Mater. Struct.* **1994**, *17*, 381–396. [CrossRef]

13.	Xiulin, Z. Prediction of probability distribution of fatigue life of 15MnVN steel notched elements under variable-amplitude loading. *Int. J. Fatigue* **1996**, *18*, 81–86. [CrossRef]

14.	Schijve, J. Statistical distribution functions and fatigue of structures. *Int. J. Fatigue* **2005**, *27*, 1031–1039. [CrossRef]

15.	Karolczuk, A.; Palin-Luc, T. Modelling of stress gradient effect on fatigue life using weibull based distribution function. *J. Theor. Appl. Mech.* **2013**, *51*, 297–311.

16.	*Rechnerischer Festigkeitsnachweis für Maschinenbauteile aus Stahl, Eisenguss-Und Aluminiumwerkstoffen*, 6th ed.; Rennert, R. (Ed.) VDMA-Verlag: Frankfurt am Main, Germany, 2012; ISBN 9783816306054.

17.	Fiedler, M.; Varfolomeev, I. *Rechnerischer Festigkeitsnachweis unter Expliziter Erfassung Nichtlinearen Werkstoff-Verformungsverhaltens, Final Report*; VDMA-Verlag: Frankfurt am Main, Germany, 2015.

18.	Fiedler, M.; Wächter, M.; Varfolomeev, I.; Vormwald, M.; Esderts, A. *Richtlinie Nichtlinear. Rechnerischer Festigkeitsnachweis unter Expliziter Erfassung Nichtlinearen Werkstoffverformungsverhaltens; für Bauteile aus Stahl, Stahlguss und Aluminiumknetlegierungen, 1. Auflage*; VDMA Verlag GmbH: Frankfurt am Main, Germany, 2019; ISBN 3816307299.

19.	Schmitz, S.; Seibel, T.; Beck, T.; Rollmann, G.; Krause, R.; Gottschalk, H. A probabilistic model for LCF. *Comput. Mater. Sci.* **2013**, *79*, 584–590. [CrossRef]

20.	Schmitz, S.; Gottschalk, H.; Rollmann, G.; Krause, R. Risk Estimation for LCF Crack Initiation. In Proceedings of the ASME Turbo Expo 2013: Turbine Technical Conference and Exposition, San Antonio, TX, USA, 3 June 2013; ASME: New York, NY, USA, 2013; p. V07AT27A007, ISBN 978-0-7918-5526-3.

21.	Hell, M.; Wagener, R.; Kaufmann, H.; Melz, T. Fatigue Design with Respect to Component Related Cyclic Material Behaviour and Considerations of Size Effects. *Procedia Eng.* **2015**, *133*, 389–398. [CrossRef]

22.	Mäde, L.; Schmitz, S.; Gottschalk, H.; Beck, T. Combined Notch and Size Effect Modeling in a Local Probabilistic Approach for LCF. *Comput. Mater. Sci.* **2017**, *142*, 377–388. [CrossRef]

23.	Maede, L.; Kumar, K. Evaluation of component-similar rotor steel specimens with a local probabilistic approach for LCF.—Under Review. *Comput. Mater. Sci.* **2019**.

24.	Siebörger, D.; Knake, H.; Glatzel, U. Temperature dependence of the elastic moduli of the nickel-base superalloy CMSX-4 and its isolated phases. *Mater. Sci. Eng. A* **2001**, *298*, 26–33. [CrossRef]

25.	Engel, B.; Beck, T.; Moch, N.; Gottschalk, H.; Schmitz, S. Effect of local anisotropy on fatigue crack initiation in a coarse grained nickel-base superalloy. *MATEC Web Conf.* **2018**, *165*, 4004. [CrossRef]

26.	Hermann, W.; Sockel, H.G.; Bertram, A. Elastic Properties and Determination of elastic Constants of Nickel-base Superalloy by Free-Free Beam Technique. In Proceedings of the 11th International Symposium on Superalloys, Champion, PA, USA, 22–26 September 1996; p. 229. [CrossRef]

27.	Schmid, E.; Boas, W. *Plasticity of Crystals. With Special Reference to Metals*; Chapman & Hall: London, UK, 1968.

28.	Gottschalk, H.; Schmitz, S.; Seibel, T.; Rollmann, G.; Krause, R.; Beck, T. Probabilistic Schmid factors and scatter of low cycle fatigue (LCF) life. *Materialwissenschaft und Werkstofftechnik* **2015**, *46*, 156–164. [CrossRef]

29.	Seibel, T. Einfluss der Probengröße und der Kornorientierung auf die Lebensdauer Einer Polykristallinen Ni-Basislegierung bei LCF- Beanspruchung. Ph.D. Thesis, Forschungszentrum Jülich, Jülich, Germany, 2014.

Metals **2019**, *9*, 813

30. Moch, N. From Microscopic Models of Damage Accumulation to Probability of Failure of Gas Turbines. Ph.D. Thesis, Bergische Universität Wuppertal, Wuppertal, Germany, 2018.

31. Engel, B. Einfluss der Lokalen Kornorientierung und der Korngröße auf das Verformungs- und Ermüdungsverhalten von Nickelbasis Superlegierungen. Ph.D. Thesis, Technische Universität Kaiserslautern, Kaiserslautern, Germany, 2018.

32. Quey, R.; Dawson, P.R.; Barbe, F. Large-scale 3D random polycrystals for the finite element method: Generation, meshing and remeshing. *Computer Methods Appl. Mech. Eng.* **2011**, *200*, 1729–1745. [CrossRef]

33. Ghazvinian, E.; Diederichs, M.S.; Quey, R. 3D random Voronoi grain-based models for simulation of brittle rock damage and fabric-guided micro-fracturing. *J. Rock Mech. Geotech. Eng.* **2014**, *6*, 506–521. [CrossRef]

34. Quey, R.; Renversade, L. Optimal polyhedral description of 3D polycrystals: Method and application to statistical and synchrotron X-ray diffraction data. *Comput. Methods Appl. Mech. Eng.* **2018**, *330*, 308–333. [CrossRef]

35. Aghaie-Khafri, M.; Farahany, S. Creep Life Prediction of Thermally Exposed Rene 80 Superalloy. *J. Mater. Eng. Perform.* **2010**, *19*, 1065–1070. [CrossRef]

36. Antolovich, S.D.; Liu, S.; Baur, R. Low cycle fatigue behavior of René 80 at elevated temperature. *Met. Trans. A* **1981**, *12*, 473–481. [CrossRef]

37. Buchholz, B. Entwicklung eines Werkstoffmodells für Isotherme und Anisotherme Ermüdung der Nickelbasislegierung René80. Ph.D. Thesis, TU Dresden, Dresden, Deutschland, 2012.

38. Neidel, A.; Riesenbeck, S.; Ullrich, T.; Völker, J.; Yao, C. Hot cracking in the HAZ of laser-drilled turbine blades made from René 80. *Mater. Test.* **2005**, *47*, 553–559. [CrossRef]

39. Österle, W.; Krause, S.; Moelders, T.; Neidel, A.; Oder, G.; Völker, J. Influence of heat treatment on microstructure and hot crack susceptibility of laser-drilled turbine blades made from René 80. *Mater. Charact.* **2008**, *59*, 1564–1571. [CrossRef]

40. Voronoi, G. Nouvelles applications des paramètres continus à la théorie des formes quadratiques. Deuxième mémoire. Recherches sur les parallélloèdres primitifs. *J. für die reine und angewandte Mathematik (Crelle's J.)* **1908**, *134*, 198–287. [CrossRef]

41. Delaunay, B. Sur la sphère vide. A la mémoire de Georges Voronoï. Bulletin de l'Académie des Sciences de l'URSS. *Classe Des Sciences Mathématiques Et Na* **1934**, *6*, 793–800.

42. León, C.A.; Massé, J.-C.; Rivest, L.-P. A statistical model for random rotations. *J. Multivar. Anal.* **2006**, *97*, 412–430. [CrossRef]

43. Engel, B.; Beck, T.; Schmitz, S. (Eds.) *High Temperature Low-Cycle-Fatigue of the Ni-Base Superalloy René80*; LCF 8: Dresden, Deutschland, 2017.

44. Rösler, J.; Harders, H.; Bäker, M. *Mechanisches Verhalten der Werkstoffe*, 5th ed.; Springer Vieweg: Wiesbaden, Germany, 2016; ISBN 978-3-8348-2241-3.

45. Siredey, N.; Denis, S.; Lacaze, J. Dendritic growth and crystalline quality of nickel-base single grains. *J. Cryst. Growth* **1993**, *130*, 132–146. [CrossRef]

46. Nitz, A.; Nembach, E. The critical resolved shear stress of a superalloy as a combination of those of its γ matrix and γ' precipitates. *Met. Mater. Trans. A* **1998**, *29*, 799–807. [CrossRef]

47. Österle, W.; Bettge, D.; Fedelich, B.; Klingelhöffer, H. Modelling the orientation and direction dependence of the critical resolved shear stress of nickel-base superalloy single crystals. *Acta Mater.* **2000**, *48*, 689–700. [CrossRef]

48. Jöchen, K. Homogenization of the Linear and Non-Linear Mechanical Behavior of Polycrystals. Ph.D. Thesis, KIT, Karlsruhe, Deutschland, 2013.

49. Gollmer, M. Schadensakkumulationsverhalten der Superlegierung René 80 unter Zweistufiger Low Cycle Fatigue Beanspruchung. Ph.D. Thesis, Technische Universität Kaiserslautern, Kaiserslautern, Germany, 2018.

Article

Transient Effects in Creep of Sanicro 25 Austenitic Steel and Their Modelling

Luboš Kloc [1,*], Václav Sklenička [1,2] and Petr Dymáček [1,2]

[1] Institute of Physics of Materials AS CR, v.v.i. Žižkova 22, CZ-61662 Brno, Czech Republic; vsklen@ipm.cz (V.S.); pdymacek@ipm.cz (P.D.)

[2] CEITEC IPM, Institute of Physics of Materials AS CR, v.v.i. Žižkova 22, CZ-61662 Brno, Czech Republic

* Correspondence: kloc@ipm.cz

Received: 15 January 2019; Accepted: 15 February 2019; Published: 19 February 2019

Abstract: Transient effects upon stress changes during creep of the new Sanicro 25 steel were investigated experimentally using the helicoid spring specimen technique. The creep behaviour was found to be qualitatively the same as that observed earlier with the creep-resistant 9% Cr ferritic-martensitic P-91 steel, but the transient strains are considerably smaller. Negative creep rate, which is strain running against the applied stress, was observed with any stress decrease. Parameters for the complex creep model were estimated and model results were compared to the creep rates measured experimentally. The model can be used for the finite element method modelling of the creep and stress relaxation effects in the components made from the Sanicro 25 steel.

Keywords: creep; transient effects; Sanicro 25; high temperature steels

---

1. Introduction

New materials capable of withstanding loading at high temperatures for long periods of time are demanded and have been developed in order to improve efficiency of the thermal power plants. The austenitic stainless steel Sandvik Sanicro 25 (UNS S31035) steel was recently developed within the Termie–AD 700 project in Europe and as a prospective material has been subjected to various mechanical testing [1–4].

Austenitic steels exhibit better creep and corrosion resistance than ferritic ones, but, due to lower thermal conductivity and higher thermal expansion, are sensitive to thermomechanical fatigue. The finite element method (FEM) is an apt tool to describe the behaviour of real components under transient conditions. A relevant mathematical description of the material's properties under a wide range of stresses and temperatures is needed to obtain valid results. The current descriptions of the creep processes are based exclusively on the creep curves measured under constant loading conditions, so the transient effects caused by stress changes are ignored. This approach is unsatisfactory [5].

Since the demanded creep lives are obviously very long, creep tests under conditions close to that used in industrial application of the material cannot be done. Extrapolation from the results of the tests accelerated by higher temperature and/or stress is used instead. An alternative approach is measurement of very small creep strains, revealing the creep behaviour at the very beginning of the creep curve. This approach is very rare [6,7], though the results are important for parts with very small tolerable strains and mainly for relaxation of stresses generated for instance by temperature gradients. The results obtained for the creep resistant steels [8] show that the extrapolation method mentioned above does not provide correct description of the small creep strains. The measurement of the small creep strains was used in this work to obtain a basis for the realistic FEM modelling of the thermomechanical fatigue of the prospective Sanicro 25 steel.

In this work, transient strains during creep of the Sanicro 25 steel are investigated under conditions close to the potential service employment of the steel. A helicoid spring specimen technique was used to get high strain sensitivity. The main aim of the work is to explore transient effects in creep of the prospective Sanicro 25 steel, and, together with our previous paper [9], to collect data for the complex phenomenological creep model [10], the only model capable of describing transient effects in creep. While the previous paper [9] was devoted to constant stress tests, this work is focused on the transient stages upon stress changes.

2. Experimental Material and Procedure

2.1. Material

Material for experiments was supplied by Sandvik Materials Technology, Sandviken, Sweden in the form of a cylindrical rod of 150 mm in diameter. Chemical composition of the material is listed in Table 1.

Table 1. Chemical composition of tested material (in wt.%).

C	Si	Mn	Cr	Ni	W	Co	Cu	Nb	N	Fe
0.1	0.2	0.5	22.5	25.0	3.6	1.5	3.0	0.5	0.23	Bal.

Specimens were tested in the as-received state of the material. The microstructure of the material was briefly described in our previous paper [9]. Grain size is generally of about 25 µm, but grains as large as 200 µm can be also observed. Grain boundaries are decorated by small Nb rich carbonitrides and some such particles occur also in the grain interiors, mainly in large grains. Microstructure of the material has been analysed in detail in previously published papers [11–14] and is not studied here, since the very small creep strains do not cause visible changes in the microstructure.

2.2. Creep Tests

The helicoid spring specimen technique were used to investigate creep responses on the stress changes at temperatures 700 and 750 °C and stresses between 25 and 90 MPa. The technique enables measurement of very small creep strains and then also very low creep rates within tests of acceptable duration. The experimental arrangement was described in detail in [15].

The shape and manufacturing process of the specimens were the same as in the previous work [9]. Special processing technology combining precision conventional machining and electroerosive cutting was used to manufacture helicoid spring specimens with an outer diameter of 34 mm, an inner diameter of 31 mm, a pitch of 3 mm and an overall length of 50 mm. The procedure was optimised to minimise any influence on the microstructure of the specimen. The creep tests were conducted in a protective atmosphere of purified argon. Testing temperature was kept within the ±1 °C interval and the homogeneity of the temperature field was in the same order. The contactless optical measurement of strain was used to avoid any friction or other disturbing factors.

3. Results

3.1. Creep Curves

Measured creep curves are shown in Figures 1 and 2. Clear transient stages are observed following all stress changes. The transient strains are not so extensive as those observed with the 9% Cr ferritic-martensitic P-91 steel [16] but are still important.

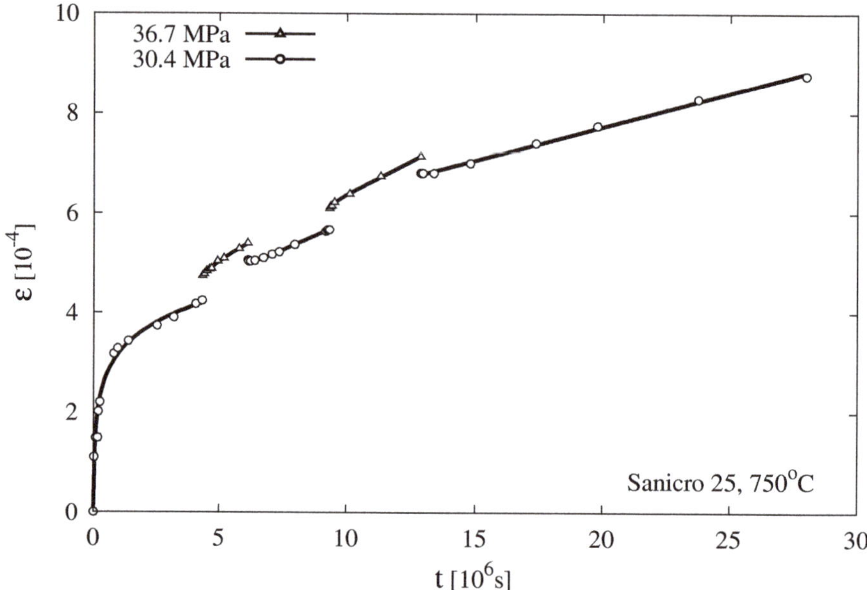

Figure 1. Creep curves at 750 °C and stresses 30.4 and 36.7 MPa. Experimental data (points) were fitted by Li equation [17] (lines).

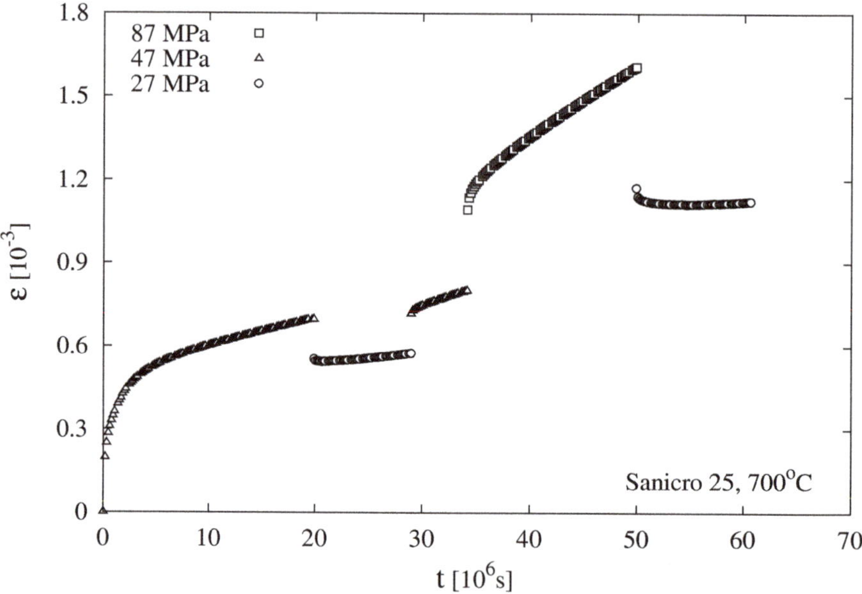

Figure 2. Creep curves at 700 °C and stresses from 27 to 87 MPa.

The short period of negative strain rate, which is strain running against applied stress, is observable even at relatively small stress reductions, indicating importance of internal stresses. This effect is visible on the creep curve detail in Figure 3.

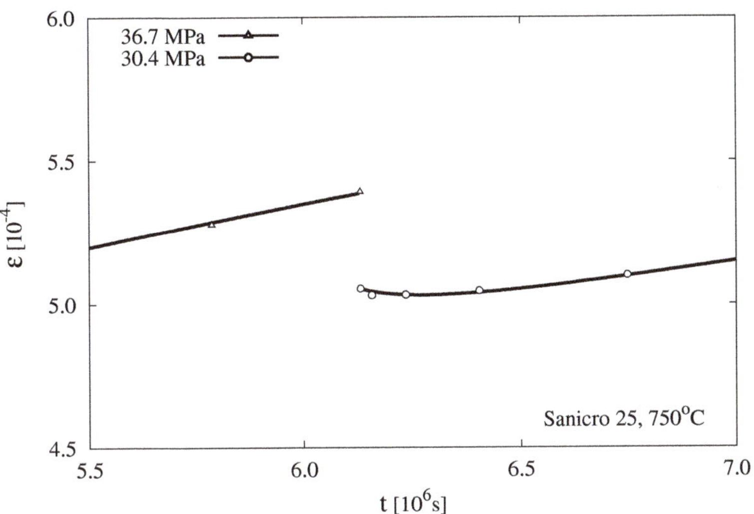

Figure 3. Detail of the creep curve at 750 °C and stress reduction from 36.7 to 30.4 MPa.

The data required to reproduce these findings are available to download from http://dx.doi.org/ 10.17632/bjw2mtgswp.1#file-a218d703-a6a8-49f7-b765-2080252c050a.

3.2. Creep Rates

Creep rates measured at the end of each segment of the creep curve are plotted in Figure 4 together with the results of constant stress tests from [9]. It is clear that, for the low-stress creep mechanism, there is no steady state creep stage, but strain rate is permanently decreasing. This is the same behaviour as it was observed with the P-91 steel [16].

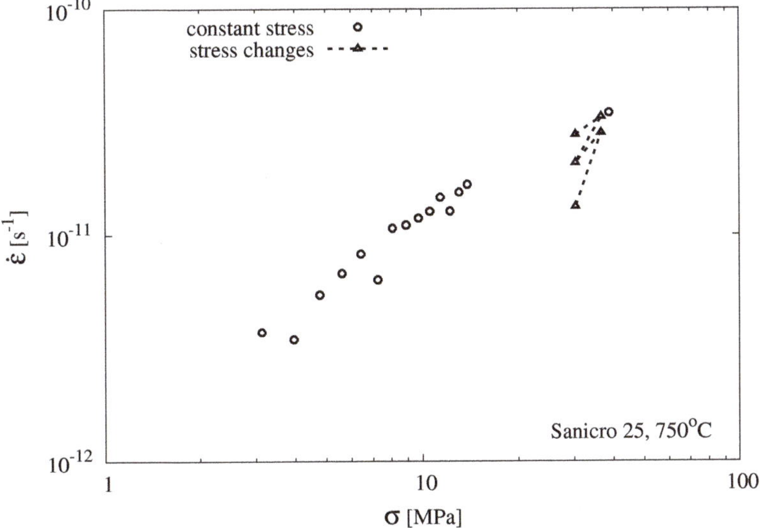

Figure 4. The creep rates $\dot{\varepsilon}$ for 750 °C, measured at the end of each creep curve segment. The creep rates from constant stress tests [9] are plotted for comparison.

4. Model

4.1. Creep Rate Description

A complex phenomenological model [10] can be used to describe observed transient effects. The model is based on an assumption of two independent deformation mechanisms, acting in parallel. The first mechanism, dominating at low stresses, is anelastic and its principle is a building of the field of internal stresses, assuming interaction of the hard elastic zones and soft elastoplastic zones. The other mechanism, dominating at higher stresses, is more obvious plastic deformation described by the modified Garofallo equation with threshold stress.

Overall creep rate $\dot{\varepsilon}$ is expressed by modified Li equation as

$$\dot{\varepsilon} = \frac{\dot{\sigma}}{E} + \dot{\varepsilon}_c + \frac{\dot{\varepsilon}_s \left(1 + r_i\right)}{1 + r_i - r_i \exp\left(-\theta\right)}, \tag{1}$$

where $\dot{\varepsilon}_s$ is given by

$$\dot{\varepsilon}_s = \mathrm{sgn}(\sigma)\, b \exp\left(\frac{-Q_h}{RT}\right) \sinh\left(p \left(|\sigma| + \sqrt{(|\sigma| - \sigma_t)^2 + \sigma_r^2} - \sqrt{\sigma_t^2 + \sigma_r^2} \right) \right), \tag{2}$$

the "creep age" θ is integrated according to

$$\theta = \frac{1}{c} \int_0^t |\dot{\varepsilon}_s|\, \mathrm{d}t, \tag{3}$$

and $\dot{\varepsilon}_c$ is obtained by the numerical solution of the equation

$$\dot{\varepsilon}_c = g \exp\left(\frac{-Q_l}{RT}\right) \left(\frac{\sigma}{\sigma_t} - \frac{(1 - k)E}{k\sigma_t}\varepsilon_c\right)^3. \tag{4}$$

In the above equations, σ is the applied stress, T is absolute temperature, and Q_h and Q_l are the apparent activation energies for the high-stress and low-stress mechanisms, respectively, E is Young's modulus, σ_t is threshold stress, σ_r, describes the residual effective stress below the threshold, r_i describes the ratio between initial and secondary creep rate, b controls the overall rate of the high-stress mechanism, p controls the transition between linear and exponential parts of the Garofallo equation, c describes the relation between the secondary stage creep rate and the primary relaxation time of the high-stress mechanism, g controls the overall rate of the low-stress mechanism and k describes the ratio between hard and soft zones.

4.2. Model Results

The parameter values used to model the creep behaviour of the Sanicro 25 steel are summarised in Table 2. The temperature dependence of the parameter σ_r must be introduced to obtain acceptable results, but with experiments at only two temperature levels, it is not possible to speculate about the character of that dependence.

Table 2. Set of model parameters for the Sanicro 25 steel.

Parameter	Value	Unit	
σ_t	130	MPa	
σ_r	3.0	MPa	@700 °C
σ_r	1.0	MPa	@750 °C
r_i	8.0		
b	3.0×10^{19}	s^{-1}	
p	0.03	MPa^{-1}	
Q_h	525	kJ/mol	
Q_l	171	kJ/mol	
g	700	s^{-1}	
E	139×10^3	MPa	
k	0.65		
c	3.5×10^{-3}		

Model results for the constant-stress creep experiments from [9] and [18] are showed in Figures 5–7, exhibiting good results for both low-stress and high-stress creep regimes.

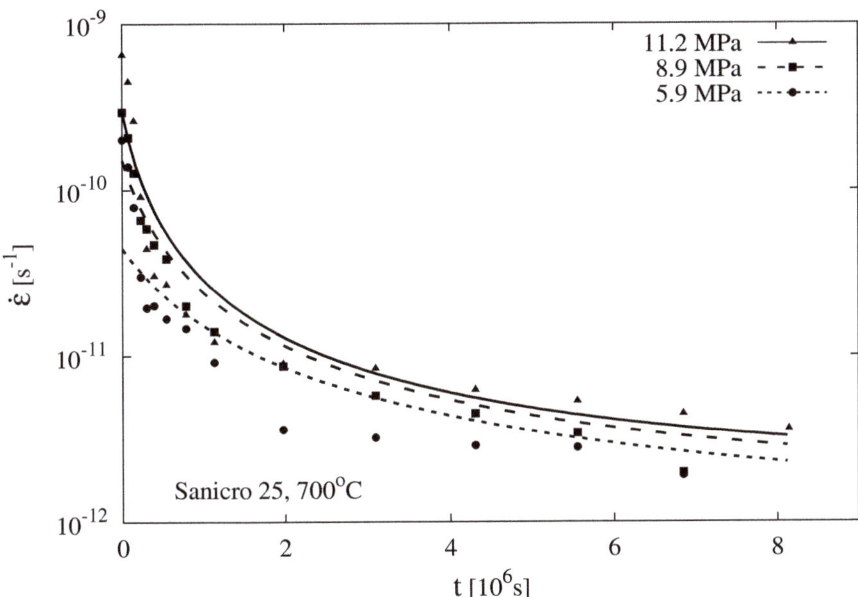

Figure 5. Creep rate dependence on time for Sanicro 25 at 700 °C and low stresses [9] (points) compared to model results (lines).

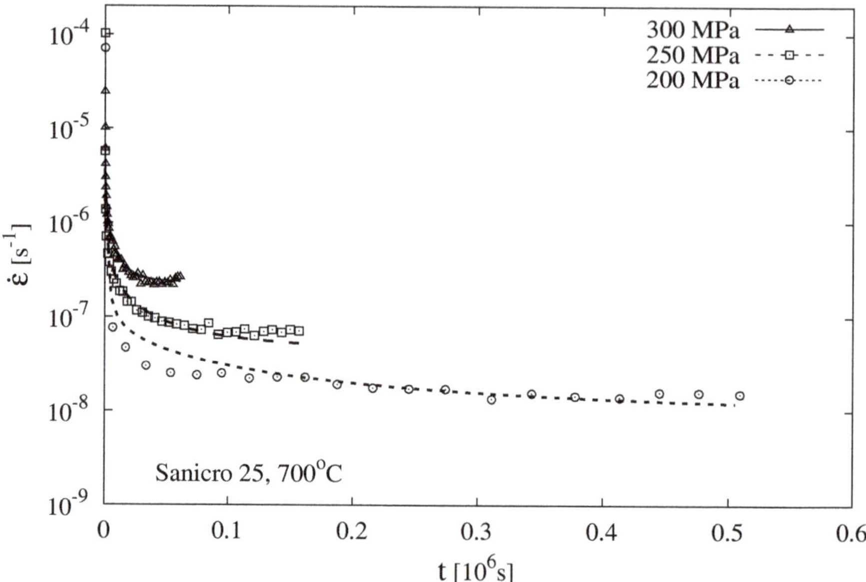

Figure 6. Creep rate dependence on time for Sanicro 25 at 700 °C and high stresses [18] (points) compared to model results (lines).

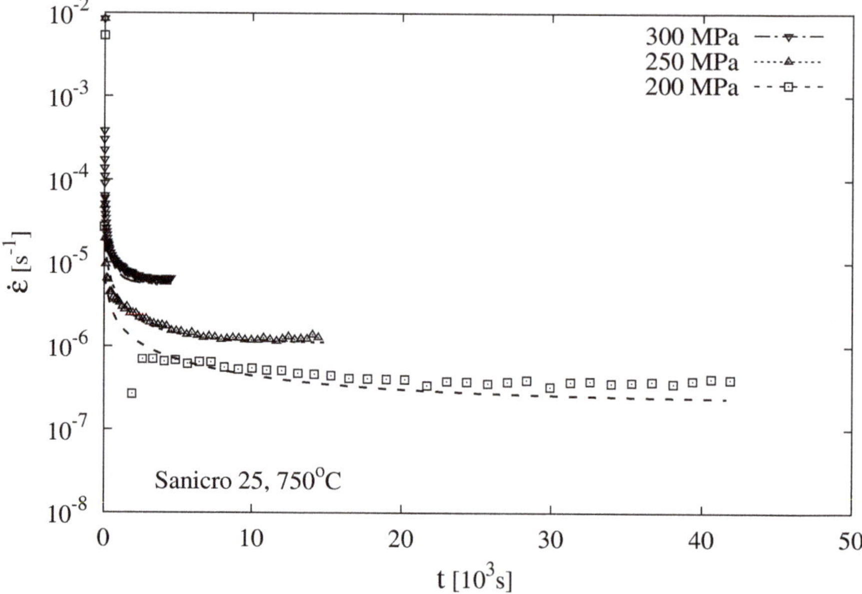

Figure 7. Creep rate dependence on time for Sanicro 25 at 750 °C and high stresses [18] (points) compared to model results (lines).

Model results for the above presented creep experiments with stress changes are plotted in Figures 8 and 9.

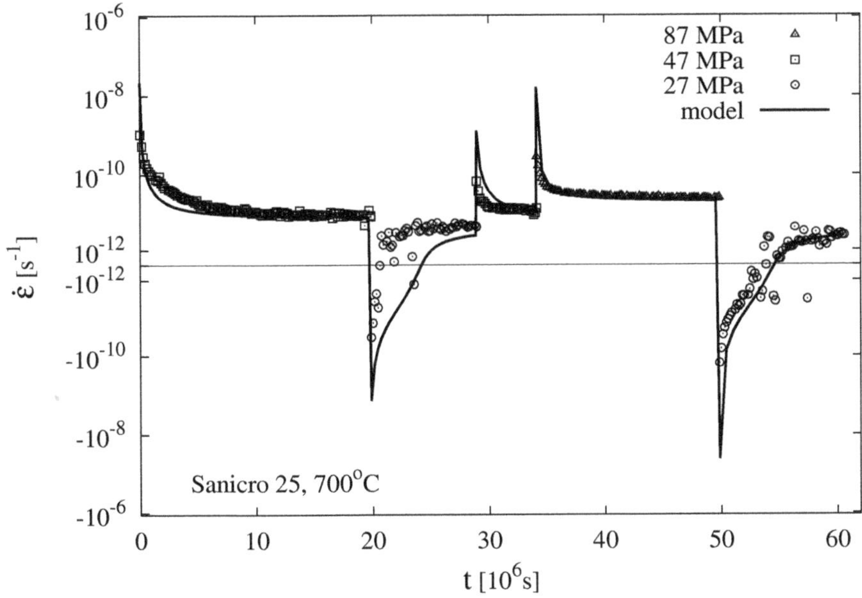

Figure 8. Creep rate dependence on time for Sanicro 25 at 700 °C and stress changes (points) compared to model results (line).

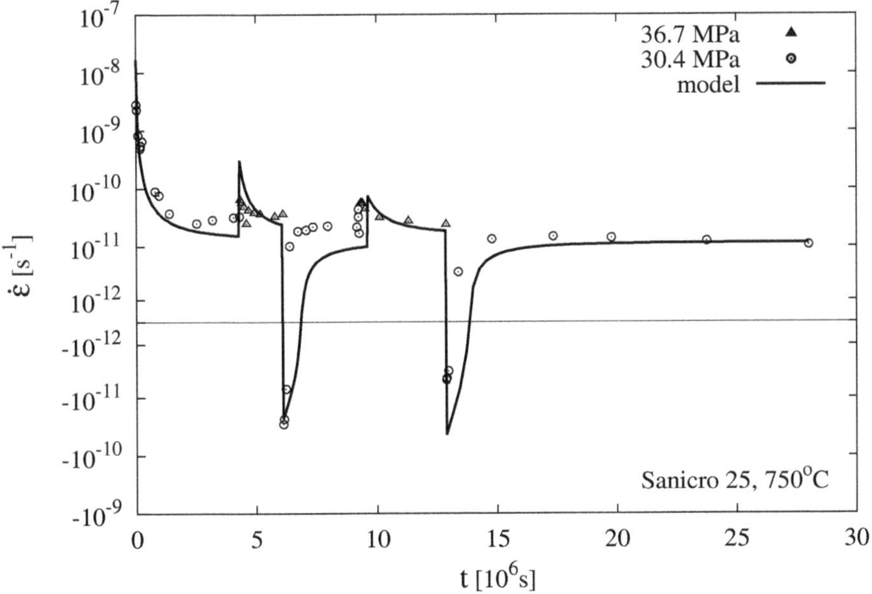

Figure 9. Creep rate dependence on time for Sanicro 25 at 750 °C and stress changes (points) compared to model results (line).

4.3. Discussion

The results are not perfect, mainly during early stress decreases and around zero strain rates. The time duration of the transient stage is slightly overestimated in the model. Some creep curves or segments of the creep curves with stress changes are fitted well, while with others the fit is considerably worse. There are two reasons for that effect. First, there is some natural scatter in the experimentally observed values. As was pointed out in the previous paper [9], the scatter in the initial creep rates is considerably higher than that of the creep rate in a secondary stage. Thus, the shape of creep curves varies and it is not possible to describe them by one set of parameters. Compromise values were then adopted for parameters r_i and k, satisfying some creep curves better than others. Second, the model was derived using many strong assumptions, which are not necessarily strictly fulfilled. All these assumptions are listed in [10]. Probably, the two deformation mechanisms are not completely independent. This fact can explain some systematic deviations of the model predictions from the experimental curves. Taking into account that other creep models ignore the transient effects completely, it can be considered as acceptable.

5. Conclusions

The transient creep behaviour upon the stress changes of the Sanicro 25 austenitic creep resistant steel was investigated at temperatures 700 and 750 °C and the applied stresses between 25 and 90 MPa.

The strain reactions to the stress changes are qualitatively the same as those observed for the P-91 ferritic steel [16], but the transient strains are considerably smaller. In the low-stress creep regime, time period of the negative creep rate is observed even with relatively small decreases in the applied stress.

The complex phenomenological creep model published in [10] was used to describe primary and secondary stages of the creep behaviour of the Sanicro 25 steel under various conditions including transient effects. All parameters of the model were estimated and the model can be used in FEM calculations, though the results are not perfect.

Author Contributions: Conceptualization, L.K.; methodology, L.K.; validation, V.S. and L.K.; investigation, L.K. and P.D.; resources, P.D. and L.K.; data curation, P.D.; writing–original draft preparation, L.K.; writing–review and editing, V.S. and P.D.; visualisation, L.K.; supervision, V.S.; project administration, L.K.; funding acquisition, L.K.

Funding: This research was funded by the Czech Science Foundation under contract no. 15-21394S and partly accomplished in CEITEC 2020 (LQ1601) with financial support from the Ministry of Education, Youth and Sports of the Czech Republic under the National Sustainability Programme II.

Acknowledgments: The authors want to thank Guocai Chai from Sandvik Materials Technology for the supply of the Sanicro 25 experimental material.

Conflicts of Interest: The authors declare no conflict of interest.

Abbreviations

The following abbreviations are used in this manuscript:

FEM   Finite Element Method

References

1. Chai, G.; Bostrom, M.; Olaison, M.; Forsberg, U. Creep and LCF Behaviors of Newly Developed Advanced Heat Resistant Austenitic Stainless Steel for A-USC. *Procedia Eng.* **2013**, *55*, 232–239. [CrossRef]
2. Polak, J.; Petras, R.; Heczko, M.; Kruml, T.; Chai, G. Evolution of the cyclic plastic response of Sanicro 25 steel cycled at ambient and elevated temperatures. *Int. J. Fatigue* **2016**, *83*, 75–83. [CrossRef]
3. Petras, R.; Skorik, V.; Polak, J. Thermomechanical fatigue and damage mechanisms in Sanicro 25 steel. *Mater. Sci. Eng. A* **2016**, *650*, 52–62. [CrossRef]

4.	Chai, G.; Forsberg, U.	Sanicro 25: An advanced high-strength, heat-resistant austenitic stainless steel. In *Materials for Ultra-Supercritical and Advanced Ultra-Supercritical Power Plants*; DiGianfrancesco, A., Ed.; Number 104 in Series in Energy; Woodhead Publishing: Cambridge, UK: 2017; pp. 391–421.
5.	Kloc, L. Critical View on the Creep Modelling Procedures. *Acta Phys. Pol. A* **2015**, *128*, 540–542. [CrossRef]
6.	Sritharan, T.; Jones, H. Creep of Type-316 Stainless-steel at Low Stresses. *Met. Sci.* **1989**, *15*, 365–368. [CrossRef]
7.	Mishra, R.S.; Jones, H.; Greenwood, G.W. Creep of a low carbon steel at low stresses and intermediate temperatures. *Acta Metall. Mater.* **1990**, *38*, 461–468. [CrossRef]
8.	Kloc, L.; Skienicka, V.; Ventruba, J. Comparison of low stress creep properties of ferritic and austenitic creep resistant steels. *Mater. Sci. Eng. A* **2001**, *319*, 774–778. [CrossRef]
9.	Kloc, L.; Dymáček, P.; Sklenička, V. High temperature creep of Sanicro 25 austenitic steel at low stresses. *Mater. Sci. Eng. A* **2018**, *722*, 88–92. [CrossRef]
10.	Kloc, L.; Sklenička, V.; Dymáček, P.; Plešek, J. New creep constitutive equation for finite element modelling including transient effects. *Mech. Mater.* **2018**, *119*, 49–55. [CrossRef]
11.	Heczko, M.; Polak, J.; Kruml, T. Microstructure and dislocation arrangements in Sanicro 25 steel fatigued at ambient and elevated temperatures. *Mater. Sci. Eng. A* **2017**, *680*, 168–181. [CrossRef]
12.	Polak, J.; Petras, R.; Heczko, M.; Kubena, I.; Kruml, T.; Chai, G.C. Low cycle fatigue behavior of Sanicro 25 steel at room and at elevated temperature. *Mater. Sci. Eng. A* **2014**, *615*, 175–182. [CrossRef]
13.	Zhou, R.Y.; Zhu, L.H.; Liu, Y.Y.; Lu, Z.R.; Chen, L.; Ma, X. Microstructural evolution and the effect on hardness of Sanicro 25 welded joint base metal after creep at 973 K. *J. Mater. Sci.* **2017**, *52*, 6161–6172. [CrossRef]
14.	Calmunger, M.; Chai, G.C.; Eriksson, R.; Johansson, S.; Moverare, J.J. Characterization of Austenitic Stainless Steels Deformed at Elevated Temperature. *Metall. Mater. Trans. A-Phys. Metall. Mater. Sci.* **2017**, *48A*, 4525–4538. [CrossRef]
15.	Kloc, L.; Marecek, P. Measurement of Very Low Creep Strains: A Review. *J. Test. Eval.* **2009**, *37*, 53–58.
16.	Kloc, L.; Sklenicka, V. Confirmation of low stress creep regime in 9% chromium steel by stress change creep experiments. *Mater. Sci. Eng. A* **2004**, *387*, 633–638. [CrossRef]
17.	Li, J. A Dislocation Mechanism of Transient Creep. *Acta Metall.* **1963**, *11*, 1269–1270. [CrossRef]
18.	Dymacek, P.; Dobes, F.; Kloc, L. Small Punch Testing of Sanicro 25 Steel and its Correlation with Uniaxial Tests. *Key Eng. Mater.* **2017**, *734*, 70–76. [CrossRef]

Article

An Explicit Creep-Fatigue Model for Engineering Design Purposes

Dan Liu and Dirk John Pons *

Department of Mechanical Engineering, University of Canterbury, Christchurch 8041, New Zealand; dan.liu@pg.canterbury.ac.nz
* Correspondence: dirk.pons@canterbury.ac.nz; Tel.: +64-3369-5826

Received: 26 September 2018; Accepted: 17 October 2018; Published: 19 October 2018

Abstract: Background: Creep-fatigue phenomena are complex and difficult to model in ways that are useful from an engineering design perspective. Existing empirical-based models can be difficult to apply in practice, have poor accuracy, and lack economy. Need: There is a need to improve on the ability to predict creep-fatigue life, and do so in a way that is applicable to engineering design. Method: The present work modified the unified creep-fatigue model of Liu and Pons by introducing the parameters of temperature and cyclic time into the exponent component. The relationships between them were extracted by investigating creep behavior, and then a reference condition was introduced. Outcomes: The modified formulation was successfully validated on the materials of 63Sn37Pb solder and stainless steel 316. It was also compared against several other models. The results indicate that the explicit model presents better ability to predict fatigue life for both the creep fatigue and pure fatigue situations. Originality: The explicit model has the following beneficial attributes: Integration—it provides one formulation that covers the full range of conditions from pure fatigue, to creep fatigue, then to pure creep; Unified—it accommodates multiple temperatures, multiple cyclic times, and multiple metallic materials; Natural origin—it provides some physical basis for the structure of the formulation, in its consistency with diffusion-creep behavior, the plastic zone around the crack tip, and fatigue capacity; Economy—although two more coefficients were introduced into the explicit model, the economy is not significantly impacted; Applicability—the explicit model is applicable to engineering design for both manual engineering calculations and finite element analysis. The overall contribution is that the explicit model provides improved ability to predict fatigue life for both the creep-fatigue and pure-fatigue conditions for engineering design.

Keywords: creep fatigue; pure fatigue; economy; engineering design

---

1. Introduction

Creep-fatigue damage is defined as the damage caused by reversed loading at elevated temperatures, hence combines the effects of fatigue and creep. This is a complex process since fatigue and creep behaviours are based on significantly different mechanisms at the microstructural level. Observationally, fatigue occurs via cracks through the grains, while creep involves the grain boundary cracking [1]. The creep-fatigue phenomenon is relevant to a wide range of industries, such as aerospace, naval, nuclear and industries [2], hence cannot be ignored in engineering design.

1.1. The Design-Based Method

To provide an easier method for engineering practitioners to evaluate fatigue behaviour, a design-based method was proposed by Marin (Equation (1)) [3]:

$$S_e = k_a k_b k_c k_d k_e k_f S_e'$$

(1)

where S_e is the endurance limit at the critical location of a machine part in the geometry and condition of use, S'_e is the rotary-beam test specimen endurance limit, k_a is the surface condition modification factor, k_b is the size modification factor, k_c is the load modification factor, k_d is the temperature modification factor, k_e is the reliability factor, k_f is the miscellaneous-effects modification factor.

Engineering designers typically use this simple equation to determine the acceptable fatigue stress in a part. This modified endurance limit is based on the endurance limit at the reference condition and several multiplicative factors for surface condition, part size, type of load, operating temperature, etc. The only mechanical property included here is the endurance limit. This property can be related to ultimate tensile strength, such as the values of the endurance limit for steels are half of the ultimate tensile strength [4]. Therefore, the benefits of this approach (Equation (1)) are ease of use since the tensile strength is readily known or easily measured. The detriments of this approach are that it only includes temperature effect when creep is active, and the fatigue evaluation is merely an approximation. Furthermore, all the modification factors have to be determined experimentally. Some degree of creep may be accommodated in the temperature factor, but the equation does not present a robust treatment for creep-fatigue.

Although the design-based method (Equation (1)) is simple enough for engineering practitioners, the poor accuracy is of concern. In addition, the consideration of multiple effects (such as shape, size and surface) is redundant if engineers merely aim to select materials. However, making further improvements to this formulation would not seem to be a viable way forward, since this numerical structure is only one of convenience rather than representing any deeper mechanics at the material science level.

There is a need for a more robust design method for creep-fatigue. Ideally such a method would have a formulation that directly related applied stress to life, included macrostructural rather than microstructural properties, and was economical to validate. The various attempts at addressing this problem are reviewed below.

1.2. The Conventional Empirical Methods

For mechanical design, a pre-evaluation of fatigue life (or damage) is normally applied at the initial stage of design to make a material selection or structural optimization. Normally, in the creep-fatigue situation, the total damage is numerically evaluated through the theory of damage accumulation and conventional-fatigue-based formulations. However, they present significant limitations.

Specifically, the creep-fatigue evaluation based on damage accumulation is normally conducted by the linear damage rule [1,5] or crack growth law [6], wherein the fatigue damage and creep damage are evaluated separately and then are numerically added. However, this is untrue to the physics of failure in that the fatigue and creep effects are not independent. Rather the effects compound each other. Existing methods based on summation of fatigue damage and creep damage ignore the interaction between fatigue and creep, and thus result in less reliable findings. Although the improved representations of creep and fatigue components have been proposed in the literature, such as the non-linear accumulated damage models for creep [7,8] and fatigue [9,10], the issue caused by ignoring the interacted effect of fatigue and creep is still not fundamentally solved.

In addition, the conventional formulations typically assume a power–law relationship between life and applied loading, as evident in the Basquin equation [11,12] and Coffin–Manson equation [13,14]. Although this approach is simple, the coefficients need to be recalculated with changed temperature and/or frequency. Hence, this makes the design process inefficient and expensive because a large number of empirical data are required and must be re-fitted for each condition. To improve this limitation, others have attempted to introduce the variables of temperature and frequency into modified models, resulting in the Coffin-Manson-based creep-fatigue models proposed by Solomon [15], Shi [16], Jing [17] and Wong & Mai [18], and the Basquin-based creep-fatigue models developed by Kohout [19] and Mivehchi [20]. However, these models may only be applied at the situations for which they were derived. They do not represent the creep-fatigue behaviour for other materials, hence the formulations cannot present a unified characteristic. Furthermore, these models are determined by curve fitting, the accuracy of which is strongly determined by the number of empirical datum points. This results in poor empirical economy. Consequently, the conventional-based creep-fatigue models are severely limited in their applicability to engineering design.

1.3. Models Based on Observation of Microstructural Damage (Mechanism-Based)

The curve-fitting method, which is applied to build the Basquin-based and the Coffin–Manson-based models (Section 1.2), provides the simplest process to construct a numerical model, and thus is well-accepted in the field of mechanical engineering. However, from the perspective of material science, the fatigue models ideally should be constructed through observations of physical phenomena (such as the crack growth, diffusion creep, and void growth). This approach has resulted in the development of several mechanism-based models. These models are variously based on micromechanical cyclic void growth [21], partition of energy and micro-crack growth [22], and multistage fatigue theory [23].

These models are attractive because they relate physically measureable microstructural properties to life or total damage. Some of these models already include the ability to accommodate multiple forms of damage (including creep, fatigue, or oxidation), and represent both creep and fatigue in one formulation. However, this class of models suffers from limitations from the perspective of an engineering designer:

- They relate to life evolution in some way, but often not in ways that are accessible to engineering design. This is a particular limitation of the damage models.
- The mechanism-based models need to be validated. They require the measurement of microstructural parameters of damage. This information is not readily available to design engineers, certainly not at the onset of design. Also, designers do not select materials based on microstructure, but rather on mechanical properties. Furthermore, microstructural data are also not easily available during the service life of the part without resorting to destructive testing.
- They have abstract mathematical formulations that are not easy to conceptualise, and are difficult to apply to design.
- They typically have multiple coefficients in power law formulations, and each equation has sub-coefficients that can only be determined empirically by fitting.
- They are not convenient for mechanical design. For example, for the material selection at the initial stage of design, it is not easy to investigate and determine the microstructural damage caused by fatigue, creep and oxidation. It is also not reasonable to assume multiple materials have the same damage. However, for the empirical-based models, the fatigue evaluation can be conveniently calculated through inputting the temperature, frequency, and applied loading.

From the perspective of mechanical engineering design, it is desirable that a creep-fatigue model should have a clear structure that is understood by engineering practitioners, include the general variables at the engineering scale (such as temperature, time, and loading), include parameters that are measureable or knowable, and be easily mathematically solved. This is not the case for the mechanism-based models. Furthermore, material standards are invariably based on assurances of mechanical properties and element composition, not on microstructural properties. Hence, while designers may be interested in microstructure, they cannot rely on in their specifications.

1.4. Extension of the Empirical Models

As mentioned in Section 1.2, the damage-accumulation-based models ignore the interaction between fatigue and creep. The microstructural interactions between creep and fatigue are beginning to be understood at a qualitative level, e.g., [24]. Various mathematical expressions for this interaction are also available, with some (albeit limited) basis in microstructure or loading partitioning, e.g., [25,26]. Hence a possible way to move the field forward from a materials and design perspective is to further improve the conventional Coffin–Manson-based creep-fatigue models.

A recent development in that direction has been the development of a model that includes temperature, cyclic time, applied loading, and with applicability to multiple (metallic) materials [27]. This 'unified' model takes the form of a mathematical representation of plastic strain with functions including empirically determined coefficients:

$$\varepsilon_p = C_0 c(\sigma, T, t_c) N_f^{-\beta_0} \tag{2}$$

with

$$c(\sigma, T, t_c) = 1 - c_1(\sigma)\left(T - T_{ref}\right) - c_2 \log\left(t_c/t_{ref}\right)$$

$$T - T_{ref} = \begin{cases} T - T_{ref} & for \quad T > T_{ref} \\ 0 & for \quad T \leq T_{ref} \end{cases} \tag{3}$$

$$t_c/t_{ref} = \begin{cases} t_c/t_{ref} & for \quad t_c > t_{ref} \ and \ T > T_{ref} \\ 1 & for \quad t_c \leq t_{ref} \ or \ T \leq T_{ref} \end{cases}$$

where ε_p is the plastic strain which reflects fatigue capacity, N_f is the creep-fatigue life, C_0 and β_0 are the fatigue ductility coefficient and fatigue ductility exponent respectively, which are related to fatigue capacity at the pure-fatigue condition, T is the temperature, t_c is the cyclic time which presents the reciprocal of loading frequency, $c_1(\sigma)$ is the stress moderating equation which reflects the creep effect caused by the applied loading, c_2 is the constant, and σ reflects the applied loading which can be related to plastic strain through the cyclic strain-stress relation.

The equation also includes the concept of a *reference condition*. Here T_{ref} is the reference temperature, which is defined as 35% of the melting temperature, t_{ref} is the reference cyclic time which is suggested as a small value of 1 s.

The limitations presented by the existing Coffin–Manson-based models are improved by this model. The improvements are that: the structure includes the parameters of typical engineering problems, is easily mathematically solved, may be applied in multiple situations on multiple metallic materials, and covers the full range of conditions from pure fatigue to creep fatigue and then to pure creep. In particular, the model provides a more economic method for fatigue-life prediction since less empirical data are required than other empirical methods such as [15,17]. In addition, the model is applicable for engineering design at the initial stage through combining with finite element analysis (FEA) [28].

Nonetheless from an engineering design perspective, the model has room for improvement. There is a need to have a representation that can predict fatigue life at a given applied loading (or can be used to evaluate the critical value of applied loading under a given life). This process of engineering calculation is applied at the early stages of engineering design, when candidate materials are being considered in relation to the functional requirements. Furthermore, it is necessary to represent the full range of fatigue, creep-fatigue, and creep conditions. From a design perspective it is essential that any model is able to be applied using the type of information available to a design engineer (which may be tentative or incomplete).

1.5. Opportunities for Modifying the Unified Model

There is something of philosophical debate between proponents of the mechanism-based models, and the empirical models. From the perspective of the mechanism-based models, design ought to be conducted by detailed examination of microstructure and the determination of multiple material parameters, some based on properties of the crystal lattice, defect sizes, oxidation factors, and curve-fitting parameters. The methods are valuable because they can relate say critical crack length to life. However, they have other limitations as described above. From the perspective of the empirical models, design ought to be conducted by performing macroscopic tests (no microstructural tests required) at various environmental conditions, and then curve-fitting to obtain coefficients for a formulation. The methods are valuable because they can be highly accurate, and they readily relate loading to life. However, they have other limitations as described above.

Both methods have strengths and weaknesses. Proponents of the different schools of thought tend not appreciate the approach taken by the other, which is strange since both rely on fitting of many coefficients, and formulations encapsulating many assumptions. In the longer term the mechanism-based models may prove to be superior, if they can eventually link the physical features of the virgin and damaged microstructure to life, using parameters that are easy to measure and available at design time. However, at present, the empirical models are superior, at least for engineering design purposes. Hence the further improvement of the existing models is still worthwhile attempting from an engineering perspective.

The development of the unified creep-fatigue model [27] was based on an assumption, which is the change rate (β_0) of applied loading to fatigue life is constant for different temperatures and cyclic times. Graphically, the curves of applied loading vs. fatigue life at the situations with different temperatures and cyclic times at the log-log coordinate are parallel. The model applied this assumption because the slopes of loading-life curves at the log-log coordinate change only slightly among the situations with different temperatures and cyclic times. Although the accuracy of fatigue-life prediction is acceptable [27], this assumption still suggests some opportunities for future improvements.

Firstly, the accuracy of the fatigue-life prediction could be further improved. Specifically, although the influence of temperature and cyclic time to the exponent (β_0) is slight, this influence may not be negligible. However, in the unified model (Equation (2)), this exponent is a constant, not a function of temperature and cyclic time. This implies that inclusion of this influence may improve accuracy of the fatigue-life prediction. In addition, the derivation of creep-fatigue-related coefficients was conducted by applying numerical optimization. This is a curve-fitting-based method, and thus the fitting quality strongly depends on the number of power series and coefficients.

Such methods generally benefit, as regards fitting accuracy, from provision of higher power series and more tunable coefficients. There are examples in the literature that specialize in this approach, and result in exceptionally good fits [15–17]. However, this comes with two significant costs: (a) parameter non-identification becomes problematic in that multiple different combinations of parameters give similar results, hence the model becomes degenerate, and (b) it becomes difficult, even impossible, to link the coefficients to any meaningful parameters of physical properties or microstructure, hence the ontological power is depleted. Therefore, it is prudent to exercise restraint when expanding the terms within predictive models. It is preferable to add parameters that have some basis in physical reality. Consequently, we propose that the unified model might be improved by introducing new parameters for temperature and cyclic time into the exponent component (β_0) (See Sections 4.3 and 5.2).

Secondly, the description of the pure-fatigue condition could be further improved. Specifically, the unified model can be restored to the Coffin–Manson equation at the pure-fatigue condition which is described by the coefficients of C_0 and β_0. These two coefficients are derived from the empirical data by numerical optimization. As mentioned above, the assumption may impact the accuracy of these two coefficients, and thus the quality of pure-fatigue description may be reduced. In this case, the modification for exponent component (β_0) may improve the accuracy of C_0 and β_0, and then a better description for pure fatigue might be obtained (see Section 4.4).

In summary, we propose that the unified creep-fatigue model [27] could potentially be further improved through introducing the parameters of temperature and cyclic time into the exponent component. This has the potential to improve the accuracy of the fatigue-life prediction for both creep-fatigue and pure-fatigue conditions.

In the present work, we propose an explicit creep-fatigue model.

2. Methodology

The present work aims to further improve the unified creep-fatigue model [27]. This new explicit model should present improved accuracy of the fatigue-life prediction for both the creep-fatigue and pure-fatigue conditions. We are also mindful of the need to make such models accessible for engineering design. This has not always been a strong feature of models in the literature. This requires consideration of the type of information available to designers, and an understanding of what they are trying to achieve.

To improve the model, we removed the assumption that β_0 in Equation (2) is constant, and then introduced the parameters of temperature and cyclic time into the exponent component. We retained from [27] the concept that the fatigue capacity is reduced due to active creep behavior, which is influenced by temperature and time [1,4]. These two elements were included into the unified model (Equation (2)) through introducing a creep moderating function to the c component in Equation (2) [27]. In the present work the additional change is the introduction of an additional creep moderating function (a function of temperature and cyclic time) to modify the fatigue ductility exponent (β_0). The numerical relationships among temperature, cyclic time and exponent component were extracted from the understanding of creep behaviour (diffusion creep). Then, to build a bridge between pure fatigue and creep fatigue, the reference condition was also introduced. By this way, the exponent component can be restored to β_0 at the pure-fatigue condition.

Creep mechanisms are normally divided into Nabarro–Herring creep, Coble creep, grain boundary sliding and dislocation creep [1,29]. Nabarro–Herring creep and Coble creep show a strong dependency on temperature, where the diffusional flow of atoms occurs under conditions of relatively high temperature. Grain boundary sliding involves displacements of grains against each other. This is a particularly important mechanism for the creep failure of ceramics at high temperature because of the glassy-phase formation which provides a good sliding condition along the grain boundary. Dislocation creep presents progressive disruption through the crystal lattice, which results from both line defects and point defects, and can occur at relatively low temperature. This process is sensitive to the applied stress on the material, with a secondary dependency on temperature [1].

Based on the brief description of these four creep mechanisms, the diffusion creep (including Nabarro–Herring creep and Coble creep), which has strong temperature dependence, is used to extract the creep effect. (In the Discussion we briefly comment on the effect of ignoring these other creep mechanisms).

Then, an explicit creep-fatigue model was developed, see Section 3.1. This model was then validated on the materials of 63Sn37Pb solder and stainless steel 316 (see Sections 4.1 and 4.2). The coefficients were determined by the empirical data (including pure-creep data and creep-fatigue data) which were extracted from the literature. Ideally, the creep-fatigue data applied to obtain the coefficients and applied to validate the model should be extracted from two different literature sources. However, in the present work the empirical data are limited so, we extracted the empirical data from one source in the literature, and then the data were divided into two groups. One group was used to extract the coefficients of this model, and the other group used to validate this model. Hence if the experiments are conducted by following the experimental standard, the data at one specific condition (temperature, loading and cyclic time) should not be impacted by the location and operator.

After this, this model was compared with the unified and other models to evaluate the accuracy of the fatigue-life prediction at creep fatigue and pure fatigue (see Sections 4.3, 4.4 and 5.2), and the economy (see Section 5.3). In this process, the accuracy of life prediction for both the creep-fatigue and pure-fatigue conditions is discussed by evaluating the average error and prediction ratio (which are defined in Sections 3.2 and 3.3). In addition, the unified and integrated characteristics of the explicit model were investigated. Although the explicit model presents better the fatigue-life prediction, introducing more parameters into a numerical representation may result in poor economy for engineering designers because more empirical data may be required. Specifically, for an economical method for creep-fatigue-life prediction, the coefficients of this model should be obtained by fewer creep-fatigue experiments, because conducting creep-fatigue test is an expensive and time-consuming process. Thus, reduced empirical effect means better economy. This potential issue of economy is discussed (see Section 5.3).

Finally, the explicit model was applied to engineering design calculation (see Appendix A.1) and finite element analysis (see Appendix A.2). We provide specific directions for how the model may be used under both approaches, and the limitations thereof.

This new explicit model was developed with engineering design in mind. In particular, the general variables at the engineering scale (such as temperature, time and loading) were introduced to this explicit model, but the variables at the microstructural level (such as crack configuration, damage size, inter-void spacing, and oxidation) were not included. Although the explicit model still relies on empirical data, it is not a purely curve-fitting-based model. Specifically, the relationships between the different variables were derived from the understanding of creep and fatigue behaviours at the microstructural level, and the formulation was constructed by harmoniously integrating these relationships. This is not simply a curve-fitting-based process, thus gives an improved method for life prediction. During the process of engineering design, the coefficients are determined from the empirical data.

3. The Explicit Creep-Fatigue Model

We introduce the parameters of temperature and cyclic time into the exponent component. The modification process is presented in Section 3.1, and the method of determining the coefficients is presented in Section 3.2.

3.1. Development of the Explicit Creep-Fatigue Model

As mentioned in Section 1.4, the previous research applied an assumption that the fatigue ductility exponent (β_0) is constant at different temperatures and cyclic times. Removing this assumption gives an opportunity to further improve the unified model (Equation (2)). The unified model aimed to be applied for engineering design, thus the general variables at the engineering scale (temperature, time and applied loading) were included. However, at one specific temperature and cyclic time, applied loading does not influence the slope of life-loading curve, thus this parameter is not included into the exponent component and only the variables of temperature and cyclic time are included. In addition, according to the concept of fatigue capacity presented in [27], the slopes of life-loading curves gradually trend to zero with an increased creep effect (elevated temperature and prolonged cyclic time).

To resolve these issues, we introduce a creep moderating function $b(T, t_c)$ to modify the fatigue ductility exponent, and then is further expanded as the form of '$1 - x$':

$$\beta_0 \rightarrow \beta_0 b(T, t_c) = \beta_0 \left[1 - b'(T, t_c)\right] \tag{4}$$

We assume that time and temperature are not convoluted with each other, and thus the overall effect caused by temperature and time are additive. Later we show that this assumption gives sufficiently accurate outcomes. Then, function $b'(T, t_c)$ is split into a thermal component and a time component:

$$\beta_0 b(T, t_c) = \beta_0 \left[1 - b_1'(T) - b_2'(t_c)\right] \tag{5}$$

Then, we determine the relationships among temperature, cyclic time and exponent component. This is achieved through investigating creep behaviour. Specifically, function $\beta_0 b(T, t_c)$ implies the rate of fatigue-capacity decreases or increases between different temperatures and/or cyclic times. This rate can be described by diffusion-creep rate, and described by Fick's law [30]:

$$J = -D\frac{d\varphi}{dx} \tag{6}$$

where J is the diffusion flux which shows that the amount of substance flowing through a unit area at a unit time (thus reflects the diffusion rate), D is the diffusion coefficient, x is the position and φ reflects the concentration of vacancies.

The diffusion process is identified as a thermodynamic system due to the strong driving force of temperature. In this system, the transfer of atoms and the formation of vacancies are numerically evaluated by free energy at atomic level [31], and then the equilibrium atomic fraction of vacancies (N_v) is given by Equation (7):

$$N_v = exp\left(-\frac{\Delta G_f}{kT}\right) \tag{7}$$

where ΔG_f is the Gibbs free energy for formation of a vacancy, k is the Boltzmann's constant and T is the temperature. In Equation (6), φ is defined as the number of vacancies per unit volume, and thus is related to the atomic fraction by Equation (8):

$$\varphi = \frac{N_v}{\Omega} \tag{8}$$

where Ω is the atomic volume. Therefore, a natural exponential relation between the diffusion flux and the temperature component can be presented:

$$J \propto exp(-1/T) \tag{9}$$

The expression of $exp(1/T)$ can be simplified to a linear dependence when the temperature is relatively high enough, which is higher than the temperature where the creep behavior is activated (normally 0.35 of melting temperature), and usually the case when creep-fatigue is being considered in an engineering application. This provides a linear relationship, but the coefficient of the temperature (the slope of this straight line) should be determined from the empirical data. Thus, a linear relationship between diffusion rate and temperature arises:

$$J \propto T \tag{10}$$

In addition, Equation (9) shows that the diffusion-creep behaviour gives a logarithmical relationship between temperature and diffusion flux. The definition of 'diffusion flux' indicates that this term measures the amount of substance flowing through a cross sectional area during a unit time. Thus, a time dependence is included in this parameter in the form of a rate function. Then, Equation (9) can be presented as:

$$J = \frac{dD_v}{dt} \propto exp(-1/T) \tag{11}$$

where D_v reflects the amount of substance flowing through a unit area. This equation gives a logarithmical relation between temperature and cyclic time:

$$T \propto \log(1/t) \tag{12}$$

Then, the linear relationship of temperature vs. exponent component and the logarithmical relationship of temperature vs. cyclic time are integrated into Equation (5). The moderating function $b(T, t_c)$ is presented by Equation (13):

$$b(T, t_c) = 1 - b_1 T - b_2 \log t_c \tag{13}$$

where b_1 and b_2 are constant and determined by empirical data.

To build a bridge between pure fatigue and creep fatigue, we introduce the thermal and cycle time reference condition into Equation (13), then this equation is modified as:

$$b(T, t_c) = 1 - b_1\left(T - T_{ref}\right) - b_2 \log\left(t_c / t_{ref}\right) \tag{14}$$

Finally, the explicit creep fatigue model is given as:

$$\varepsilon_p = C_0 c(\sigma, T, t_c) N_f^{-\beta_0 b(T, t_c)} \tag{15}$$

with

$$
\begin{aligned}
c(\sigma, T, t_c) &= 1 - c_1(\sigma)\left(T - T_{ref}\right) - c_2 \log\left(t_c / t_{ref}\right) \\
b(T, t_c) &= 1 - b_1\left(T - T_{ref}\right) - b_2 \log\left(t_c / t_{ref}\right) \\
T - T_{ref} &= \begin{cases} T - T_{ref} & for \quad T > T_{ref} \\ 0 & for \quad T \le T_{ref} \end{cases} \\
t_c / t_{ref} &= \begin{cases} t_c / t_{ref} & for \quad t_c > t_{ref} \ and \ T > T_{ref} \\ 1 & for \quad t_c \le t_{ref} \ or \ T \le T_{ref} \end{cases}
\end{aligned}
\tag{16}
$$

3.2. The Method of Determining the Coefficients

The coefficients of the explicit model (Equation (15)) are determined by the empirical data, including pure-creep data and creep-fatigue data.

3.2.1. Selecting the Reference Condition

The creep damage is assumed to be active above the reference temperature and the reference cyclic time. The reference temperature is defined as 35% of the melting temperature [32], and the reference cyclic time is suggested as a small value, nominally 1 s.

3.2.2. Deriving the Coefficients of Function $c(\sigma, T, t_c)$

The method to derive the coefficients of $c(\sigma, T, t_c)$ proposed in [27] is extended to the present work. In this case, function $c_1(\sigma)$ and constant c_2 are presented by Equations (17) and (18):

$$c_1(\sigma) = -\frac{c_2}{P_{MH}(\sigma)} \tag{17}$$

$$c_2 = \frac{1}{\log\left(t_a / t_{ref}\right)} \tag{18}$$

In Equation (17), $P_{MH}(\sigma)$ is a function which represents the relationship between the Manson–Haferd parameter and applied stress (σ). The Manson–Haferd parameter under one specific stress is numerically presented as:

$$P_{MH} = \frac{T - T_a}{\log t - \log t_a} \tag{19}$$

where T is the absolute temperature, t is the creep-rupture time, and ($\log t_a$, T_a) is the point of convergence of the $\log t$-T lines. In particular, T_a is defined as the reference time (T_{ref}) below which creep is dormant.

Both Equations (17) and (18) are obtained by the empirical data of pure creep. Specifically, during creep-rupture tests, the temperatures (T), stresses (σ) and creep-rupture times (t) are recorded. Then, the relationships between T and $\log t$ under different stresses are plotted (Figure 1), wherein the temperature at the point of convergence is identified as the reference temperature, and the value of $\log t$ at this convergence point ($\log t_a$) is given by the average value of the $\log t(T_{ref})$ at different stresses. The value of $\log t$ then gives c_2.

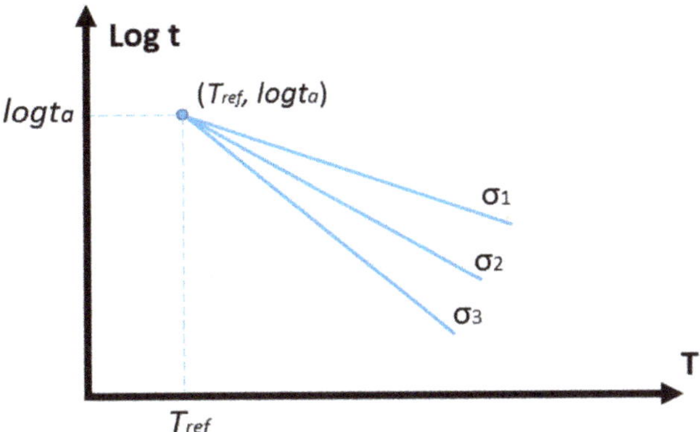

Figure 1. Plotting of T vs. $\log t$.

According to Figure 1, the Manson–Haferd parameters under different stresses are given, then the relationship between the Manson–Haferd parameter and applied stress ($P_{MH}(\sigma)$) can be obtained through curve fitting. Then, function $c_1(\sigma)$ is expanded.

3.2.3. Deriving the Coefficients

The remaining coefficients in the explicit model are determined by the empirical data of creep fatigue. Specifically, during the creep-fatigue tests, the temperatures (T), cyclic time (t), stresses (σ), plastic strain (ε_P) and fatigue life (N) are recorded. In particular, with the empirical data of plastic strain vs. stress, the coefficients (K' and n') of the cyclic strain–stress relation under different temperature-cyclic time conditions are obtained. In the present work, these two coefficients are applied to describe the engineering quantities-based relationship, and a power-law-based transition between strain and stress is included. They then are involved in the function $c_1(\sigma)$ for transforming stress into plastic strain (Equation (20)), and a moderating factor (f_m) is introduced to compress the stress effect on creep-related damage. We did not separate the whole of applied loading (σ) into two components. This is because we cannot say one part the applied loading contributes to creep, and another part contributes to fatigue. Therefore, we defined that the whole of applied loading works for both fatigue and creep damage.

$$c_1(\sigma) = -\frac{c_2}{P_{MH}(\sigma)} \to c_1(\varepsilon_P) = -\frac{c_2}{P_{MH}\left[f_m \cdot K'(T,t_c) \cdot \varepsilon_P^{n'(T,t_c)}\right]} \tag{20}$$

In the present work, we define f_m as a stress-moderating factor which is applied to compress the cyclic stress to an equivalent constant stress. This moderating factor is related to the shape of the loading wave, and presents the average level of the cyclic loading. Illustratively, the area below the contour of the cyclic loading along the time dimension should be equal to the area below the contour of the equivalent constant loading at the same time period. This is based on an assumption that creep makes the same contribution to tensile and compressional portions. Although this assumption may be not appropriate for some materials [33,34], it simplifies the method of extracting this factor. For example, f_m is defined as 0.6366 for the sinusoidal wave and as 0.5 for the triangular wave.

Then, numerical optimisation was applied to derive the coefficients of C_0, β_0, b_1 and b_2 by minimizing the average difference (δ_a) between the predicted fatigue life ($N_{pre,ij}$) and the experimental results ($N_{exp,ij}$) (Equation (21)).

$$\delta_a = \sum \left(\log N_{pre,ij} - \log N_{exp,ij}\right)^2 / n \tag{21}$$

where n is the number of data, and N_{ij} presents the fatigue life obtained at multiple conditions of $(T, t_c)_j$ and strain amplitude i.

3.3. Evaluation of the Explicit Model

The quality of fatigue-life prediction is evaluated by the prediction ratio. Specifically, the prediction ratio (Equation (22)) gives the ratio of predicted creep-fatigue life to experimental creep-fatigue life:

$$H_{ij} = \frac{N_{pre,ij}}{N_{exp,ij}} \tag{22}$$

In the present work, we define that:

An acceptable prediction ratio should be between 0.75 and 1.25.

This range is narrower (more conservative than the range shown in other literature, wherein a factor of 2 or 1.5 is normally given [25,35,36]. This also can be shown illustratively, where all data points of N_{pre} vs. N_{exp} under multiple temperatures and cyclic times should fall between the upper bound (+25%) and the lower bound (−25%) relative to ideal correlation ($H = 1$).

4. Validation

The explicit model is validated on the materials of 63Sn37Pb solder and stainless steel 316. The coefficients are determined by using the method proposed in Section 3.2, where the empirical data are extracted from the literature. The quality of fatigue-life prediction is evaluated by the method proposed in Section 3.3.

4.1. Validation on 63Sn37Pb Solder

4.1.1. Deriving the Coefficients

The reference temperature for 63Sn37Pb solder was chosen as 160 K and the reference cyclic time was defined as 1 s. The creep-rupture data [37] are plotted in Figure 2, and the point of convergence $(T_{ref}, \log t_a)$ is evaluated as (160 K, 8.232). This gives

$$c_2 = \frac{1}{\log\left(t_a / t_{ref}\right)} = \frac{1}{\log(10^{8.232}/1)} = 0.1215 \tag{23}$$

and the relationship between stress and the Manson–Haferd Parameter:

$$-\frac{1}{P_{MH}(\sigma)} = 8.1979 \times 10^{-3} + 8.3244 \times 10^{-4}\sigma + 6.6651 \times 10^{-6}\sigma^2 \tag{24}$$

Then, substituting into Equation (24), function $c_1(\sigma)$ is expressed as:

$$c_1(\sigma) = -\frac{c_2}{P_{MH}(\sigma)} = 9.9586 \times 10^{-4} + 1.01122 \times 10^{-4} \cdot f_m \cdot \sigma + 8.09657 \times 10^{-7} \cdot f_m{}^2 \cdot \sigma^2 \tag{25}$$

and the magnitude of f_m is given as 0.6366 for the sinusoidal wave.

The creep-fatigue coefficients [16] obtained from the literature are tabulated in Table 1. Minimizing the difference between the predicted creep-fatigue life ($N_{pre,ij}$) and the experimental creep-fatigue life ($N_{exp,ij}$) yields $C_0 = 7.790$, $\beta_0 = 0.858$, $b_1 = 0.000234$ and $b_2 = 0.00596$, and returns an average error (δ_a) (Equation (21)) of 0.000509.

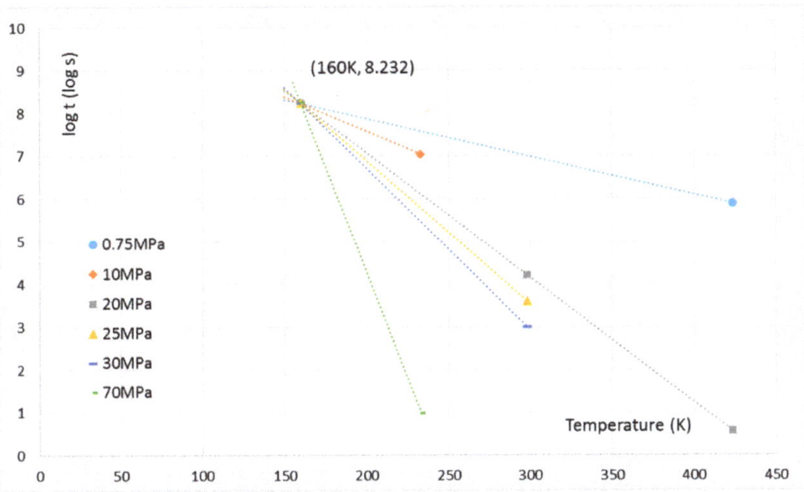

Figure 2. Creep-rupture characteristics of 63Sn37Pb solder.

Table 1. Creep-fatigue data for 63Sn37Pb solder.

Temperature (K)	Frequency (Hz)	Cyclic Time (s)	Creep-Fatigue Coefficients $\varepsilon_p=\varepsilon_f' N_f^{-\beta}$		Strain-Stress Coefficients $\sigma/2=K'(\varepsilon_p/2)^{n'}$	
			ε_f'	β	K'	n'
233	1	1	2.98	0.773	129.5	0.0652
398	1	1	1.45	0.723	84.026	0.1199
298	0.001	1000	1.01	0.708	90.437	0.1438

Consequently, the coefficients of the explicit creep-fatigue equation for 63Sn37Pb solder are collected in Table 2:

Table 2. The coefficients of the explicit formulation for 63Sn37Pb solder.

C_0	β_0	C_2	T_{ref} (K)	T_{ref} (s)	b_1	b_2	f_m	δ_a (log(cycle)^2)
7.790	0.858	0.1215	160	1	0.000234	0.00596	0.6366	0.000509
$c_1(\sigma, f_m)$		$9.9586 \times 10^{-4} + 1.01122 \times 10^{-4} \cdot f_m \cdot \sigma + 8.09657 \times 10^{-7} \cdot f_m^2 \cdot \sigma^2$						

4.1.2. Evaluation

To evaluate the explicit creep-fatigue model, another groups of creep-fatigue data (Table 3) [16] are used to compare with predicted fatigue life which is supported by the results shown in Section 4.1.1.

Table 3. Creep-fatigue data for 63Sn37Pb solder.

Temperature (K)	Frequency (Hz)	Cyclic Time (s)	Creep-Fatigue Coefficients $\varepsilon_p=\varepsilon_f' N_f^{-\beta}$	
			ε_f'	β
298	1	1	2.28	0.756
348	1	1	1.86	0.743
298	0.1	10	1.57	0.719
298	0.01	100	1.28	0.712

The prediction ratio (N_{pre}/N_{exp}) under multiple temperatures and cyclic times are plotted in Figure 3, where all data points fall between the upper bound (+25%) and the lower bound (−25%). The upper bound and the lower bound present the prediction ratios are 0.75 and 1.25 respectively. This implies that the explicit creep-fatigue equation provides a high quality of fatigue-life prediction, specifically, a relatively high correlation between predicted and experimental creep-fatigue life.

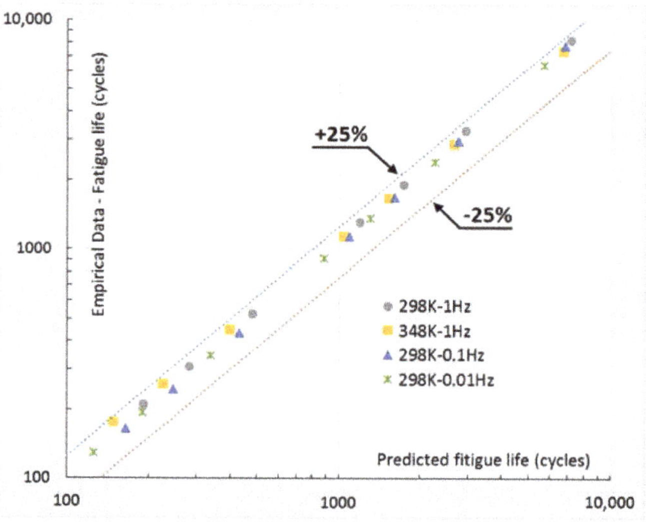

Figure 3. Prediction ratio for 63Sn37Pb solder.

4.2. Validation on Stainless Steel 316

4.2.1. Deriving the Coefficients

The reference temperature for stainless steel 316 was chosen as 585 K and the reference cyclic time was defined as 1 s. The creep-rupture data [38] are plotted in Figure 4, and the point of convergence (T_{ref}, $\log t_a$) is evaluated as (585 K, 10.783). This gives

$$c_2 = \frac{1}{\log\left(t_a/t_{ref}\right)} = \frac{1}{\log(10^{10.783}/1)} = 0.09274 \tag{26}$$

and the relationship between stress and the Manson–Haferd Parameter:

$$-\frac{1}{P_{MH}(\sigma)} = 0.006011 + 7.0286 \times 10^{-5}\sigma - 1.1429 \times 10^{-7}\sigma^2 \tag{27}$$

Then, substituting into Equation (27), function $c_1(\sigma)$ is expressed as:

$$c_1(\sigma) = -\frac{c_2}{P_{MH}(\sigma)} = 5.575 \times 10^{-4} + 6.5184 \times 10^{-6} \cdot f_m \cdot \sigma - 1.0599 \times 10^{-8} \cdot f_m^2 \cdot \sigma^2 \tag{28}$$

and the magnitude of f_m is given as 0.5 for the triangular wave.

The creep-fatigue coefficients [39] obtained from the literature are tabulated in Table 4. Minimizing the difference between the predicted creep-fatigue life ($N_{pre,ij}$) and the experimental creep-fatigue life ($N_{exp,ij}$) yields $C_0 = 7.768$, $\beta_0 = 0.571$, $b_1 = -0.000225$ and $b_2 = -0.0223$, and returns an average error (δ_a) (Equation (21)) of 0.00255.

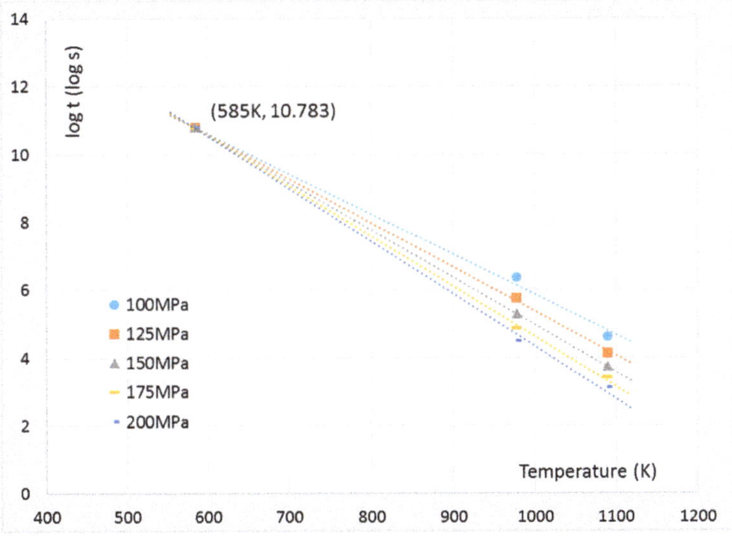

Figure 4. Creep-rupture characteristics of stainless steel 316.

Table 4. Creep-fatigue data for stainless steel 316.

Temperature (K)	Strain Rate (%/min)	Creep-Fatigue Coefficients $\varepsilon_p = \varepsilon_f' N_f^{-\beta}$		Strain-Stress Coefficients $\sigma/2 = K'(\varepsilon_p/2)^n$	
		ε_f'	K'	K'	n
723	0.4	0.279	0.522	444	0.338
873	4	0.347	0.578	175	0.173
973	40	0.425	0.578	150	0.211

Consequently, the coefficients of the explicit creep-fatigue equation for stainless steel 316 are collected in Table 5:

Table 5. The coefficients of the explicit formulation for stainless steel 316.

C_0	β_0	C_2	T_{ref} (K)	T_{ref} (s)	b_1	b_2	f_m	δ_a (log(cycle)^2)
0.768	0.571	0.0927	585	1	−0.000225	−0.0223	0.5	0.00255
$c_1(\sigma, f_m)$		$5.575 \times 10^{-4} + 6.5184 \times 10^{-6} \cdot f_m \cdot \sigma - 1.0599 \times 10^{-8} \cdot f_m^2 \cdot \sigma^2$						

4.2.2. Evaluation

To evaluate the explicit creep-fatigue model, another groups of creep-fatigue data (Table 6) [39] are used to compare with predicted fatigue life which is supported by the results shown in Section 4.2.1.

The prediction ratio (N_{pre}/N_{exp}) under multiple temperatures and cyclic times are plotted in Figure 5, where all data points fall between the upper bound (+25%) and the lower bound (−25%). The upper bound and the lower bound present the prediction ratios are 0.75 and 1.25 respectively. This implies that the explicit creep-fatigue equation provides a high quality of fatigue-life prediction, specifically, a relatively high correlation between predicted and experimental creep-fatigue life.

Table 6. Creep-fatigue data for stainless steel 316.

Temperature (K)	Strain Rate (%/min)	Creep-Fatigue Coefficients $\varepsilon_p=\varepsilon'_f N_f^{-\beta}$	
		ε'_f	β
723	4	0.369	0.521
873	40	0.408	0.563
973	0.4	0.246	0.555
973	4	0.470	0.615

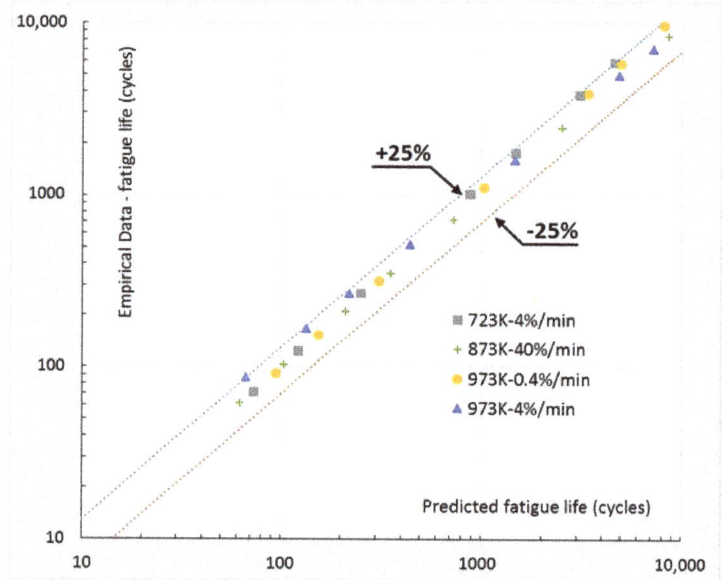

Figure 5. Prediction ratio for stainless steel 316.

4.3. Accuracy Comparison: Explicit vs. Unified Models

The present work aims to improve the accuracy of the creep-fatigue-life prediction through further modifying the unified model. Thus, the ability of life prediction by applying the explicit model should be better than applying the unified model. This is proved through comparing the explicit model with the unified model on the materials of 63Sn37Pb and stainless steel 316.

Specifically, the explicit formulation removes the assumption applied in the unified creep-fatigue model (Equation (2)), and then a creep moderating function was introduced into the exponent component. In this way, the explicit model has better ability to describe creep fatigue. To prove this, we applied the creep-fatigue data (Table 1 for 63Sn37Pb solder and Table 3 for stainless steel 316) to extract the coefficients of the unified model (Equation (2)) and the explicit model (Equation (15)). Then, these coefficients were applied to predict the fatigue life for the situations shown in Table 4 for 63Sn37Pb solder and Table 6 for stainless steel 316. The empirical data, and predicted life given by the unified model and the explicit model are illustrated in Figure 6 for 63Sn37Pb and Figure 7 for stainless steel 316.

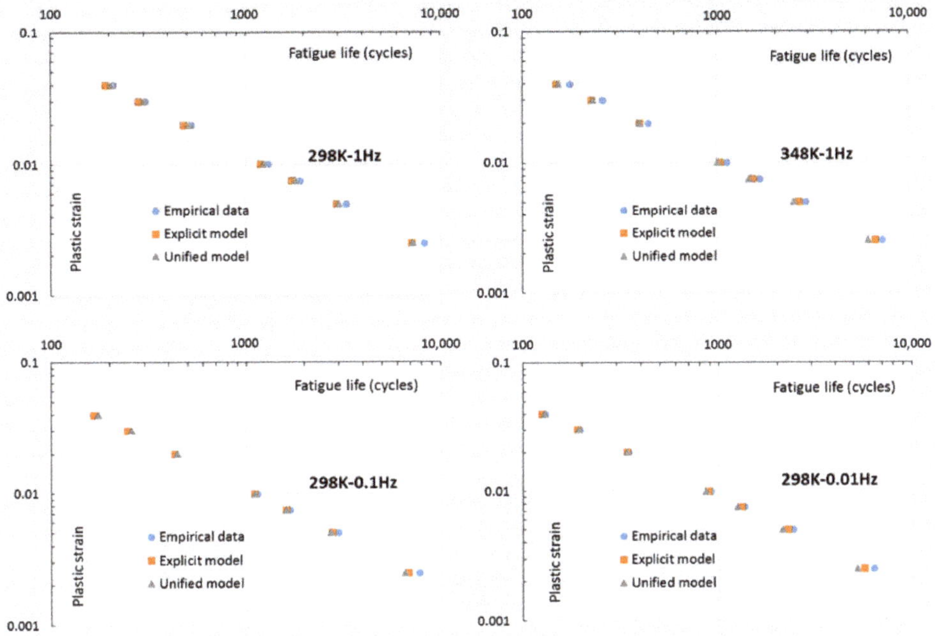

Figure 6. Empirical data and predicted life for 63Sn37Pb.

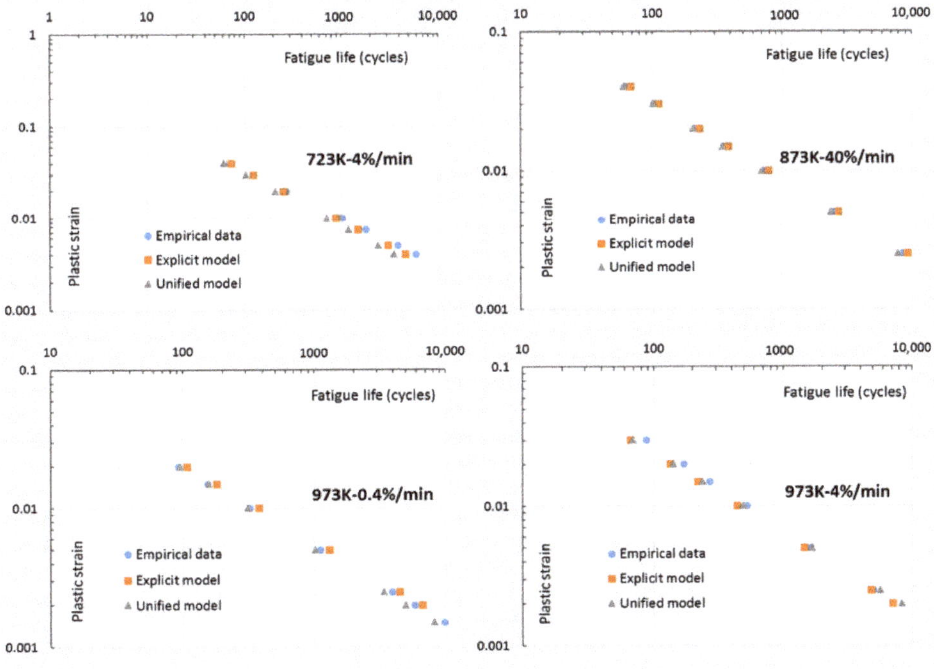

Figure 7. Empirical data and predicted life for Stainless steel 316.

Figures 7 and 8 show that life-loading curves given by the explicit model are closer to the empirical data, thus we conclude that the explicit model has better ability to predict fatigue life at the creep-fatigue condition. This is also proved by the average errors and prediction ratios (see Equations (21) and (22) in Sections 3.2.3 and 3.3 for the definitions of the average error and prediction ratio respectively) in Table 7. In particular, the value of the prediction ratio in Table 7 is represented by a range which is given by the maximum and minimum prediction ratios of the whole results. This representation is also shown in Table 8.

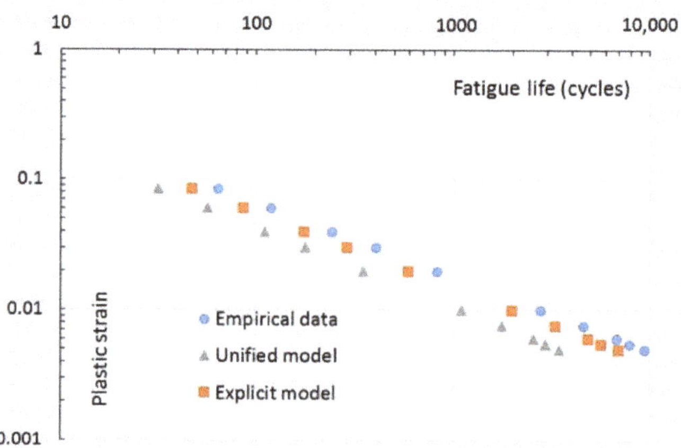

Figure 8. Predicted life and empirical data of stainless steel 316 for pure fatigue.

Table 7. The average errors and prediction ratios given by the unified model and the explicit model.

Materials	Average Errors (log($cycle$)^2)		Prediction Ratios Range	
	Unified Model	Explicit Model	Unified Model	Explicit Model
63Sn37Pb solder	0.00177	0.00139	0.80–1.10	0.85–1.05
Stainless steel 316	0.0110	0.00883	0.75–1.20	0.75–1.00

Table 8. Accuracy of prediction regarding the empirical-data number.

Number of Data Groups to Derive Coefficients	Average Errors for Predicting Fatigue Life (log($cycle$)^2)		Prediction Ratios Range	
	Wong & Mai's Model	Explicit Model	Wong & Mai's Model	Explicit Model
Six groups of data	0.002481	0.001176	0.85–1.00	0.80–1.00
Three groups of data	0.006608	0.01547	1.00–1.10	0.60–1.15

Table 7 shows that the explicit model provides smaller average errors and narrower ranges of prediction ratio for both the materials of 63Sn37Pb solder and stainless steel 316. This demonstrates that the explicit model has better accuracy for quantitatively representing creep fatigue.

4.4. The Ability to Describe Pure Fatigue

Both the unified creep-fatigue model and the explicit model can be restored into the Coffin–Manson equation at pure fatigue. This loading condition is numerically presented by the coefficients of C_0 and β_0, thus the accuracy of pure-fatigue description is determined by them. In the present work, this ability was evaluated through comparing the predicted life with the empirical data on stainless steel 316 [39] (Figure 8).

Figure 8 shows that the loading-life curve formulated by the coefficients of C_0 and β_0 in the explicit model is closer than the unified model to the empirical data. This is also described by the prediction ratio. The prediction ratios for these two models is presented in Figure 9. The dotted lines (bounds) which are labeled by 1, -25%, -50% and -75% in Figure 9 represent the prediction ratios of 1, 0.75, 0.5 and 0.25 respectively.

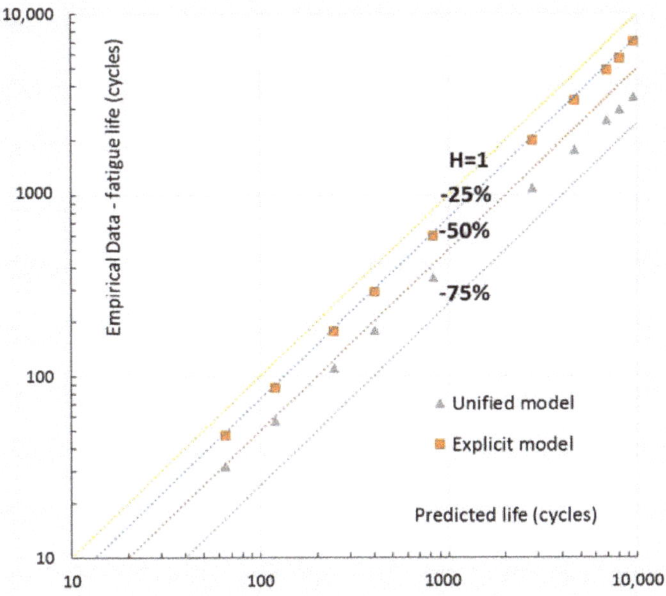

Figure 9. Prediction ratio for stainless steel 316 at pure fatigue.

Figure 9 shows that the pure-fatigue prediction ratios are around 0.75 for applying the coefficients of the explicit model, but the prediction ratios are lower, between 0.5 and 0.25, for using the coefficients of the unified model.

Both Figures 8 and 9 imply that the errors between the empirical data and the predicted life given by the explicit model are smaller. Therefore, we conclude that the explicit model has better ability to describe pure fatigue than the unified model.

4.5. General Process of Validation for Other Materials

The explicit model could be further validated through involving more empirical data on more materials. The general process of validation can be summarised as follows:

(1) Obtain the empirical data for one specific material. The data include pure-creep data and creep-fatigue data, which could be extracted from the literature, or collected by performing testing. In particular, the creep-fatigue data under multiple temperatures and cyclic times are divided into two groups (3 to 4 sub-group data at different temperatures and cyclic times for each group). One group data (named Group1) are applied to determine the coefficients of the explicit model, and other group data (named Group2) are applied to compare with the predicted life (evaluate accuracy of the fatigue-life prediction).

(2) Determine the coefficients of the explicit model. The coefficients are determined by the method presented in Section 3.2.

(3) Predict fatigue life. With the coefficients obtained in step 2, the predicted life under the situations presented in 'Group2' are calculated through using the explicit model.

(4) Evaluate the explicit model. The evaluation of the explicit model is conducted by the method given in Section 3.3. This process is numerically and illustratively presented by the prediction ratios. If the predicted data satisfy the range of acceptation given in Section 3.3, we can conclude that the explicit model can be applied on this material to predict creep-fatigue life.

5. Discussion

5.1. The Characteristics of the Explicit Creep-Fatigue Model

The explicit creep-fatigue model was validated on the materials of 63Sn37Pb solder and stainless steel 316 (see Section 4). This implies that this model has ability to be applied at multiple temperatures and cyclic times, and the relationships between different variables (temperature, cyclic time, applied loading and life) in the explicit model are applicable for different materials.

In addition, at the reference (the pure-fatigue) condition (where $T = T_{ref}$ and $t_c = t_{ref}$), the explicit creep-fatigue model can be restored to the Coffin–Manson equation. At the pure-creep condition (where $c(\sigma, T, t_c) = 0$), the explicit creep-fatigue model can be reformed as the Manson–Haferd parameter for creep. Consequently, the explicit formulation recovers both of the standard fatigue and creep formulations.

Consequently, the explicit model (Equations (15) and (16)) has the following features:

1. Provides one formulation that covers the full range of conditions from pure fatigue, to creep fatigue, then to pure creep.
2. Recovers the mathematical formulation of both of the standard fatigue and creep formulations (Coffin–Manson and Manson–Haferd respectively).
3. Accommodates multiple temperatures. Specifically, the explicit model can be applied to predict fatigue life at situations with different temperatures.
4. Accommodates multiple cyclic times. The explicit model is applied at the cyclic loading without hold time, thus the cyclic time refers to the period of one cycle of this loading condition. This is a limitation of this model, which will be discussed in Section 5.4.
5. Accommodates multiple materials. The explicit model was not a purely empirical-based model because the physical meaning was indirectly introduced into the explicit model. This process is quite different from the curve-fitting method. Thus, we conclude that the explicit model is potential able to be applied for multiple materials: we have demonstrated validation for 63Sn37Pb solder and stainless steel 316. Further validation on different materials is needed: this will be discussed in Section 5.4.
6. Provides a physical basis for the structure of the formulation. The basis of the c term has been explained previously [40]. Specifically, diffusion-creep behaviour gives a linear relationship of temperature vs. loading and a logarithmical relationship of temperature vs. cyclic time. Plastic zone around the crack tip gives a power-law relation between life and loading. The new b term is justified on principles of diffusion-creep rate and represents Fick's law (see Section 3.1). Both c and b terms were built on the concept of fatigue capacity, which was formulated as '$1 - x$'. This formulation numerically presents the negative effect of creep on fatigue. In addition, the introduction of the reference condition gives an opportunity to connect pure fatigue with creep fatigue.

Attributes 1–2 may be considered 'integrated' attributes, 3–5 'unified' attributes, and 6 a 'natural origin' attribute. Regarding integration, the models based on microstructural features, e.g., the integrated creep-fatigue theory [41,42], also offer an integrated characteristic. However, the determination of microstructural variables (such as crack, damage size and inter-void spacing) is challenging in the engineering situation. The explicit model is potentially easier to use in the engineer case.

Furthermore, the explicit model can be restored to the Coffin-Mason equation at pure fatigue and can be reformed as the Manson–Hefard parameter at pure creep. Both of these two formulations are conventional engineering models, and can be applied for engineering design without the need of observations at the microstructural level. This is a positive feature.

By natural origin we do not necessarily mean that the model has a physical basis traceable to microstructure and mechanical properties. Rather that the formulation of the model is consistent with existing representations of principles of physics (e.g., laws). We acknowledge that a full connection of all parameters in the explicit model to measureable variables of microstructure remains elusive. This limitation applies to all creep and fatigue models.

While there are other creep-fatigue formulations that also have high accuracy, they lack one or more of the features of the explicit model: they do not have the integrated characteristic; they are typically accurate only for specific cases (poor unified attribute); or they rely on the inclusion of many coefficients (typically into power series) which have no natural origin. Many of the competing models are so over-endowed with coefficients, e.g., [16,18], that they also have the risk of parameter non-identifiability.

5.2. The Ability to Describe Creep Fatigue

The unified model (Equation (2)) presents better ability to predict life at the creep-fatigue condition. This was proved through comparing the unified model with the existing creep-fatigue models. For example, the unified model was compared with Solomon's model [15], Jing's model [17], and Wong & Mai's model [18].

Both Solomon's model and Jing's model use fixed coefficients. When they are applied to other situations, Solomon's model results in a poor average error (23.96), and Jing's model cannot give any numerical solution. Thus, they only can be used in the situations where they were derived, and cannot be extended to other situations and other materials where there are no empirical data. This is because these models determine their coefficients by numerical optimisation across all variables (including temperature, frequency, fatigue life and applied loading). Hence when changing to a different material it is necessary to recalculate all the coefficients: it is not possible to simply change only some of the coefficients. However, Wong & Mai's model has potential to be applied to multiple materials. This is because this model has seven independent coefficients which are required to be recalculated for different materials. The accuracy of these models comes at the cost of high specificity of the coefficients, and the risk of parameter non-identifiability. Also, the coefficients in the power series terms have no physical identity, but only exist to provide improve mathematical fit.

To allow a comparison with the explicit model, we re-calculated the coefficients for Solomon's model, Jing's model, and Wong & Mai's model, as follows.

Solomon's model (Equation (29)) is:

$$\varepsilon_p = C_1(T)\left(N_f f^{k-1}\right)^{-\beta_0} \tag{29}$$

with

$$C_1(T) = c_1 - c_2 T - c_3 T^2 - c_4 T^3 \tag{30}$$

where T is temperature in °C, f is the frequency, N_f is the fatigue life, and c_1, c_2, c_3, c_4, k and β_0 are constants derived from the empirical data.

Jing's model (Equation (31)) is:

$$\varepsilon_p = C_2(T) N_f^{\beta(T)} \tag{31}$$

with

$$\begin{aligned} C_3(T) &= c_4 - c_5 T + c_6/\sqrt{T} \\ \beta(T) &= b_3 - b_4 T + b_5/\sqrt{T} \end{aligned} \tag{32}$$

where c_4, c_5, c_6, b_3, b_4 and b_5 are constant derived from the empirical data.

Wong & Mai's model (Equation (33)) is:

$$\varepsilon_p = C_0 s(\sigma) c(T, f) N_f^{-\beta_0 b(T, f)} \tag{33}$$

with

$$s(\sigma) = \begin{cases} 1 & \text{when creep is dormant} \\ exp\left[-\left(\sigma_{yield}\varepsilon_p^{n'}\right) / A'\right] & \text{when creep is active} \end{cases}$$
$$c(T, f) = 1 - c_1\left(T - T_{ref}\right) - c_2 \log\left(f / f_{ref}\right)$$
$$b(T, f) = 1 - b_1\left(T - T_{ref}\right) - b_2 \log\left(f / f_{ref}\right) \tag{34}$$

where n' is cyclic hardening index, σ_{yield} is the yield stress, T is the temperature in Kelvin, f is the frequency, T_{ref} is the reference temperature below which creep becomes dormant, f_{ref} is the reference frequency, and C_0, β_0, A', c_1, c_2, b_1 and b_2 are constant.

We recalculated all these constants, for these three models, for stainless steel 316, and plotted the results in Figure 10 to compare with the explicit model and empirical data.

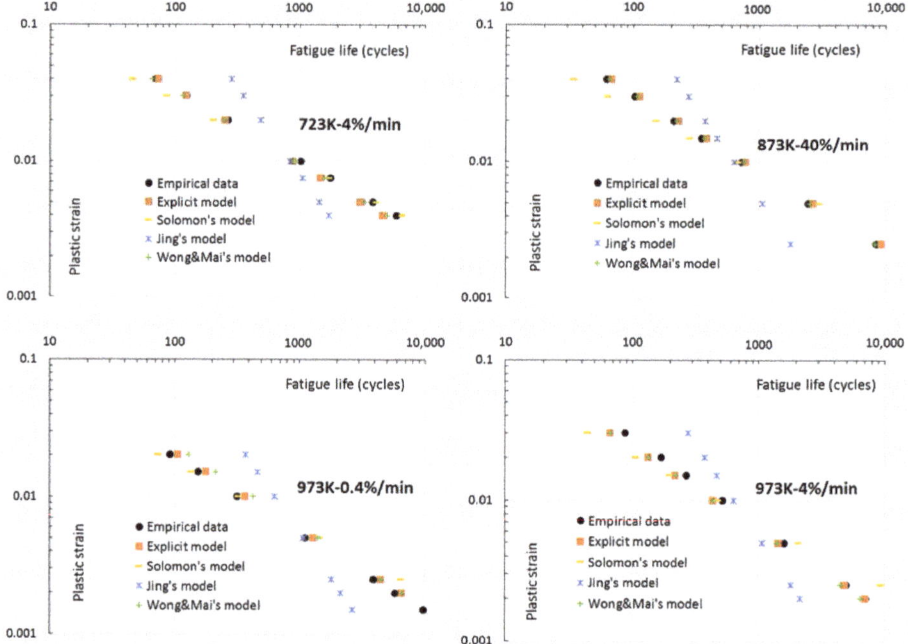

Figure 10. Empirical data for stainless steel 316, and predicted life given by the explicit model, Solomon's model, Jing's models, and Wong & Mai's model.

The average errors calculated by Equation (21), for the explicit model, Solomon's model, Jing's model, and Wong & Mai's model, for stainless steel 316, are plotted in Figure 11:

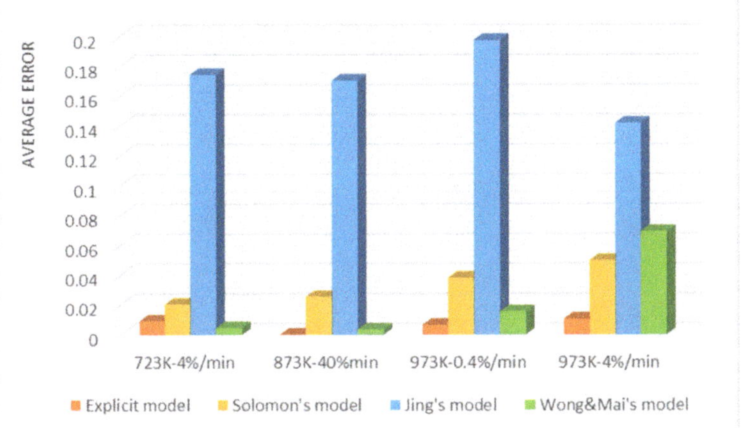

Figure 11. Average error for predicted life given by the explicit model, Solomon's model, Jing's model, and Wong & Mai's model.

Both Figures 10 and 11 show that the explicit model (Equation (15)) has better numerical accuracy for describing creep fatigue than the models of Solomon, Jing, and Wong & Mai models. For Solomon's and Jing's models, the average error given by these two models are much higher than the explicit model. This is because the relationships between different variables in these two models were completely derived from the empirical data of one specific material, thus they cannot be extended to other materials. However, Wong & Mai's model presents better ability for life prediction than Solomon's and Jing's models (except the situation of 973 K-4%/min). This may be because Wong & Mai's model involves a material property (yield stress), and it also applied a concept that would later be referred to as fatigue capacity. However, this model has seven independent coefficients which are required to be determined by empirical data. Consequently, this leads to another issue, that of economy.

5.3. The Economy

The economy is an important factor which is considered during the process of engineering design [4]. An economical method should provide a good balance between the accuracy and cost. Although the mechanism-based models may not need any creep-fatigue tests, observing and measuring microstructure is not a simple and economic process for engineering practitioners. The empirical-based models are more suited for engineering purposes, hence are the point of comparison for the explicit model. We selected Wong & Mai's equation [18] to compare with the explicit model regarding to the economy, since the Wong & Mai's equation shows better life-prediction ability than other existing models (see Section 5.2).

In the present work, empirical data under different temperatures and cyclic times was taken from the literature (Table 4). The first stage took seven groups of data and split this into a group of six and one. The six groups of empirical data were applied to derive the coefficients of Wong & Mai's equation and the explicit model. Then, these coefficients were used to predict fatigue life at the condition of the remaining data set (named 'predicted condition'). The discrepancy was noted.

The second stage repeated this analysis but with three and four groups respectively, and again the discrepancy was noted. Finally, the average errors and prediction ratios obtained in these two situations were compared. Thus, it becomes possible to infer how sensitive each model is to the available quantity of data. A model with better economy would be one where the degradation in accuracy was less sensitive to the quantity of data.

The comparison between Wong & Mai's equation [18] and the explicit model for the accuracy of the fatigue-life prediction regarding the empirical-data number is shown in Table 8 and Figure 12 for stainless steel 316.

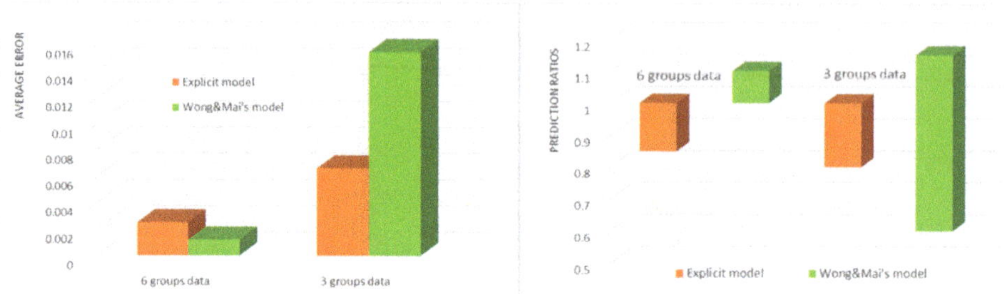

Figure 12. Average error and prediction ratios regarding the empirical-data number.

Table 8 and Figure 12 show that the Wong & Mai equation gives better accuracy for the fatigue-life prediction (at the condition of 973 K-0.4%/min) than the explicit model when six groups of creep-fatigue data are available. This is because the Wong & Mai equation has more independent coefficients which are extracted through the numerical optimisation (a curve-fitting-based method), which results in a better fitting quality when enough empirical data are involved [43].

When only three groups of creep-fatigue data were selected to obtain the coefficients, then the situation changes. The Wong & Mai equation experiences more severe degradation against both measures: average error and prediction ratio.

In both cases the numerical optimization still yields quite small average errors for fitting for the Wong & Mai equation. However, poor accuracy results when Wong & Mai's equation with the coefficients obtained at this stage is extended to predict fatigue life at the 'predicted condition' (973 K-0.4%/min). Specifically, for the Wong & Mai equation, the average error worsens from 0.001176 to 0.01547, and the range of prediction ratio also widens.

The explicit model also degrades, but not to the same extent. Specifically, the average error is 0.002481 for the coefficients obtained from six groups of creep-fatigue data, and 0.006608 for the coefficients obtained from three groups of data. Meanwhile, the range of prediction ratio only slightly changes between these two situations.

Thus, the explicit model shows greater robustness for smaller datasets. This is significant is it indicates that fewer creep-fatigue experiments are necessary to obtain the coefficients of the explicit model. As a result, we conclude that the explicit model is the more economical method because less empirical data are required.

The explicit model was developed through introducing the parameters of temperature and cyclic time into the exponent component, and thus two more coefficients (b_1 and b_2) were included. This leads to a risk that more empirical data are required to obtain high-fitting accuracy. However, Table 8 implies this risk is not significant. Although the accuracy is reduced if less empirical data are applied to determine the coefficients, the difference of errors between these two situations is small based on a log-scale calculation, and the reduced accuracy still provides good ability for life prediction (this was proved in Section 4).

We suggest that the robustness of the explicit model arises because the parameters and corresponding coefficients were introduced into the exponent component (modifying the fatigue ductility exponent) rather than the coefficient component (modifying the fatigue ductility coefficient). In this case, the process of numerical optimisation for the coefficient and exponent components was conducted in two relatively separate directions. While this reduced the accuracy somewhat, it also reduced the risk of parameter non-identifiability. We expect that the economy would worsen if the modification was conducted for the coefficient component, since this already has a several tunable coefficients.

Consequently, we conclude that the explicit creep-fatigue model presents an economical method for the fatigue-life prediction, and introducing two more coefficients into the exponent component does not significantly impact the economy.

5.4. Limitations and Implications for Future Research

Limitations that designers need to note are that strictly speaking the method has only been validated for the materials of 63Sn37Pb and stainless steel 316. However, this explicit model is potentially usable for other metallic materials. We anticipate difficulties applying this model for plastics and composites because they present totally different material characteristics and failure mechanisms. However, it is not impossible that this explicit model may be further improved and extended to other material categories as more empirical data are included. In particular, nylon is widely used in the engineering industry for load bearing parts. Thus, it may be an interesting future project to check and adapt this model to engineering nylon (such as nylon 6).

The explicit model is not yet ideal regarding natural origin as it does not include quantifiable microstructural properties. To achieve this, it would be necessary to better understand the microstructural processes of fatigue & creep—especially their interactions—and how those affect plastic strain and life. Some work is available in this area, e.g., [24], but there is still a long way to go before the values of coefficients in a creep-fatigue model can be predicted ab initio from microstructural inspection. In addition, as mentioned in Section 2, the explicit model ignores the dislocation creep and grain boundary sliding. In this case, introducing these two behaviours to reflect creep effect at high stress may be beneficial.

Another potential avenue of future research is to continue the process of extending existing models towards a more complete theory, as has been illustrated here with the redevelopment of the unified model into the explicit. During the process of development, more microstructural-level-based parameters may be included, with a corresponding inclusion of new terms into the model. We suggest that it is worthwhile designing these extensions to include other well-established phenomena, as we demonstrated in Section 3, rather than merely chasing better accuracy by adding more power terms and coefficients.

The situation of cyclic loading without hold time (dwell-fatigue) is not covered by the explicit model. In this loading condition, fatigue makes more of a contribution than creep, because the total time is too small to produce marked creep damage. However, for cyclic loading with hold time, the creep effect gradually intensifies as the hold time increases. Then more creep damage is produced than fatigue damage, and the failure finally occurs due to the creep effect. We could imagine that in the situation with a relatively short hold time, the explicit model may still present a reasonable prediction of fatigue life, but the accuracy of this prediction may become worse when the hold time is prolonged. This implies that the explicit formulation has an opportunity to be further improved to cover the situation with hold time or relatively long cyclic time. To achieve this, it would seem necessary to modify the formulation (especially, the creep component in this explicit model) to include new terms of as yet-unknown mathematical form. Conceptual works, e.g., [24], may be useful in identifying the basic form of these relationships.

At elevated temperature, the crack surface is oxidized, and then the material becomes more brittle. This results in further crack propagation. Therefore, the oxidation effect should ideally be included. The class of models based on observation of microstructure, e.g., [41,42], are superior in this regard because they can measure the voids and internal damage.

The class of models based on macroscopic empirical testing, to which the explicit model belongs, lack the microstructural parameters of crack length, oxidation, etc. At least not as primary variables, but the effects are partly accommodated through other means. In the explicit model, the coefficients are determined from the empirical data through numerical optimization, thus any oxidation effects are incorporated into the fitting process. Although the accuracy of fatigue-life prediction may be impacted, the results still show acceptable accuracy (see Sections 4.3 and 5.2).

In future work it might be possible to include oxidization in the explicit equation. Superficially this might involve simply including a power series term. However this may not be entirely successful, because our observation is that simply adding more terms and coefficients does improve accuracy, but at the cost of introducing model degeneracy. This has the further consequence of making the model more highly dependent on the specific situation, i.e., reduces the ability of the model to generalize to other materials and situations. The challenge is to include the oxidation effect, in a way that is coherent with how the effect operates physically, and to do so using parameters that are identifiable by the engineering designer. This opens an opportunity to further improve this engineering-based model.

5.5. Application to Engineering Design and Structural Mechanics

The present work aims to develop a creep-fatigue model for engineering design, thus this section is included to briefly explain how this model is applied by manual engineering calculations and finite element analysis at the engineering design process.

Fundamentally, the explicit creep-fatigue model can be used to predict fatigue life at a given applied loading, or can be used to evaluate the critical value of applied loading under a given life. This process of engineering calculation is normally applied at the initial stage of engineering design. For example, the explicit model can be applied to select a material.

In addition, the explicit model can also represent the pure fatigue condition. This is because it can be restored to the Coffin–Manson equation at the reference condition ($T = T_{ref}$ and $t_c = t_{ref}$), which represents pure fatigue wherein the creep effect is dormant. In this case, this restored equation can be used to predict the fatigue life or critical value of applied loading at pure fatigue. The accuracy of pure-fatigue description was demonstrated in Section 5.3, which implies the coefficients of C_0 and β_0 obtained in creep fatigue can be extended to predict fatigue life at pure fatigue.

The method may be applied to manual calculation or finite element analysis, as shown in Appendix A.

6. Conclusions

The present work modified the unified creep-fatigue model by introducing the parameters of temperature and cyclic time into the exponent component. In this way, the accuracy of the fatigue-life prediction for both the creep-fatigue and pure-fatigue conditions are improved. The explicit model has the following beneficial attributes: Integration—it provides one formulation that covers the full range of conditions from pure fatigue, to creep fatigue, then to pure creep. The inclusion of the reference condition gives an opportunity to connect pure fatigue with creep fatigue. It also recovers the mathematical formulation of both of the standard fatigue and creep formulations (Coffin–Manson and Manson–Haferd respectively); Unified—it accommodates multiple temperatures, multiple cyclic times, and multiple metallic materials; Natural origin—it provides a physical basis for the structure of the formulation, in its consistency with diffusion-creep behaviour, the plastic zone around the crack tip, and fatigue capacity; Economy—although two more coefficients were introduced into the explicit model, the economy is not significantly impacted; Applicability—the explicit model is applicable to engineering design. This was demonstrated by its application to manual engineering calculations, and also to finite element analysis.

The overall contribution is that the explicit model provides improved ability to predict fatigue life for both the creep-fatigue and pure-fatigue conditions.

Author Contributions: The explicit model was developed and validated by D.L., the discussion and application of this new model were conducted by D.L. and D.J.P.

Funding: This research received no external funding.

Conflicts of Interest: The authors declare no conflict of interest. The research was conducted without personal financial benefit from any funding body, and no such body influenced the execution of the work.

Nomenclature

A	cross-sectional area
A'	constant
b_1, b_2, b_3, b_4 and b_5	constants
C_0	fatigue ductility coefficient
c_1, c_2, c_3, c_4, c_5 and c_6	constants
D	diffusion coefficient
D_v	amount of substance flowing through a unit area
E	Young's modulus under consideration
F	applied force
f	frequency of applied force cycles
f_m	stress moderating factor
f_{ref}	reference frequency
ΔG_f	Gibbs free energy for formation of a vacancy
H_{ij}	prediction ratio
J	diffusion flux
K'	cyclic strength coefficient
K	Boltzmann's constant
k_a	surface condition modification factor
k_b	size modification factor
k_c	load modification factor
k_d	temperature modification factor
k_e	reliability factor
k_f	miscellaneous-effects modification factor
L	length
ΔL	change of the length
$N_{exp,ij}$	experimental results of fatigue life
N_f	creep-fatigue life
$N_{pre,ij}$	predicted fatigue life
N_v	equilibrium atomic fraction of vacancies
n	number of data
n'	cyclic strain hardening exponent
P_{MH}	Manson–Haferd parameter
S_e	endurance limit
S_e'	rotary-beam test specimen endurance limit
T	temperature
T_{ref}	reference temperature
t	creep-rupture time
t_c	cyclic time
t_{ref}	reference cyclic time
x	position
β_0	fatigue ductility exponent
δ_a	average difference
ε_e	elastic strain
ε_p	plastic strain which reflects fatigue capacity
ε_t	total strain
σ	applied loading
σ_{yield}	yield stress
φ	concentration of vacancies
Ω	atomic volume

$b(T,\, t_c)$	creep moderating function
$b_1'(T)$	temperature moderating function
$b_2'(t_c)$	cyclic time moderating function
$c_1(\sigma)$	stress moderating equation
$s(\sigma)$	stress function
$(\log t_a,\, T_a)$	point of convergence of the $\log t$-T lines

Appendix A. Application for Engineering Design

Appendix A.1. Manual Calculation Method for Design

The design engineer needs to determine the coefficients of the explicit equation for the material of interest. This is achieved by the following steps:

(a) Obtain empirical data for the material of interest. The data needed are pure creep, and creep fatigue. The pure creep data are generally commonly available in the literature, and if not then the empirical test is not onerous. The creep-fatigue data may also exist, but if not then a more extensive testing regime is necessary. This is where the economy of the explicit model is advantageous. In particular, in this process of creep-fatigue testing, the cyclic strength coefficients (K') and cyclic strain hardening exponents (n') under different temperatures and cyclic times are extracted. Then, the parameters of cyclic strain-stress relation (K' and n') are formulated in functions of temperature and cyclic time ($K'(T, t_c)$ and $n'(T, t_c)$) through curving fitting. This is conventional practice; see Equation (19).

(b) Extract the coefficients of the explicit model using the method shown in Section 3.2. This process involves numerical optimisation, which may be achieved by using a spreadsheet (e.g., MS Excel® 2013) or other solver. Numerical optimisation is applied to determine the coefficients of C_0, β_0, b_1 and b_2 by minimizing the average difference (δ_a) (Equation (21)).

This method involves the following six steps and is shown in Figure A1:

1. Opening Excel® solver: DATA → Solver
2. Selecting objective: selecting optimum cell in the spreadsheet. In the present work, the cell which shows average error is selected.
3. Defining optimized condition: defining the criterion of numerical optimization for the value of the optimum cell. In the present work, the option of 'Min' is selected to find the minimum average error.
4. Selecting variables: selecting adjustable cells in the spreadsheet. In the present work, the cells which give the values of C_0, β_0, b_1 and b_2 are selected.
5. Selecting solving method: selecting the numerical optimization algorithm. In the present work, we select the method of 'GRG Nonlinear' which is applied to the smooth-nonlinear situation.
6. Clicking 'Solve' to get results.

After this, designers get all the coefficients of the explicit equation. Then this equation can be applied to predict fatigue life under consideration. This is conducted by the following steps:

(c) Determine loading conditions. They are force (F), temperature (T) and cyclic time (t). These will be known, or able to be estimated, by the designer.

(d) Determine plastic strain in the geometry under consideration. The plastic strain is determined by a stress-centric approach (Equation (A1)):

$$\varepsilon_p = \varepsilon_t - \varepsilon_e = \frac{\Delta L}{L} - \frac{\sigma}{E} = \frac{\Delta L}{L} - \frac{F}{A \cdot E} \tag{A1}$$

where ε_t is the total strain, ε_e is the elastic strain, L is the length, ΔL is the change of the length, σ is the applied stress, E is the Young's modulus under consideration, F is the applied force, and A is the cross-sectional area.

(e) Calculate fatigue life. With the cyclic strain-stress relation extracted in step (a), the coefficients obtained in step (b), and the values of temperature, cyclic time and plastic strain given by steps (c) and (d), the fatigue life is calculated by Equation (A2):

$$N_f = \left[\frac{\varepsilon_p}{C_0 c(\sigma, T, t_c)} \right]^{-1/\beta_0 b(T, t_c)} \tag{A2}$$

Overall, we conclude that this simple process can readily be used to provide a more detailed and accurate representation of life under creep-fatigue conditions.

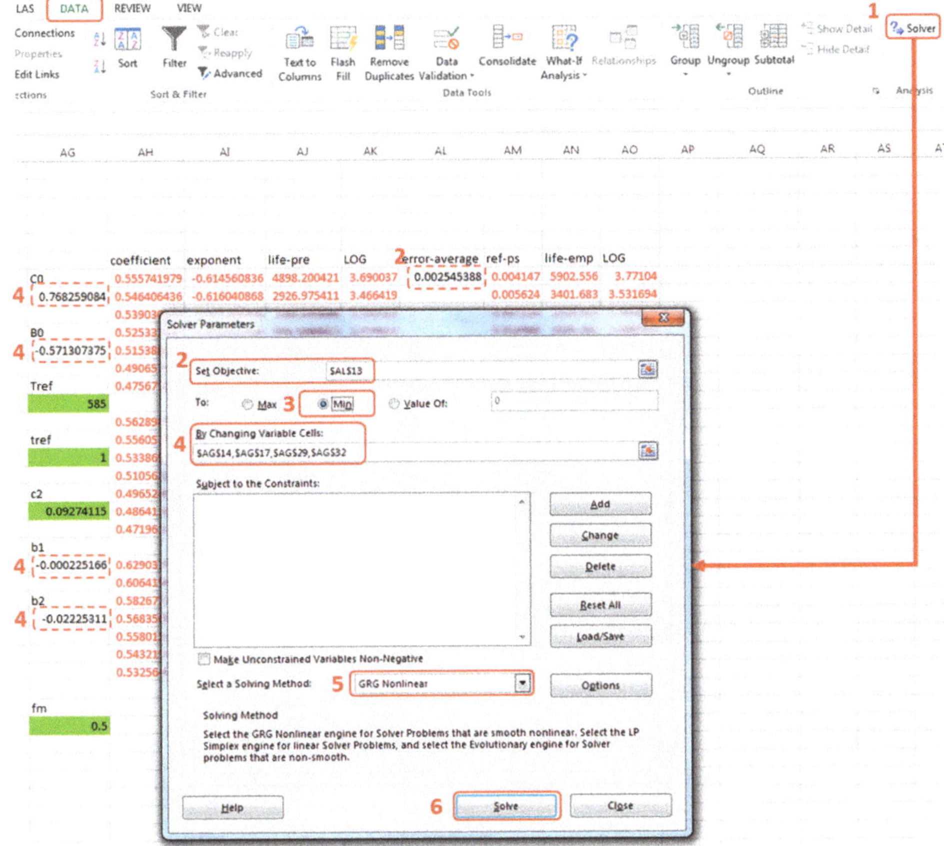

Figure A1. Spreadsheet method for numerical optimization, using Excel®.

Appendix A.2. Using the Explicit Equation for Finite Element Analysis (FEA)

The conventional method for creep-fatigue in FEA is for the algorithms to determine the plastic strain in the part, based on the stress and the pure creep loading (strain dependency on applied temperature). The Coffin–Manson equation (Equation (A3)) is then used to determine the life.

$$\Delta\varepsilon_p = \varepsilon_f' N_f{}^c \tag{A3}$$

where $\Delta\varepsilon_p$ is the plastic strain amplitude, ε_f' is the fatigue ductility coefficient and c is the fatigue ductility exponent.

Conventional finite element creep-fatigue simulation is conducted under cyclic loading and elevated temperature. Generally, this is a complex process. There are two areas where the explicit equation simplifies the process.

First, the conventional FEA process requires empirical data for the specific loading case. This comprises a creep test (strain vs. time) for the applied loading. The creep parameters are determined using curve-fitting, and input to software. There is a need to redo the creep test when the applied loading is changed. In contrast the explicit method has the following advantages: (a) It merely requires data from a creep rupture test—this is a simpler test to perform; (b) If the loading changes, say due to modifications in the design conditions (geometry, temperature, stress, etc.), then there is no need to perform another empirical test nor to recalculate the coefficients. The explicit equation already includes all the variables for multiple different loading conditions.

Second, the conventional FEA process treats the creep effect as independent to the fatigue induced strain. Consequently this may result in non-convergence for FEA, and then the analysis settings may need to be repeatedly modified to attempt convergence. In contrast the explicit method has the advantage of removing the creep parameters. The creep strain is instead included in an integrated manner with the fatigue formulation. Consequently non-convergence is less of an issue.

The benefits of the explicit model is that it provides a method for representing the interaction between creep and fatigue. Thus, the creep effect may be removed from the simulation and the simpler process of creep-fatigue life prediction can be applied. Ideally the explicit model would be formulated within the FEA software, but this is not currently the case. However there is a way to circumvent this problem, which involves adapting the coefficients of the Coffin–Manson equation to represent the full creep-fatigue behaviour. We illustrate this using ANSYS® 17.0 (ANSYS, Inc., Canonsburg, PA, USA). We do not show the other parts of the FEA workflow, such as the setting up of the model and the convergence as we assume the analyst will be familiar with those.

The process is as follows:

(a)   Obtain empirical data for the material of interest. See step (a) in Appendix A.1.
(b)   Extract the coefficients using the method shown in Section 3.2. See step (b) in Appendix A.1.
(c)   Determine loading conditions. They are stress, temperature and cyclic time.
(d)   Determine parameters imported into ANSYS® as engineering data. These parameters include the general material properties (such as yield stress, Young's modulus and Poisson's ratio) and the strain-life parameters under consideration. The general material properties are commonly available in the design handbook. The strain-life parameters are determined by the following method:

The ductility coefficient (ε'_f) and ductility exponent (c) of the Coffin–Manson equation are given by the coefficient component ($C_0 c(\sigma, T, t_c)$) and the exponent component ($-\beta_0 b(T, t_c)$) of the explicit equation respectively.

The cyclic strength coefficient (K') and cyclic strain hardening exponent (n') under consideration are determined by the relation obtained in step (a).

The strength coefficient (σ'_f) and strength hardening exponent (b) are given by the compatibility equations (Equations (A4) and (A5)):

$$\sigma'_f = K' \varepsilon'^{n'}_f \tag{A4}$$

$$b = n'c \tag{A5}$$

(e)   Operate FEA. In this process, *the thermal effect is removed* and the finite element simulation is performed under the pure-fatigue condition. Finally, FEA gives the fatigue life.

Note that this method may only be applied to evaluate the creep-fatigue life of the part. Specifically, the resulting stress/strain distribution and deformation shown by the FEA is unreliable, because its algorithms will not have modelled the creep effect. Nonetheless we believe the life prediction should be robust. However, this opens an opportunity for future research, where the combination between the explicit model and FEA may be further improved.

The accuracy of life prediction using the method above is demonstrated by the following example. In this example, we evaluated the fatigue life for a cylindrical specimen by two different FEA processes. On the one hand, we conducted simulation with creep effect and thermal condition. During this process, creep-related parameters were included as the engineering data, and the simulation was operated under cyclic loading at elevated temperature.

On the other hand, we conducted simulation using the explicit model (using the method above). In this process, the coefficient and exponent of the explicit model were imported into ANSYS® (Revision 17.0) as the fatigue parameters in engineering data. Since they already contain the creep effect, the creep-related parameters and thermal condition are removed from FEA. The fatigue life given by these two methods are shown in Figures A2 and A3.

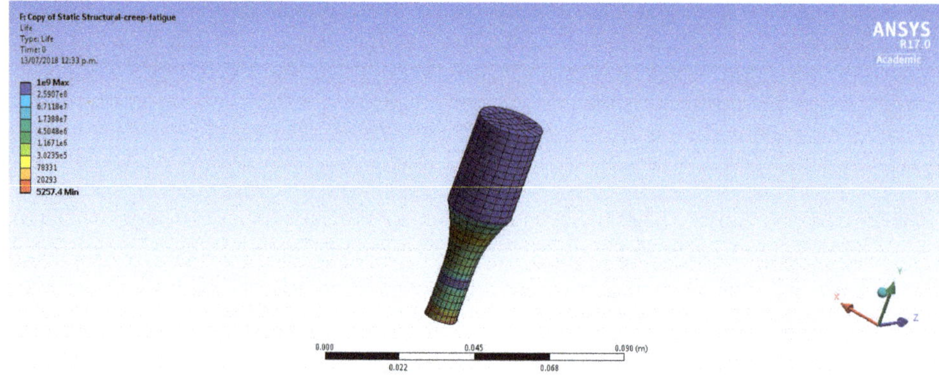

Figure A2. Fatigue life obtained at the creep-fatigue condition using conventional method.

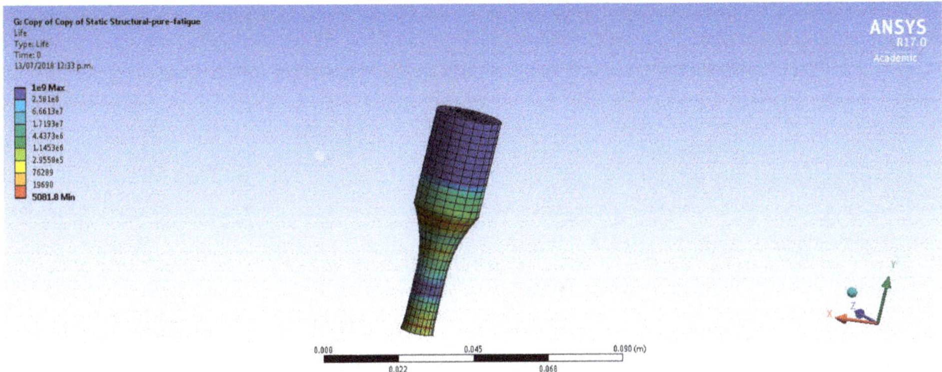

Figure A3. Fatigue life obtained at the pure-fatigue condition using explicit method.

This example is also summarised in Table A1.

Table A1. Fatigue evaluation by FEA.

	Simulation Using Conventional FEA Methods	**Simulation Using Explicit Method**
Temperature	977 K	977 K but model parameters set to room temperature
Loading amplitude	100 MPa	100 MPa
Creep parameters	Were imported	Were removed
Mechanical properties	Values at pure fatigue	Values at the running condition
Fatigue parameters	Parameters at pure fatigue	Parameters were obtained from the explicit model at the running condition
Fatigue life	5257	5081

Table A1 shows that these two life-evaluation processes give similar results (fatigue life), with the explicit method being slightly more conservative. The explicit method is faster to implement even for a single pass through the design, and has further time advantages when there are revisions and loops in the design process. We conclude that combining the explicit model with FEA can reduce the difficulty and complexity of analysis regarding fatigue evaluation, and speed up the design process. These benefits are particularly attractive early design stages, when the design is still experimental and the loading conditions are not finalized.

References

1. Dowling, N.E. *Mechanical Behavior of Materials: Engineering Methods for Deformation, Fracture, and Fatigue*; Pearson: London, UK, 2012; p. 954.
2. Chowdhury, P.; Sehitoglu, H. Mechanisms of fatigue crack growth—A critical digest of theoretical developments. *Fatigue Fract. Eng. Mater. Struct.* **2016**, *39*, 652–674. [CrossRef]
3. Marin, J. *Mechanical Behavior of Engineering Materials*; Prentice Hall: Upper Saddle River, NJ, USA, 1962.
4. Joseph Edward Shigley, C.R.M. *Mechanical Engineering Design*; McGraw-Hill Education: New York, NY, USA, 2001; p. 1248.
5. Taira, S. Lifetime of structures subjected to varying load and temperature. In *Creep in Structures*; Springer: Berlin/Heidelberg, Germany, 1962; pp. 96–124.
6. Ainsworth, R.; Ruggles, M.; Takahashi, Y. Flaw assessment procedure for high-temperature reactor components. *J. Press. Vessel Technol.* **1992**, *114*, 166–170. [CrossRef]
7. Grell, W.; Niggeler, G.; Groskreutz, M.; Laz, P. Evaluation of creep damage accumulation models: Considerations of stepped testing and highly stressed volume. *Fatigue Fract. Eng. Mater. Struct.* **2007**, *30*, 689–697. [CrossRef]
8. Pavlou, D. Creep life prediction under stepwise constant uniaxial stress and temperature conditions. *Eng. Struct.* **2001**, *23*, 656–662. [CrossRef]

9.   Lv, Z.; Huang, H.-Z.; Zhu, S.-P.; Gao, H.; Zuo, F. A modified nonlinear fatigue damage accumulation model. *Int. J. Damage Mech.* **2015**, *24*, 168–181. [CrossRef]

10. Shang, D.-G.; Yao, W.-X. A nonlinear damage cumulative model for uniaxial fatigue. *Int. J. Fatigue* **1999**, *21*, 187–194. [CrossRef]

11. Basquin, O. The exponential law of endurance tests. *Am. Soc. Test. Mater.* **1910**, *10*, 625–630.

12. Wöhler, A. Theorie rechteckiger eiserner brückenbalken mit gitterwänden und mit blechwänden. *Z. für Bauwesen* **1855**, *5*, 121–166.

13. Coffin, L.F., Jr. *A Study of the Effects of Cyclic Thermal Stresses on a Ductile Metal*; Knolls Atomic Power Lab.: New York, NY, USA, 1953.

14. Manson, S. Behavior of Materials Under Conditions of Thermal Stress. Available online: https://ntrs.nasa.gov/archive/nasa/casi.ntrs.nasa.gov/19930092197.pdf (accessed on 8 March 2018).

15. Solomon, H. Fatigue of 60/40 solder. *IEEE Trans. Compon. Hybrids Manuf. Technol.* **1986**, *9*, 423–432. [CrossRef]

16. Shi, X.; Pang, H.; Zhou, W.; Wang, Z. Low cycle fatigue analysis of temperature and frequency effects in eutectic solder alloy. *Int. J. Fatigue* **2000**, *22*, 217–228. [CrossRef]

17. Jing, H.; Zhang, Y.; Xu, L.; Zhang, G.; Han, Y.; Wei, J. Low cycle fatigue behavior of a eutectic 80au/20sn solder alloy. *Int. J. Fatigue* **2015**, *75*, 100–107. [CrossRef]

18. Wong, E.; Mai, Y.-W. A unified equation for creep-fatigue. *Int. J. Fatigue* **2014**, *68*, 186–194. [CrossRef]

19. Kohout, J. Temperature dependence of stress–lifetime fatigue curves. *Fatigue Fract. Eng. Mater. Struct.* **2000**, *23*, 969–977. [CrossRef]

20. Mivehchi, H.; Varvani-Farahani, A. Temperature dependence of stress–fatigue life data of frp composites. *Mech. Compos. Mater.* **2011**, *47*, 185–192. [CrossRef]

21. Kiran, R.; Khandelwal, K. A micromechanical cyclic void growth model for ultra-low cycle fatigue. *Int. J. Fatigue* **2015**, *70*, 24–37. [CrossRef]

22. Maurel, V.; Rémy, L.; Dahmen, F.; Haddar, N. An engineering model for low cycle fatigue life based on a partition of energy and micro-crack growth. *Int. J. Fatigue* **2009**, *31*, 952–961. [CrossRef]

23. Xue, Y.; Horstemeyer, M.; McDowell, D.; El Kadiri, H.; Fan, J. Microstructure-based multistage fatigue modeling of a cast ae44 magnesium alloy. *Int. J. Fatigue* **2007**, *29*, 666–676. [CrossRef]

24. Liu, D.; Pons, D. Crack propagation mechanisms for creep fatigue: A consolidated explanation of fundamental behaviours from initiation to failure. *Metals* **2018**, *8*, 623. [CrossRef]

25. Halford, G.; Hirschberg, M.; Manson, S. *Creep Fatigue Analysis by Strain-Range Partitioning*; National Aeronautics and Space Administration: Washington, DC, USA, 1971.

26. Neu, R.; Sehitoglu, H. Thermomechanical fatigue, oxidation, and creep: Part II. *Life prediction. Metall. Trans. A* **1989**, *20*, 1769–1783. [CrossRef]

27. Liu, D.; Pons, D.J.; Wong, E.-H. The unified creep-fatigue equation for stainless steel 316. *Metals* **2016**, *6*, 219. [CrossRef]

28. Liu, D.; Pons, D.J. A unified creep-fatigue equation with application to engineering design. In *Creep*; Tomasz, T., Marek, S., Zieliński, A., Eds.; InTechOpen: Rijeka, Croatia, 2018.

29. Finnie, I.; Heller, W.R. *Creep of Engineering Materials*; McGraw-Hill: New York, NY, USA, 1959; p. 341.

30. Evans, R.W.; Wilshire, B. *Introduction to Creep*; The Institute of Materials, University of Michiganb: Arbor, MI, USA, 1993; p. 115.

31. Poirier, J.-P. *Creep of Crystals: High-Temperature Deformation Processes in Metals, Ceramics and Minerals*; Cambridge University Press: Cambridge, UK, 1985; p. 276.

32. Ashby, M.F.; Shercliff, H.; Cebon, D. *Materials: Engineering, Science, Processing and Design*; Butterworth-Heinemann: Oxford, UK, 2013; p. 784.

33. Tsuno, N.; Shimabayashi, S.; Kakehi, K.; Rae, C.; Reed, R. In Tension/compression asymmetry in yield and creep strengths of ni-based superalloys. In Proceedings of the International Symposium on Superalloys, Champion, PA, USA, 14–18 September 2008; pp. 433–442.

34. Yamashita, M.; Kakehi, K. Tension/compression asymmetry in yield and creep strengths of Ni-based superalloy with a high amount of tantalum. *Scr. Mater.* **2006**, *55*, 139–142. [CrossRef]

35. Jahed, H.; Varvani-Farahani, A.; Noban, M.; Khalaji, I. An energy-based fatigue life assessment model for various metallic materials under proportional and non-proportional loading conditions. *Int. J. Fatigue* **2007**, *29*, 647–655. [CrossRef]

36. Zhu, S.-P.; Yang, Y.-J.; Huang, H.-Z.; Lv, Z.; Wang, H.-K. A unified criterion for fatigue–creep life prediction of high temperature components. *Proc. Inst. Mech. Eng. G J. Aerosp. Eng.* **2017**, *231*, 677–688. [CrossRef]
37. Shi, X.Q.; Wang, Z.P.; Yang, Q.J.; Pang, H.L.J. Creep behavior and deformation mechanism map of sn-pb eutectic solder alloy. *J. Eng. Mater. Technol.* **2002**, *125*, 81–88. [CrossRef]
38. Fritz, L.J.; Koster, W. *Tensile and Creep Rupture Properties of (1) Uncoated and (2) Coated Engineering Alloys at Elevated Temperatures*; Lewis Research Center: Cleveland, OH, USA, 1977.
39. Kanazawa, K.; Yoshida, S. In Effect of temperature and strain rate on the high temperature low-cycle fatigue behavior of austenitic stainless steels. In Proceedings of the International Conference on Creep and Fatigue in Elevated Temperature Applications, Philadelphia, PA, USA, 23 September 1973; Institution of Mechanical Engineers: Philadelphia, PA, USA.
40. Liu, D.; Pons, D.J. Physical-mechanism exploration of the low-cycle unified creep-fatigue formulation. *Metals* **2017**, *7*, 379.
41. Wu, X.; Quan, G.; MacNeil, R.; Zhang, Z.; Liu, X.; Sloss, C. Thermomechanical fatigue of ductile cast iron and its life prediction. *Metall. Mater. Trans. A* **2015**, *46*, 2530–2543. [CrossRef]
42. Wu, X.; Quan, G.; Sloss, C. Mechanism-based modeling for low cycle fatigue of cast austenitic steel. *Metall. Mater. Trans. A* **2017**, *48*, 4058–4071. [CrossRef]
43. Chapra, S.C.; Canale, R.P. *Numerical Methods for Engineers*; McGraw-Hill: New York, NY, USA, 2010; p. 960.

MDPI
St. Alban-Anlage 66
4052 Basel
Switzerland
Tel. +41 61 683 77 34
Fax +41 61 302 89 18
www.mdpi.com

Metals Editorial Office
E-mail: metals@mdpi.com
www.mdpi.com/journal/metals